Semimagnetic Semiconductors and Diluted Magnetic Semiconductors

ETTORE MAJORANA
INTERNATIONAL SCIENCE SERIES
Series Editor:
Antonino Zichichi
European Physical Society
Geneva, Switzerland

(PHYSICAL SCIENCES)

Recent volumes in the series:

Semimagnetic Semiconductors and Diluted Magnetic Semiconductors

Edited by

Michel Averous

Université de Montpellier II
Montpellier, France

and

Minko Balkanski

Université Pierre et Marie Curie
Paris, France

Plenum Press • New York and London

Library of Congress Cataloging in Publication Data

International School of Materials Science and Technology (1990: Erice, Italy)
    Semimagnetic semiconductors and diluted magnetic semiconductors / edited by
Michel Averous and Minko Balkanski.
        p.   cm.—(Ettore Majorana international science series. Physical sciences; v.
55)
    "Proceedings of the 20th course of the International School of Materials Science and
Technology on semimagnetic semiconductors and diluted magnetic semiconductors,
held July 1—15, 1990, in Erice, Sicily, Italy"—T.p. verso.
    Includes bibliographical references and index.
    ISBN 0-306-43931-X
    1. Diluted magnetic semiconductors—Congresses. 2. Materials—Magnetic proper-
ties—Congresses. I. Averous, M. II. Balkanski, Minko, 1927–   . III. Title. IV. Series.
QC611.8.M25I58   1990   91-20120
537.6′22—dc20   CIP

Proceedings of the 20th Course of the International School of
Materials Science and Technology on Semimagnetic Semiconductors
and Diluted Magnetic Semiconductors, held July 1—15, 1990,
in Erice, Sicily, Italy

ISBN 0-306-43931-X

© 1991 Plenum Press, New York
A Division of Plenum Publishing Corporation
233 Spring Street, New York, N.Y. 10013

PREFACE

Semimagnetic semiconductors (SMSC) and diluted magnetic semiconductors (DMS) have in the past decade attracted considerable attention because they confer many new physical properties on both bulk materials and heterostructures. These new effects are due either to exchange interactions between magnetic moments on magnetic ions, or to exchange interactions between magnetic moments and the spin of the charge carrier. These effects vary with the transition metal (Mn, Fe, Co) or rare earth (Eu, Gd, etc) used and thus provide a range of different situations. The field is very large (zero gap, small gap, wide gap), and the magnetic properties also are very rich (paramagnetic spin glass, antiferromagnetism). These materials are very convenient for studying the magnetism (the magnetism is diluted) or the superlattices (SL) with a continuous change from type II SL to type III SL.

This Course attempted to provide a complete overview of the topic. The participants of this summer school held in Erice came from ten countries and were from various backgrounds and included theoreticians, experimentalists, physicists, and chemists. Consequently, an attempt was made to make the Course as thorough as possible, but at the same time attention was devoted to basic principles. The lecturers, drawn from all the groups in the world involved in the field, were asked to be very didactic in their presentation.

After two introductory lectures, Dr. Triboulet gave an overview of the recent developments in crystal growth and the phase diagrams of the various materials with diluted magnetic ions. Several lectures were then devoted to the specific theory involved in the type of semiconductors with diluted magnetism (see for example Hass' lectures), or to the problem of exchange interaction between magnetic ions (pair triangle, etc) presented by Dr. Shapira. Dr. P Wolff, Co-director of the Course, developed the idea of the magnetic polaron along both theoretical and experimental lines, and Dr. Awschalom developed the concept of spin dynamics and dimensional crossover in DMS. Dr. Dietl and Dr. Lascaray spoke on the magneto-transport and magnetooptics of the DMS (SMSC) respectively, and both presented very new and interesting results. Dr. Voos demonstrated the interest of DMS (SMSC) in the physics of superlattices, and how by applying energy in the form of heat or a magnetic field the energy levels can change due to the exchange interaction. The lectures by Dr. Benoît à la Guillaume and Dr. Twardowski were devoted to the special case of iron (Fe) which gives very different effects when compared with manganese (Mn). Dr. Bauer presented a new family of SMSC, namely the IV-VI compounds, which are of interest in terms of potential applications.

Space does not allow us to mention all the work presented each day during the school by the participants involved in the various fields.

We would like to thank Dr. P Wolff who not only organized the round table but who played an active role in it, as well as all the sessions. He also gave the closing address in which he drew together the overall ideas developed in this school, and also proposed many promising research directions for DMS.

Our hope is that this Course was followed successfully by the graduate and postgraduate participants present at the summer school, and also that this book may prove useful as a basic text for semiconductor researchers.

The Editors
December 1990

CONTENTS

BACKGROUND ON SEMIMAGNETIC SEMICONDUCTORS

M. AVEROUS

Groupe d'Etudes des semiconducteurs
Université de Montpellier Place Eugène Bataillon
34060 MONTPELLIER - FRANCE

INTRODUCTION

Diluted Magnetic Semiconductors have in past decade attracted considerable attention of the Physicists due to interesting physical properties. All these properties are due to the exchange interactions induced by the magnetic moments localized at the magnetic ions.

Diluted Magnetic Semiconductors or Semimagnetic-semiconductors could be II VI , IV VI, II V or III V compounds in which fraction of non magnetic cations has been substituted by Magnetic Transition Metal or Rare earth ions. They differ from Magnetic Semiconductors in which one of the two sublattice is constitued by magnetic ions (i.e. Eu O for ex.).

The incomplete d-shell (Transition Metals) or 4 f-shell (rare earths) of the magnetic atoms gives rise to a variety of properties in which their localized magnetic moments play important roles, either individually or collectively through their mutual interactions, or through the exchange interaction between these localized moments and the spin of the charge carriers.

CRYSTAL STRUCTURE

Most of the research performed so far on these materials have been devoted to Mn - based S.C.S.M. which represents a rather simple magnetic case. [1] to [4]. The Mn^{++} ground stale is a magnetically active spin sextet GA1. The Mn - base S.C.S.M. are the exemple of a permanent magnetic moments system (so - called Brillouin - type paramagnetism).

The family of $II_{1-x}Mn_x$ VI alloys, along with their crystal structures, is presented in schematic form in Fig 1. [5].

The bold lines indicates ranges of the molar fraction x which homogeneous crystal phases form . However since several years new S.M.S.C., with Fe, Gd, Eu and very recently Co as magnetic ions appear.

A lot of new phenomena are discovered every day as well in bulk materials as heterostructures, quantum wells, Superlattice, depending of the position of the magnetic ions (barrier, well...etc).

Semimagnetic Semiconductors and Diluted Magnetic Semiconductors
Edited by M. Averous and M. Balkanski, Plenum Press, New York, 1991

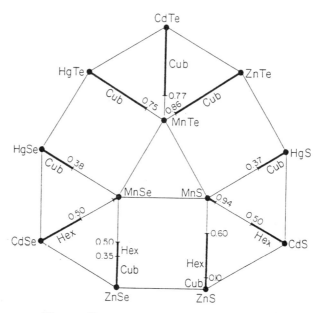

Fig 1. $A^{II}_{1-x} Mn_x B^{VI}$ alloys and their crystal structure

The purpose of these lectures, is to give the background which is necessary to follow with good profit the specialized lectures in different fields.

So the contents will be :

      -A) Crystal Structure and composition
      -B) Band Structure
      -C) Magnetic properties
      -D) Narrow gap Semimagnetic-Semiconductors

Magneto optics in Narrow gap are developped here because these is no special lecture devoted to this subject.

The range of solid solutions is very wide although the crystal structure of MnTe, MnSe and MnS differ from the crystal structure of II VI compounds. [6].

The crystal structures of II VI compounds are Zinc blende or wurtzite. They are formed with tetrahedral bonding (s-p^3), involving the six valence p electrons of the element VI and the two valence s electrons of the element II. Manganese is a metal transition with valence bound electrons corresponding to 4 s^2 orbital. So it can contribute to the s - p^3 bonding by these 4 s^2 electrons and therefore replace the II element in the II VI wurtzite are Zinc Blende structure, despite the fact that it differs from this element by its 3 d shell half filled.

However the exactly half filled 3 d orbitals play a role in the existence of very wide solid solutions.

For example Iron, is about an order of magnitude less miscible, and the other metals transition are less than that (Co). The reason is that following the Hund's rule, in Mn all five spins are aligned, and it would required large energy to add an electron with antiparallel spin :

In this way, the 3 d^5 orbit behave as a complete one, and thus Mn is like a groupe II element. It is quite different for Iron or Cobalt.

Except the rôle of the exchange interaction, the magnetic properties and the intra magnetic ions transitions, the physical properties are deduced, from the properties of the starting II VI compound.

So it is interesting to have a brief review of structural properties of II VI compounds.

Particularly their crystallographic properties are systematic except HgS, whose stable phase is cinnabar. They all form in either the Zinc Blende or the Wurtzite phase. Both structures form as a result of tetrahedral s - p^3 bonding.

Zinc Blend is a cubic structure with two fcc sublattices interpenatrating, (Fig 2).

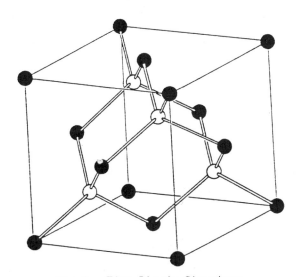

Fig 2. *Zinc Blende Structure*

Anions and cations, shifted with respect to each other by 1/4 of the body diagonal of the fcc cubic unit cell.

The wurtzite structure is formed by two interpenatrating fcc sublattices shifted along the c axis.

Obviously the distant between nearest "cation - cation" d and the bond length b : distant cation - anion are the same in the two structures from geometrical considerations one have

$$d = \sqrt{\frac{8}{3}} \, b$$

In the case of ZB the lattice parameter

$$a = \sqrt{2} \, d = \frac{4}{\sqrt{3}} \, b$$

3

In the case of wurtzite there are two lattice parameters a and c
- a is the separation between like atom of the sublattices in the
hexagonal plane.
- c is the distance between repeating planes. It is easy to show :

$$a = d \; ; \; c = \sqrt{\frac{8}{3}} \, d$$

CRYSTAL STRUCTURE OF $II_{1-x}Mn_xVI$

The crystal structure of ternary alloys with Mn is shown in Fig
I. Generally, these SCSM have the structure of the II VI parent. Except
ZnMnS and ZnMnSe which are cubic (Zinc Blende) at low x then above a
critical value of x they are wutzite. For each ternary compound an upper
limit of x is imposed by the fact that MnTe, MnSe or MnS do not
crystallize in the Zinc Blende or wurtzite structure.
    The systematic behaviour of lattice parameters in the II VI compounds
are conserved in SCSM. For example in SCSM crystallizing in wurtzite
structure the a/c relationship predicted by the ideal hexagonal close
packing of spheres is the same. Both a and c are determined by the nearest
neigbor distance d of the anion (or cation) sublattices. so d is the key
parameter which determines the structure dimension of the ternary SCSM
lattice.
    -the lattice parameters for all SCSM obey Vegard's law very closely
(Fig 3). [7], [8].

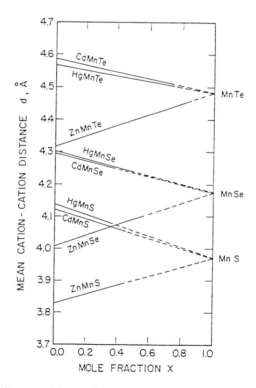

Fig 3. Mean cation-cation distance vs Mn mole fraction

The microscopic situation is more complicated (local structure) it is expected that the microscopic structure is similar to the one of the parent binary compound, with the scale adjusted by the Vegard law behaviour, with the cations sites occupied ramdonly by two species of ions. (This is the base of the virtual crystal approximation).

Recent EXAFS experiments show a very different picture. [9].

(EXAFS : Extended X-Ray Absorption Fine Structure).

The result is, for ex in the case of $Cd_{1-x}Mn_xTe$, that the CdTe and MnTe bond lengths remain nearly constant throughout the entire range of composition (x < 0,7) (Fig. 4)

These results are not in contradiction with x-ray diffraction data (Vegard law), because the x-ray diffraction data give the lattice parameter averaged over all anion and cation sites.

It follows that the anion sublattice will be locally distorded, to adjust to CdTe and MnTe constant bonds.

Since the MnTe bond length is shorter than the CdTe bond, the Te ion will be displaced from its equilidrum position towards Mn and away from Cd (voir Fig 5).

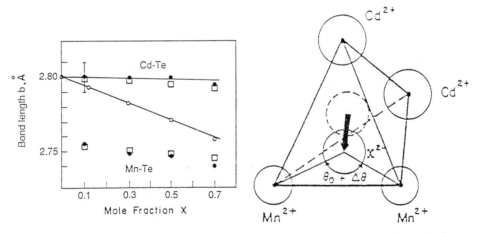

Fig 4.   EXAFS Results on CdTe and MnTe bond length

Fig 5.   Distorded position of the anion in mixed CdMnX

The Mn^{++} ions have a $3d^5$ configuration and $^6S_{5/2}$ ground state (S = 5/2, L = 0). A spherical symetric S state cannot be affected by a cristalline electric field. However a crystal field splitting of the ground state of Mn^{++} appears because it is not a pure ^6S state but contains admixture of higher lying levels. Generally these admixtures are small and at the present step neglected. In specialized lectures they will be taken into account.

Consequently the Mn^{++} ground state is a magnetically active sextet G_A. In that respect Mn - base SCSM are the example of a permanent magnetic moments system (so called Brillouin - type paramagnetism).

In the case of Fe^{++} the situation is very different because Fe^{++} possesses both spin and orbital moments (S = 2, L = 2) (see Benoît à la Guillaume's lectures).

A tetrahedral crystal field splits the ^5D (d^6) iron ground term into an orbital doublet ^{5}E and an orbital triplet ^{5}T. The spin orbit interaction splits these terms : ^{5}E into a singlet A1, a triplet T1 a doublet E, a triplet T2 and a singlet A2. [10].

Consequently the situation for Fe is essentially different from Mn.

All the ground term of magnetic ions must be review as well metal transition as rare earth, and then the role of the different interactions : S.O., C.F.,...etc.

If the crystal field interaction is large the electronic level energy of the magnetic ions depend of the crystal host.

I don't give here the theory of magnetic interaction. I recall simply the behaviour when the magnetic ions could be considered as isolated (i.e. at very low values of x). In the lectures of Dr. Shapira the complete theory, when there exists cluster (pairs, triangles and so on... with different environment) will be treated in details.

For the simplest case (Mn^{++} diluted in CdTe) from specific heat and susceptibility, Dr. Galazka suggest the following magnetic diagram (Fig 7). [3]. :

These diagram is roughly the same for $II_{1-x}Mn_xVI$.

A simplified spin Hamiltonian for Mn^{2+} in a cubic host that include the hyperfine interaction is :

$$\mathbb{H} = g \, \mu_B \, H.S + \frac{a}{6} \, (\tilde{S}_x^{\,4} + \tilde{S}_y^{\,4} + \tilde{S}_z^{\,4}) + A \, I \, S \qquad (1)$$

- a is a crystal field parameter,
- $\tilde{S}x$, $\tilde{S}y$ and $\tilde{S}z$ are the projection of the spin operators.

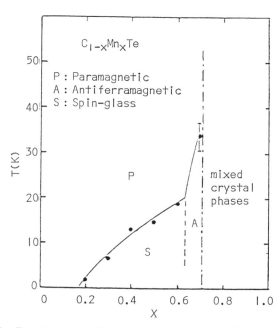

Fig 7. Magnetic "phase diagram" for $Cd_{1-x}Mn_xTe$

Where the first term accounts for the electronic Zeeman interaction (see Dr. Lascaray's lectures), the second, for the cubic crystal field, and the third for the hyperfine interaction.

When the energy of the Mn levels are known, the magnetization M, susceptibility x and specific heat c follow directly :

$$\chi = \frac{N_{Mn} \; g^2 \mu_B^2 \; S \; (S + 1)}{3 \; k_B \; T} \qquad (2)$$

Where N_{Mn} is the number of Mn ions/unit volume :

- g is the Lande factor
- μ_B is the Bohr magneton
- S is the total spin
- k_B the Boltzmann constant

When $x \simeq 0,005$ the probability that Mn ions interaction increases rapidly. This part will be treated by Shapira.

Let us Recall that :

$$M = \chi H \qquad (3)$$

Concerning the specific heat in the case of pure II VI host (x = 0) only the lattice vibrations contribute to the specific heat and an extrapolation of the low temperature specific heat to T = 0 determines the Debye temperatures.

In SMSC, the low temperature specific heat is substantially greater than in corresponding pure II VI.

The excess specific heat has its origine mainly in the magnetic properties of Mn ions. Small changes in the lattice vibrations spectrums could be neglected.

In particular for $x > 0,2$ the specific heat behaves very similarly to the magnetic contribution found in metallic spin glasses. This point will be developped in the lecture on magnetic interaction.

BAND STRUCTURE

WIDE GAP SCSM ALLOYS

In the crystal structure we have shown, that there exist for $Cd_{1-x}Mn_xTe$, $Cd_{1-x}M_xSe$...etc a single phase either Zinc blend or Wurtzite in a large range of x (0,7 for $Cd_{1-x}Mn_xTe$, 0,5 for $Cd_{1-x}Mn_xSe$...). [5].

All the experimental data (optical absorption excitonic emission...) suggest that as well in the Zinc blende phase as in the Wurtzite one the gap is always direct as in the host compound. The host compound band structure could be understood in terms of s - p^3 orbital bonding configuration.

The highest-lying valence band is three fold degenerate (without S.O. interaction), and the states at the center of the Brilloun zone are bonding combinations of functions which are p - like about the nuclei. The lowest conduction band is s - like. The f
fig 8 give the energy band structure of cubic ZnTe as calculated by the empirical pseudo potential method at k = 0. [11].

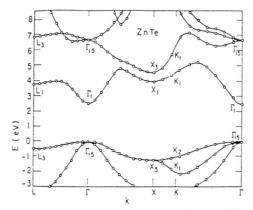

Fig 8. *Band Structure of cubic ZnTe*

When the S.O. interaction is included, the degeneracies are lifted. The Γ_{15v} state sixfold degenerated (threefold if we don't consider the two spin states), splits into a fourfold Γ_8 ($J = 3/2$) state and a lower, two fold Γ_7 ($J = 1/2$) state.

Fig 9 show the band structure of the wurtzite structure at $k = 0$. The valence band states are considered to be derived from the Γ_{15} Zinc blende valence band state through the actions of two perturbations : Crystal Field splitting and S.O. splitting. It is a reasonable zeroth order approximation.

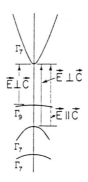

Fig 9. *Band Structure of wurtzite at k = 0*

The Γ_8 level is splitted into a Γ_9 and Γ_7 level when a hexagonal crystal field is applied : Γ_9 and Γ_7 level are two fold degenerate.

The energy gap varies with the magnetic ions concentration. There is several effects.

- a) An effect due to magnetic interactions see elsewhere,
- b) An effect due to the linear dependance with x (Vegard law), because the two compounds at $x = 0$ and $x = 1$ have different energy gap.

Fig 10 show the variation of the energy gap with the Mn mole fraction (x), for $Cd_{1-x}Mn_xTe$.

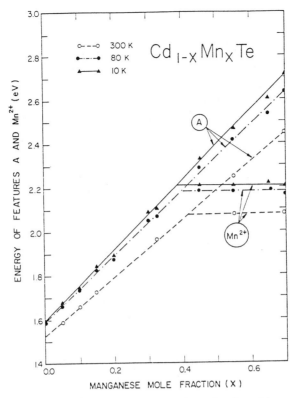

Fig 10.   Energy gap vs Mn mole fraction
A lines : Free exciton. Mn^{2+} lines : 6A1→41T1 absorption band

VARIATION OF THE BAND GAP WITH T

In usual S.C., the temperatures variations of the band gap are well described by followning equation. [12]. :

$$Eg \ (T) = Eg \ (0) - \frac{aT^2}{T+b} \qquad (4)$$

- Eg (T) band gap at T
- Eg (0) band gap at T = 0

These equation predicts that $\frac{dE_g}{dT})_{T \to 0} \to 0$

and $\frac{dE_g}{dT}$) -> Cte at high T

In SMSC there is departure from the above law near the magnetic ordering temperatures [13].

Dr Lascaray analyses this effect due to magnetic interaction in his lectures.

The Mn ions not only change the band structure but also introduce new optical transitions. These include absorption and emission features which can be correlated with the expected splittings of Mn multiplets in a crystal field.

Fig 6 gives the splitting of states of the d^5 configuration by a crystal field of Oh symmetry. Similar splitting is expected for a crystal field of TD symmetry. Details of Mn^{2+} optical transitions in II VI compounds have been extensevely studied. In the regime of very dilute Mn systems, vibrational interactions, Jahn-Teller effects give a lot of structure in the spectrum. When x increases only broad emission or excitation feature remain.

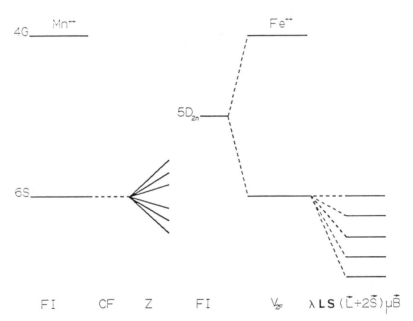

FI   CF   Z   FI V_{2e} $\lambda \mathbf{L S} (\vec{L} + 2\vec{S}) \mu \vec{B}$

Fig 6. States of Mn^{++} and Fe^{++} splitted by crystal field Spin Orbit and Zeeman splitting

The problem is that the energy of the transitions intra Mn between 6A1-4T1 are depending of the x value of the same order of magnitude lower than the energy gap and could be seen only for high manganese concentrations or under hydrostatic pressure, when Eg is higher than the energy transition. [15].

For Eg $\simeq$ E6A1 - 4T1 (for x $\simeq$ 0.7) There is a mixing of these two transitions (fig 11) :

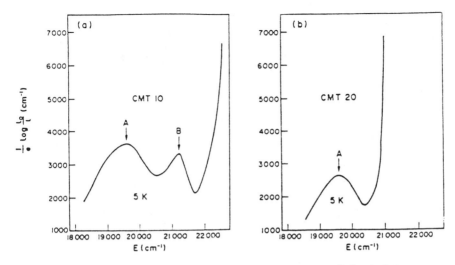

Fig 11.   *Absorption Spectra of Cd$_{0,3}$Mn$_{0,6}$Te (b) at 5 k*
            A is 6A$_1$→4T$_1$ absorption Band
            B is 6A$_1$→4T$_2$ absorption Band

    All the phenomena described above are not correlated to the magnetic interactions which exist near the magnetic transition temperature and treated elsewhere.

SMALL GAP SCSM ALLOYS

CASE OF Hg$_{1-x}$MnTe

    HgTe is a semimetal. The six p - like states are decoupled by the S.O. interaction into a quadruplet Γ$_8$ (J = 3/2) and a doublet Γ$_7$ (J = 1/2) as in the case of wide gap seen previously. The Γ$_7$ forming a lower lying valence band, with the energy Δ = Γ$_8$ - Γ$_6$, determined by the S.O. splitting. The s - like band, transforming as Γ$_6$ (J = 1/2) lies below Γ$_8$ quadruplet, leading to a symmetry induced zero gap semiconductor. [15].
    The Γ$_8$ levels then form the degenerate valence and conduction band edges, separate from the Γ$_6$ band by the interaction gap

$$E_0 = \Gamma_6 - \Gamma_8 = 0,3 \text{ eV at } 4,2 \text{ k}$$

CdTe and MnTe in cubic phase (virtual) are wide gap semiconductors. We have seen their band structure at the centre of the Brillouin Zone.
    The band structure of Hg$_{1-x}$Cd$_x$Te or Hg$_{1-x}$Mn$_x$Te varies continously from zero gap situation to open gap one.
    The interaction gap Eo, increases with x thus there exists a

semimetal-semiconductor transition for x = 0,165 in Hg₁₋ₓ CdₓTe and x = 0,075 in Hg₁₋ₓMnₓTe at 4,2K, (fig 12).

In the absence of magnetic field the electronic states may be descrited in terms of the Kane band structure model.[19].

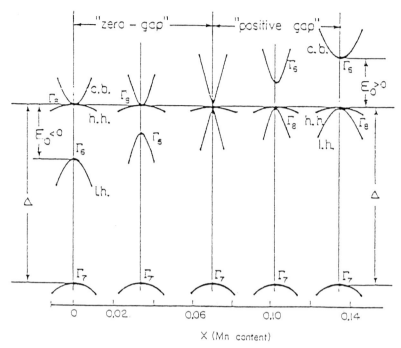

Fig 12. Band Structure of $Hg_{1-x} Cd_x Te$ or $Hg_{1-y} Mn_y Te$

MAGNETOOPTICS IN SMALL GAP SMSC

Magnetooptics in wide gap SMSC will be treated in the Lascaray's lectures, the transport phenomena including small gap SMSC in the Dietl's lectures and the exchange interactions in the Shapira's lectures.

So I think it is important to treat here the magnetooptics in small gap or zero gap SMSC for two main reasons :

The first one is the very nice physic which is possible to find in this situation.

The second one is in vue of potentiel application : I.R detectors, I.R lasers, Heterostructures, Superlattices...etc.

To understand the magnetooptics in small gap SMSC we have to explain the behaviour of the Landau levels Let us recall the background of the theory developped in the Pidgeon-Brown Model.[21].

The proper effective mass Hamiltonian Ho is provided by the eight-band Pidgeon-Brown Model.[21].

In their model the k-p interactions between closely Γ6, Γ7, Γ8 bands are treated exactly and the interaction with higher bands are included up to order k^2. The magnetic Hamiltonian Ho without magnetic interactions, is a 8x8 matrix, that reduces to two 4x4 matrices Da and Db, at the Γ point when inversion assymetry and warping in Γ8 bands are neglected.

$$D = \begin{pmatrix} D_a & 0 \\ 0 & D_a \end{pmatrix} \tag{5}$$

On the basis of the band edge Bloch functions $u_i(J,M_j)$ associated with :

$$a \begin{cases} u_1(\tfrac{1}{2}, \tfrac{1}{2}) = |\ S \uparrow > \\[2ex] u_3(\tfrac{3}{2}, \tfrac{3}{2}) = \dfrac{1}{\sqrt{2}} |\ (X + iY)\uparrow > \\[2ex] u_5(\tfrac{3}{2}, -\tfrac{1}{2}) = \dfrac{1}{\sqrt{6}} |\ (X - iY)\ \uparrow\ + 2Z \downarrow > \\[2ex] u_7(\tfrac{1}{2}, -\tfrac{1}{2}) = \dfrac{i}{\sqrt{3}} |\ - (X - iY)\ \uparrow\ + Z \downarrow > \end{cases} \tag{6}$$

$$b \begin{cases} u_2(\tfrac{1}{2}, -\tfrac{1}{2}) = i|\ S \downarrow > \\[2ex] u_4(\tfrac{3}{2}, \tfrac{1}{2}) = \dfrac{i}{\sqrt{6}} |\ (X + iY)\ \downarrow\ - 2Z \uparrow > \\[2ex] u_6(\tfrac{3}{2}, -\tfrac{3}{2}) = \dfrac{i}{\sqrt{2}} |\ (X - iY)\ \downarrow > \\[2ex] u_8(\tfrac{1}{2}, \tfrac{1}{2}) = \dfrac{1}{\sqrt{3}} |\ (X + iY)\ \downarrow\ + Z \uparrow > \end{cases} \tag{7}$$

S, X, Y, Z are the Luttinger - Kohn periodic amplitudes that transform as atomic s - and p - functions, respectively, under the operations of the tetrahedral group at the Γ point in the spherical approximation H_0 could be written.
H_0 contains the following band parameters :
The interaction bound gap $E = \Gamma_6 - \Gamma_8$.
The spin orbit splitting energy $\Delta = E\Gamma_8 - E\Gamma_6$
The energie of the s - p interaction :

$$E_p = \frac{2}{m_0} |\ < S\ |\ P_x\ |\ X >\ |^2 \tag{8}$$

and modified Luttinger parameters γ_1, $\bar{\gamma}$ and κ which described the interaction between the Γ_8 band and higher bands.
Now we must take into account the exchange interaction by adding H_{exch} to H_0.
H_{exch} is an Heisenberg type Hamiltonian.

$$H_{exch} = \sum_{R_i} J\ (r - R_i)\ S_i\ \sigma \tag{9}$$

S_i is a localized spin at the site R_i ($S_i = 5/2$), σ is the spin of the free carriers ;

13

$J (r - R_i)$ is an integral exchange rapidly varying over a unit cell.

Since the free carrier has an extended wave function, it interacts with a large number of localized spin. So we can replace The Mn spin operators by their thermal average. $\langle S \rangle$.

Without magnetic field $\langle S \rangle = 0$.

Under magnetic field along the z direction $\langle S \rangle$ has only a non zero component $\langle S_z \rangle$.

With the VCA and molecular field approximation :

$$H_{exc} = x \, \sigma_z \langle S_z \rangle \sum_R J (r - R). \tag{10}$$

R denotes the sites of the cations sublattice.

$\langle S_z \rangle$ is directly related to the magnetization M through the relation :

$$M = -N_0 \times g_{Mn} \, \mu_B \, \langle S_z \rangle \tag{11}$$

No is the number of unit cells per unit volume, x is the Mn molar fraction and μ_B is the Bohr magneton.

for Mn : $\langle S_z \rangle = -S \, f_{norm}$ with $S = \dfrac{5}{2}$

f_{norm} the normalized magnetization

$$f_{norm} = \frac{2}{5} \, B_{\frac{5}{2}} \left(\frac{g_{Mn} \, \mu_B \, H}{k_B \, T} \right) \tag{12}$$

For isolated magnetic ions $B_{5/2}$ is the Brilloum fonction for a spin $S = 5/2$.

$$B_S(x) = \frac{2S + 1}{2} \coth \frac{2S + 1}{2} x - \frac{1}{2} \coth \frac{x}{2} \tag{13}$$

In the basis of the Bloch fonctions u_1 (J, M_J) the exchange Hamiltonian consits of two 4x4 matrices.

$$M = \begin{pmatrix} M_a & 0 \\ 0 & M_b \end{pmatrix} \tag{14}$$

$$\text{with } M_a = \begin{pmatrix} -3A\alpha & 0 & 0 & 0 \\ 0 \, \overline{\beta} & 3A & 0 & 0 \\ 0 & 0 & -A & -2A\sqrt{2} \\ 0 & 0 & -2A\sqrt{2} & A \end{pmatrix} \tag{15}$$

$$M_b = \begin{pmatrix} -3A\alpha & 0 & 0 & 0 \\ 0 \, \overline{\beta} & 3A & 0 & 0 \\ 0 & 0 & -A & -2A\sqrt{2} \\ 0 & 0 & -2A\sqrt{2} & A \end{pmatrix} \tag{16}$$

α is the s - d integral
β is the p - d integral

$$\alpha = \langle S / J (r) \, S \rangle, \quad \beta = \langle X | J(r) | \, x \rangle = \langle Y | J(r) | \, Y \rangle = \langle Z | J(r) | \, Z \rangle$$

The exchange parameter A = 1/6 No βx < S2> is proportional to the magnetization M.

The landau levels are obtained by solving the two decoupled matrix equations :

$$(D_a - E_a I) \psi_a = 0 \qquad (D_b - E_b I) \psi_b = 0 \qquad (17)$$

$$\text{with } D_a = D_a + M_a$$

$$D_b = D_b + M_b$$

ψ_a and ψ_b the waves functions given in terms of envelope fonctions f_n (r) and band edge Bloch functions.

$$\psi_a (n) = a_{1n} f_n u_1 + a_{3n} f_{n-1} u_3 + a_{5n} f_{n+1} u_5 + a_{7n} f_{n+1} u_7$$

$$\psi_b (n) = b_{2n} f_n u_2 + b_{4n} f_{n-1} u_4 + b_{6n} f_{n+1} u_6 + b_{8n} f_{n-1} u_8$$

for n > 0, f (r) is the Landau level function

$$f_n(r) = \exp (i(k_y y + k_z z) \, \emptyset_x \, (x + \lambda^2 ky).$$

where $\emptyset_x$ is the x - th harmonic oscillator eigen function of characteristic length $\Lambda = (\hbar c /eH)^{1/2}$. If n < 0, f_n = 0

For each quantum level n ≥ 1, there exists four solutions in each a and b ladder that describe Γ6, Γ7 (light and heavy) and Γ7 landau levels. The eigenvalues and eigenvectors are obtained by solving numerically eq. 17 for proper band and exchange parameters.

For the levels n = - 1. The solutions are analytical

$$E_{b(-1)} = - \hbar\omega_0 \left(\frac{1}{2} (\gamma_2 + \bar{\gamma}) - \frac{3 \kappa}{2} \right) - 3 A \qquad (18)$$

$$\psi_{b(-1)} = f_0(r) \times u_6$$

$$E_{a(-1)} = - \hbar\omega_0 \left(\frac{1}{2} (\gamma_1 + \bar{\gamma}) - \frac{\kappa}{2} \right) - A \qquad (19)$$

$$\psi_{a(-1)} = f_0(r) u_5$$

$\hbar\omega$ is the free electron cyclotron energy.

The exchange contribution is included in A and gives a shift of the level which is three time larger for b(-1).
In the Γ band, the energies of a and b ladders corresponding to n↑ (or a) and n↓ (or b) levels are : (Parabolic approximation).

$$E_{a_b} (n) = (n + \frac{1}{2}) \frac{eH\hbar}{m_{\Gamma_6}^* c} \pm \frac{1}{2} g_{\Gamma_6}^* \mu_B H \mp \frac{5}{4} \alpha N_0 x \, f \, norm \qquad (20)$$

$m_{\Gamma_6}^*$ and $g_{\Gamma_6}^*$ are the band edge effective mass and gyromagnetic factor deduced from k - p interaction.

$$\frac{m_0}{m_{\Gamma_6}^*} = 1 + \frac{E_p}{3\varepsilon_0} \frac{3 \varepsilon_0 + 2 \Delta}{\varepsilon_0 + \Delta} \qquad (21)$$

$$g^*_{\Gamma_6} = 2 \left(1 - \frac{E_p}{3\varepsilon_0} \frac{\Delta}{\varepsilon_0 + \Delta} \right) \qquad (22)$$

One can write the Eq. (20)

$$E_{a_b}(n) = (n + 1/2) \frac{e H \hbar}{m^*_{\Gamma_6} c} \pm \frac{1}{2} \bar{g}^*_{\Gamma_6} \mu_B H \qquad (23)$$

with :

$$\bar{g}^*_{\Gamma_6} = g^*_{\Gamma_6} - \frac{5}{2} \alpha N_0 x \qquad (24)$$

For the Γ bands we obtain :

$$\frac{m_0}{m^*_{\Gamma_8}\text{light}} = - \left(\gamma_1 + 2 \bar{\gamma} + \frac{2E_p}{3\varepsilon_0} \right) \quad ; \quad \frac{m_0}{m^*_{\Gamma_8}\text{heavy}} = 2 \bar{\gamma} - \gamma_1 \qquad (25)$$

$$g^*_{\Gamma_6}\text{light} = 2 \left(3\bar{\gamma} - \kappa + \frac{E_p}{3\varepsilon_0} \right) - \frac{5}{6} \frac{\beta N_0 x f_{\text{norm}}}{\mu_B H} \qquad (26)$$

$$g^*_{\Gamma_6}\text{heavy} = - 6 K - \frac{5}{2} \frac{\beta N_0 x f_{\text{norm}}}{\mu_B H} \qquad (27)$$

It is worth to note that the effective masses are not changed by the exchange interaction, but an additional term due to the exchange appears in the gyromagnetic factors, that depend on T and H through f_{norm}.

Thus the effect of exchange on the landau levels gives a temperature and field dependant shift. The orbital motion in quantizing magnetic field remains unaffected.

This results give interesting physical properties which appeared through magnetooptic experiments.

INTERBAND MAGNETOOPTICS

In the case of zero gap or small gap S.M.S.C. the interband magnetooptic experiments are mainly made by IR magneto absoption on very thin samples for Faraday and Voigt configurations, using σ and π polarisations.

The interband $\Gamma_6 - \Gamma_8$ magneto absorption spectra in HgMnTe in zero gap situation are given in fig 13.

16

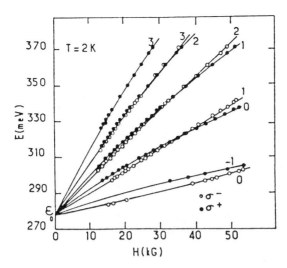

Fig 13. *Γ6- Γ8 transitions vs magnetic field for* $Hg_{1-x}Mn_xTe$
(x = 4 °/oo) at 2 k

The experimental lines are identified with the dominant transitions according to the selection rules :

$$\varepsilon^- \ (\text{or } \sigma^-) : \quad a_{\Gamma_6}(n) \rightarrow a_{\Gamma_8}(n + 1) ; \qquad (28)$$

$$\varepsilon^+ \ (\text{or } \sigma^+) : \quad b_{\Gamma_6}(n + 1) \rightarrow b_{\Gamma_8}(n) ; \qquad (29)$$

$$\varepsilon // H : \quad b_{\Gamma_6}(n + 1) \rightarrow a_{\Gamma_8}(n)$$

For very low x (x ≃ 0.001) the relative ordering of the transitions is identical to that observed in the HgTe or $Hg_{1-x}Cd_xTe$. For x > 0,4 % an inversion of the relative positions of the circular polarization transition $a_{\Gamma_6}(n) \rightarrow a_{\Gamma_8}(n + 1)(\sigma^-)$ and $b_{\Gamma_6}(n) \rightarrow b_{\Gamma_8}(n - 1)(\sigma^+)$ appears.

If we compare with the same interaction gap we can conclude that this anomalous effect is due to the exchange interaction. The energy separation between consecutive transitions for a given polarization (i.e. the period of oscillations of a given spectrum) remains identical for both compounds.

These features are illustrated in fig 14 by plots of $\hbar\Omega$ and the spin sphitting $S_{c(1)} = E_{b(1)} - E_{a(1)}$ of the n = 1 conduction Landau level, directly measured from the energy difference at a given field, between the σ^+ and $\varepsilon // $ transitions. [17].

The quantity $\hbar\Omega$ in the parabolic limit which represents the sum of the cyclotron energies in Γ_6 and Γ_8 conduction bands, has the same values for HgMnTe and HgCdTe due to the same value of band parameters in both compound for the same interaction gap. The strong enhancement of the spin splitting in the semimagnetic compounds provides direct evidence of electron spin sublevels induced by the exchange interaction.

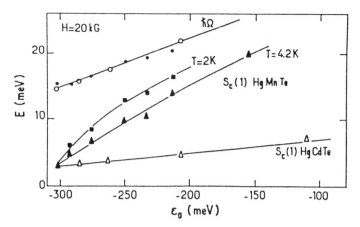

Fig 14. Spin and cyclotron splittings Sc(1) and ℏΩ vs ε0
(○, △→ HgCdTe), (●, ■, ▲→ HgMnTe)

On fig 15 which represents the magnetoabsorption spectrum at ℏΩ = 243.3 meV for Hg$_{1-x}$Mn$_x$Te (x = 1,5%) in the Faraday configuration and σ* polarization a new and intense line appears due to bΓ6(o) → bΓ8(-1) this is a strong evidence for the emergence of the bΓ8(-1) heavy hole level above the Fermi energy. [16].

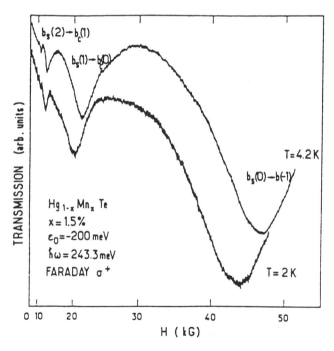

Fig 15. Magneto transmission spectra of Hg$_{1-x}$Mn$_x$Te (x = 1,5 %)

From fig 14 it could be noticed in the case of S.M.S.C. a strong dependance with T of the magnetooptic transitions. This effect provides from exchange interaction which induces an $< S_z >$ dependant terme in the gyromagnetic factor.

The fig 16 and 17 show the band edge effectives masses and the g_e factor for HgCdTe and HgMnTe at the same ε_0. There is no effect of the exchange m_e/m_o, and a large effect on the g^* factor due to the term which depend on $<S_z>$. [16].

All the above results magnetooptical spectra could be interpreted in terms of Pidgeon Brown model generalized to include the s – d and p – d exchange contributions.

The fig 18 shows the Γ_8 landau levels versus magnetic field. The point is the non linear dependance, and so the non equivallent distance giving rise a Semiconductor-Metal transition at 35 kG for x = 0,4 % and T = 2k. [16].

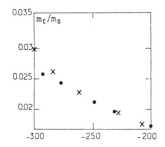

Fig 16.   Band edge effective masses for   Fig 17.   gyromagnetic factors for
          HgCdTe (x) and HgMnTe alloys   HgCdTe (x) and HgMnTe
          (4,2 k)   alloys (x, $\bullet \rightarrow$ 4,2 k)
                                                      ($0 \rightarrow$ 2 k)

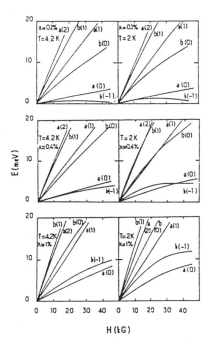

Fig 18. Γ_8 landau level energies versus H for $Hg_{1-x}Mn_xTe$
          (various x) at 4,2 k and 2 k.

19

NARROW GAP Hg₁₋ₓ MnₓTe (100 < Eg < 300 meV positive band gap).

The Γ₈ → Γ₆ magnetooptical spectra again show interesting features, that are very different of the results obtained in HgCdTe with the same gap. A reduction of the energy gap appears with magnetic fields, since the lowest energy transition, observed for σ^- polarization - decreases in energy when H increases. It could be remark a near coincidence of σ^- and ε //H transitions, except for the two lowest lying transitions (labelled -1 and 0). This fact is illustrate in the magneto transmission spectra shown in Fig (19) ε_g = 219 meV T = 4.2K. [18].

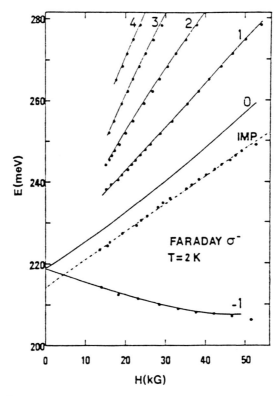

Fig 19. *Γ₆ - Γ₈ transitions for Hg₁₋ₓMnₓTe*
(x = 0,128), (ε₀ = 219 met)

CONCLUSION

This course is an introduction to the summer school on Diluted Magnetic Semiconductors. That means : we tried to give the basic concepts which are necessary to follow the other courses. In the first part we restrict our topic to the case of non interacting magnetic ions (isolated case). In the last part devoted to the magnetooptics of small gap materials we tried to show the role of exchange interactions on the transitions energy. This exchange interaction could produce with the magnetic field a S.C.-M. transition, due to the cross over of the lowest landau levels.

ACKNOWLEDGMENTS

I would like to thank Dr G. Bastard, who was in Erice for the Low Dimensions Summer School at the same time as our school. He had accepted very kindly to give a lecture on the small gap and zero gap S.M.S.C.

REFERENCES

[1] Galaska R.R. Inst. Phys. confiser. 43, 133, 1979.

[2] Galaska R.R. and Kossut J. Lecturers notes in physics 132, 245, 1980, Ed. : Spinger Verlag, Berlin.

[3] Galaska R.R. Proc. Int. Conf. an Narrow Gap Semiconductors, Linz, September 1981, Ed. : Spinger Verlag, Berlin.

[4] Furdyna J.K. J. Appl. Phys. 53, 7637, 1982.

[5] Giriat Wand J.K. Furdyna Semiconductors and Semimetals vol 25, 1, 1988.

[6] Pajaczkowska A. Prog. Crystal Growth Charad 289, 1978.

[7] Furdyna J.K., Giriat W., Mittchell D.F. and Spronle G., J. Solid State Chem. 46, 349, 1983.

[8] Delves R.J. and Lewis B., J. Phys. Chem. Solids 24, 549, 1963.

[9] Balzarotti A., Czyzyk M., Kisiel A., Motta N., Podgorny M. and Zimmal-Starnawska M., Phys. Rev. B 30, 2295.

[10] Villeret M., Rodriguez S. and Kartheuser E., Phys. Rev. B 41, 14, 10028, 1990.

[11] Cohen M.L. and Bergstresser T.K., Phys. Rev. 141, 789, 1966.

[12] Varshni Y.P., Physics 34, 149, 1967.

[13] Diouri J., Lascaray J.P., El Amrani M., Phys. Rev. B 31, 7995, 1985.

[14] Lascaray J.P., Piouri J., El Amrani M., Coquillat D., Solid State Commun 47, 709, 1983.

[15] Rigaux C., Semiconductors and Semimetals, vol 25, 229, 1988. Edited by Academic Press, Inc.

[16] Bastard G., Rigaux C., Guldner Y., Mycielski J. and Mycielski A., J. Physique 39, 87, 1978.

[17] Bastard G., thèse de Doctorat, Paris 1978.

[18] Bastard G., Rigaux C., Guldner Y., Mycielski A., Furdyna J.K. and Mullin D.P., Phys. Rev. B 24, 1981.

[19] Kane E.O., J. Phys. Chem. Sol. 1, 249, 1951.

[20] Luttinger J.M., Phys. Rev. 102, 1030, 1953.

[21] Pidgeon C.R. and Brown R.N., Phys. Rev. 146, 575, 1966.

[22] Pastor K., Grynberg M. and Galazka R.R., Sol. Stat. Com. 29, 739, 1979.

THERMODYNAMICS AND CRYSTAL GROWTH OF DILUTED MAGNETIC SEMICONDUCTORS:

"FROM THE ATOMS TO THE CRYSTALS"

R. TRIBOULET

C.N.R.S.,
Laboratoire de Physique des Solides de
Bellevue
1, Place Aristide Briand - F 92195 MEUDON-CEDEX
(France)

I. INTRODUCTION

Ternary $A^{II} XB^{VI}$, and to a very little extent, quaternary $A^{II}_1 A^{II}_2$ XB^{VI} alloys, with X being essentially Mn, and with a small insight to other transition metals and some rare earths, will be considered in this review. These alloys have drawn considerable attention during the past decade because of their magnetic properties which will be emphasized in other courses.

It is also not useless to repeat this evidence that the properties of DMS, their thermodynamics properties (solubility range, phase equilibra, structure...) their crystal growth, stem from the nature of their constituting atoms ! The approach of this review will be then to follow as close as possible the characteristics of the atoms and of their chemical bonds in view of studying the properties of the alloys made with them.

We will thus first focus our attention on the Mn characteristics and their consequences on the ranges and structures of solid solutions, on the possibilities of solubility of other transition metals and rare earths, on the phase diagrams shapes, on the perfection of the crystals of alloys, on the DMS microscopic structure.

The crystal growth of bulk DMS crystals will be then presented in view of the specific characteristics of "metallurgical" Mn and of the pseudobinaries phase diagrams $A^{II}B^{VI}$ - MnB^{VI}.

Some insight will be provided on the thin layers growth by such topical techniques as Molecular Beam Epitaxy (MBE) and Metal Organic Vapour Phase Epitaxy (MOVPE).

A lot will be borrowed to two remarkable reviews (Pajaczkowska, 1978, Giriat and Furdyna, 1988) separated from ten years, from which the author has extracted most of the content of this course, while adding his own experience, conception and results in the domain. In annexes will be

Semimagnetic Semiconductors and Diluted Magnetic Semiconductors
Edited by M. Averous and M. Balkanski, Plenum Press, New York, 1991

23

presented in some details (i) some crystallographic properties of the $A^{II}B^{VI}$ semiconductors, following the approach of Yoder-Short, Debska and Furdyna, 1985, (ii) the thermodynamics of the CdSe-MnSe system, after Sigai and Wiedemeier, 1974, (iii) an example of the thermodynamics modelling of the chemical vapour transport in the case of the simple system Zn-Se, (iiii) the segregation within three cases : simple normal freezing with reference to (Pb,Mn)Te, normal freezing in solution, with reference to $(Zn_{1-x}Mn_x)_{1-y}Te_y$, solution zone melting, with reference to the THM growth of (Pb,Sn)Te. These annexes and examples are intended to help the crystal grower in the analysis of his results.

II. THE Mn CHARACTERISTICS

a) The Mn-chalcogen bond

Mn belongs to the 7th column of the periodic chart of elements. It is a transition metal with valence electrons corresponding to the $4s^2$ orbitals. It differs from the group II elements by the fact that its 3d shell is only half-filled with 5 electrons instead of 10 in the group II elements. Its $4s^2$ electrons can nevertheless participate in the s-p^3 tetrahedral bonding. Mn can therefore substitutionaly replace the group II elements in the $A^{II}B^{VI}$ structures.

According to Hund's rule, all five spins of the half-filled Mn^{++} 3d orbitals are parallel. Adding an electron with opposite spin to the atom would require considerable energy. In this sense, the $3d^5$ orbits act as a complete shell (like the $4d^{10}$ shell in Cd) and the Mn atom is thus more likely to resemble a group II element in its behavior than other transition-metal elements. That is why Mn forms stable ternary phases over a wide range of compositions while, following Giriat and Furdyna, 1988, other transition metals, even those, as iron, which readily form divalent ions and which are of ionic radius comparable to those of their A^{II} counterpart, are much less miscible in $A^{II}B^{VI}$ compounds.

The stable phase of all the Mn monochalcogenides is of rocksalt structure with octahedral coordination which needs d^2sp^3 orbitals. Since two metastable modifications of the Mn monochalcogenides with sphalerite and wurtzite structures, and tetrahedral coordination, exist, this means that Mn can quite easily change its atomic coordination from octahedral with d^2sp^3 orbitals to the tetrahedral one with s-p^3 orbitals.

In contrast, since all lower d orbitals of Cd are filled, higher d orbitals must be used in order to form a rocksalt type configuration with octohedral environment requiring the use of d^2sp^3 orbitals. Such rocksalt configuration is energetically unfavorable and is only promoted for pure Cd chalcogenides by higher pressure. One can thus expect Cd chalcogenides in the Mn chalcogenides rich rocksalt phase to present higher deviations from ideality than Mn chalcogenides in the Cd chalcogenides rich wurtzite phase where Mn with its lower unfilled d orbitals can easily be found in both octahedral and tetrahedral configurations. Small temperature dependence can thus been expected for the solid solutions of Mn chalcogenides in Cd chalcogenides, and, in contrast, distinct temperature dependence of Cd chalcogenides in Mn chalcogenides.

This approach is consistent with the shape of the three pseudo-binary Cd, Mn (Te, Se, S) phase diagrams where the liquidus on the Cd-rich side are extremely few temperature dependent while they are much more on the Mn side, taking account of the fact that the Mn chalcogenides have always higher melting points than the corresponding Cd chalcogenides, with nevertheless an always moderate temperature difference.

This arguing does not hold any more in the case of the (Zn, Mn) chalcogenides systems where the Zn chalcogenides have always higher melting points than their Mn counterparts, or in the case of the Hg chalcogenides with melting points always much smaller than their Mn counterparts. The (Cd,Mn) (S, Se, Te) phase diagrams present not only a very weak variation of the liquidus temperature as a function of composition over the solid solution range, but also extremely weak two phases fields, that means quasi superposed liquidus and solidus. Mn is found to replace Cd quite ideally, and the mixing entropy in these systems to compensate quite exactly the configuration entropy.

The Mn chalcogenides have been found to be highly ionic under their NaCl structure. A statistical steady increase of the ionicity of the bond of the ternary $A^{II}_{1-x}Mn_xB^{VI}$ alloys when increasing the Mn content has been found by Perkowitz et al, 1988, from infrared phonon data in the $Cd_{1-x}Mn_xTe$ alloys, while the extrapolated values for MnTe is apparently anomalous. This increase of the ionicity, proportional to G, the change in Gibbs free energy, shows that the addition of Mn destabilizes the Cd-Te bond, and the zinc-blende $Cd_{1-x}Mn_xTe$, as proposed by Maheswaranathan et al, 1985, from ultrasonic transit times measurements allowing to determine the elastic constants of (Cd,Mn) Te crystals and confirmed by Guergouri et al from plastic deformation experiments. Maheswaranathan et al attribute the influence of Mn to Mn3d orbitals hybridizing easily with sp^3 Mn-Te bonding orbitals because the Mnd levels lie within the valence band mainly in an energy range slightly higher than that of the sp^3 orbitals, as infered from photoemission data (Webb et al, 1981). The Mn3d - containing hybrids should have somewhat higher energy and less bond strength than the sp^3 orbitals. In contrast, the Cd or Zn d levels do not hybridize with the sp^3 bonding orbitals because their d states have much lower energies than the sp^3 states in CdTe, (Cd,Mn)Te and ZnTe.

We will see later, in the crystal growth part, that this increase of ionicity and destabilization of the Cd-Te bond when adding Mn in CdTe is found to favour an hexagonal phase and to initiate twinning. This influence of ionicity on twinning can be simply physically understood. Along the <111> direction, the zinc blende structure is described by a stacking sequence ABCABC... of lattice sites (Fig. 1). In a purely covalent compound, the positive charges of the cores of the atoms create a repulsive force between atoms situated in A and C sites. The charge transfer between atoms, resulting from an increase of the bond ionicity, induces an attractive coulombic force between atoms located in A and C sites and a tendancy towards a ABABAB... stacking sequence corresponding to the hexagonal structure if systematic, or to a twinning plane if affecting just a plane or a limited number of planes (Fig. 2).

b) **The Mn purity**

The needs of electronic grade Mn are so reduced that commercially available Mn presents only a poor purity of $\sim$ 3N. Mn is only electrolytically extracted under the γ form, which is extremely oxydizable, and not purified further. For the γ Mn, even a fast transfer from chemical etching with HNO_3, to remove the oxide layer, to the crucible gives rise to prohibitive oxidation.

Mn can be purified by sublimation, which allows to get rid of the oxides (especially MnO which is impossible to reduce under H_2 flux even at high temperature) and converts the γ Mn phase to the α Mn which is more resistant to oxidation (Kaniewski et al, 1978). Sublimation, as described by these authors, is carried out under high dynamic vacuum at a temperature about 1000°C. The efficiency of such a process for the impurities particularly troublesome for II-VI's has nevertheless not be

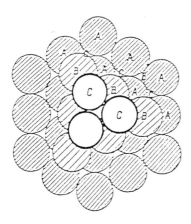

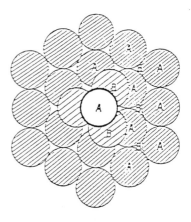

Fig. 1. Stacking sequence of tangent spheres in the cubic (zinc blende) structure.

Fig. 2. Stacking sequence of tangent spheres in the hexagonal (wurtzite) structure.

proved so far. It will be shown later that, in some specific cases as for (Cd,Mn)Te, the problem of the purity and oxidation of Mn can be overcome by making use of particular growth processes.

III. THE DMS STRUCTURE AND MICROSTRUCTURE

A very illustrative "diagrammatic overview" of the $A^{II}_{1-x}Mn_xB^{VI}$ alloys and their crystal structure has been proposed by Giriat and Furdyna, 1988, and is reproduced here Fig. 3. The heavy lines indicate ranges for which ternary alloys with homogeneous crystal phases can be formed, that means cooled to room temperature without precipitation of an extra phase. "Cub" and "hex" indicate the crystal structure, either zinc-blence or wurtzite respectively. The remarkable extension of the $(A^{II}B^{VI}_{1-x})$ - Mn_x solid solution range can be explained from the electronic configuration of Mn as seen above.

The magnetic exchange interactions, as the Mn^{++} - Mn^{++} one, can be deeply affected by the microscopic environment of the atoms, that means by the Mn - anion distance and the angle sustended by the Mn^{++} pair and the anion as well. It is stated by Giriat and Furdyna, 1988, following the approach of Balzarotti et al, 1984, related to (Cd,Mn)Te, that the distortions of the lattice induced by alloying affect essentially the anion sublattice. The Te atoms can see various environments, according to the different possible metals configurations (namely 4Cd, 3Cd + 1Mn, 2Cd - 2Mn, 1Cd - 3Mn, 4Mn) and thus should have to move while the metals are always surrounded by four Te atoms. In contrast with this approach, it has been shown by Baudour et al, 1989, from X diffraction studies, that in (Hg,Zn)Te, the atoms were statistically shifted from their average ideal crystallographic positions with a mean displacement amplitude about 0.08 Å for all the atoms, anions or cations.

IV. PREPARATION OF $A^{II}_{1-x}Mn_xB^{VI}$ ALLOYS

4.1. Bulk crystal growth.

4.1.a. $Cd_{1-x}Mn_xTe$.

The pseudo-binary CdTe-MnTe phase diagram (Fig. 4) (Triboulet et al,

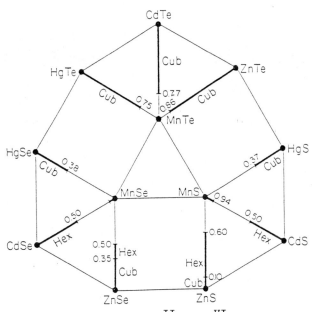

Fig. 3. A diagrammatic overview of the $A_{1-x}^{II}Mn_xB^{VI}$ alloys and their crystal structures. The bold lines indicate ranges of the molar fraction x for which homogeneous crystal phases form. "Hex" and "Cub" indicates wurtzite and zinc blende, respectively, $Cd_{1-x}MN_xS$ AND $Cd_{1-x}Mn_xSe$ also form homogeneous crystalline phases near $x \simeq 1.0$ in the NaCl structure, but these are not relevant to the present article.

1970) presents several unique features which have made (Cd,Mn)Te probably the most extensively studied system :

        (i) the solid solutions range extends from x = 0 to $\sim$ 0.75. A gap of miscibility occurs for x exceeding this limit

        (ii) liquidus and solidus merge over almost the whole solid solution range, expressing a distribution coefficient of Mn close to one : homogeneous crystals of the same composition as the starting liquid phase can be grown from classical normal freezing techniques as Bridgman one

        (iii) the "melting point" decreases only slightly as the Mn content increases and stays in the range of easy use of silica tubes.

        A phase transition in the solid state is found within the solid solution range, with a transition temperature decreasing with increasing Mn content, from a high temperature α' phase, which is thought likely of hexagonal structure, down to a the low temperature ZB α one. The crossing of the $\alpha' - \alpha$ transition line during the cooling process is probably at the origin of the important twinning found in (Cd,Mn)Te crystals grown by Bridgman (Fig. 5) or by High Pressure Liquid Encapsulation Czochralski (HP - LEC) (Hobgood et al, 1987) as studied by Wu and Sladek, 1982, from X diffraction, etching and optical transmission measurements.

        A growth process at a temperature lower than the transition one, by the Travelling Heater method, from source materials grown by the vertical Bridgman technique in vitreous carbon crucibles contained in silica

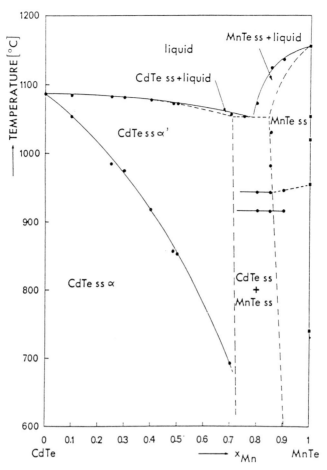

Fig. 4. Pseudo binary CdTe-MnTe phase diagram.

ampoules preventing any reaction between Mn oxide and silica, allows not only to get rid of twinning but also to compensate for the poor purity of commercially available Mn.

The THM crystals are much more transparent than the Bridgman ones inside which a laser beam is clearly seen to diffract indicating a high amount of internal scattering, contrary to the THM ones. The THM crystals benefit also from the distribution coefficient of Mn close to one and although the segregation which should occur during THM should be added up to the Bridgman one, they are found to be very homogeneous (0.005 mole/cm for x ~ 0.5). They are highly purified by the THM solution zone refining process beside the Bridgman charges as shown in table 1 in which are compared results of emission spectroscopy between a Bridgmam ingot and the THM one grown using it as source material. The amount of impurities brought by commercial Mn appears striking with hundreds of ppm of various elements. THM purification is siginificant for most of the impurities detected, except for Cu, the distribution coefficient of which is probably close to one. Finally, the THM crystals appear to be almost

Table 1. Emission spectroscopy analysis of elements (im ppm) detected in
a Bridgman ingot and in the THM ingot using it as source material.

elements	Bridgman	T.H.M.
Ca	50	4
Fe	100	3
Mo	100	0,3
V	100	0,1
Cr	20	
Mg	20	
As	50	0,1
Na	20	
Cu	5	3
Sn	20	0,3
B	50	0,1
Be	1	
Al	50	4
Pb	10	
Ti	50	0,1
Li	1	0,1
K	1	
Sb	10	0,1
Ni	50	3
Co	50	0,1

twin-free as confirmed by Laue back reflection spectra (Fig. 6) taken
normal to the (110) cleavage plane. THM crystals of any composition just
show the 2 mm symmetry expected for the ZB structure while Laue patterns
taken on Bridgman crystals contain, in addition to the reflections
expected for the ZB structure, some extrareflections which have been
attributed to twinning.

Although microhardness measurements over the whole CdTe-MnTe solid
solution range (Fig. 7) (Triboulet et al, 1990) confirm that Mn
destabilizes the Cd-Te bond, as shown from ultrasonic transit times as a
function of pressure on (Cd,Mn)Te crystals (Maheswaranathan et al, 1985)
and effective charge and ionicities measurements (Perkowitz et al, 1988),
the THM (Cd,Mn)Te crystals are found of good crystalline quality with
rocking curve halfwidths in the 30 arcsec range. Such crystals could be
used as substrate for the epitaxic overgrowth of the Hg-based ternary
alloys as (Hg,Cd)Te or (Hg,Zn)Te. They cover a wide range of lattice
constants according to the expression $a (\overset{\circ}{A}) = 6.482 - 0.150 \ x_{Mn}$

4.1.b. $Cd_{1-x}Mn_xSe$.

The pseudo-binary CdSe - MnSe phase diagram, investigated by Cook
(1968) presents great similarities with the CdTe - MnTe one : a solid
solution range which extends from x = 0 to x $\sim$ 0.5, but with a crystal
structure of wurtzite type, with a miscibility gap for x exceeding this
limit, a near coïncidence of the liquidus and solidus curves, with a
melting point starting from 1260°C for pure CdSe and decreasing very
slightly with increasing Mn content (Fig. 8).

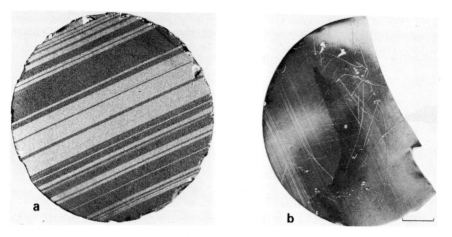

Fig. 5. Sectional view of (Cd,Mn)Te Bridgman ingots (a) x_{Mn} = 0.25; (b) x_{Mn} = 0.50 (X ray topography).

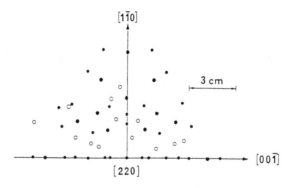

Fig. 6. One half of a Laue pattern of the (110) plane in single crystals of $Cd_{1-x}Mn_xTe$, x = 0.25 ● THM crystal ; ● + o Bridgman crystal. Film to sample distance, d = 5.5 cm.

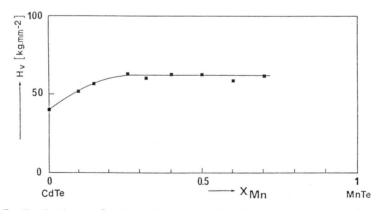

Fig. 7. Variation of the Vickers microhardness as a function of composition in the CdTe-MnTe system.

1260° makes still possible the use of thick walled silica ampoules. The vertical Bridgman technique can still be used successfully to grow $Cd_{1-x}Mn_xSe$ crystals. To avoid any violent reaction during synthesis from the elements, the use of prereacted CdSe and MnSe compounds is recommended. They are prepared by slowly heating stoichiometric amounts of high purity elements, loaded in silica ampoules, up to its melting point for CdSe and up to 1000°C during several days for MnSe, with a "cold point" (cooler end of the tube) at about 700°C to avoid any excessive and detrimental increase of the pressure in the tube during the synthesis reaction.

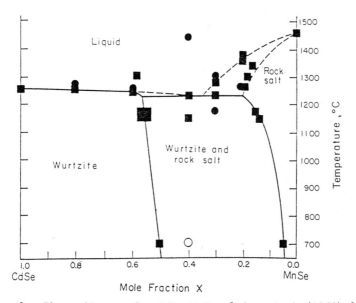

Fig. 8.   Phase diagram for CdSe–MnSe. [After Cook (1968).]

The wurtzite structure is favourable to grow routinely large (Cd,Mn)Se macrocrystals, while the near-coincidence of liquidus and solidus makes them uniform along the direction of growth.

The THM growth of (Cd,Mn)Se which should allow to solve the problem of the Mn purity, as in the case of the (Cd,Mn)Te alloys, has not so far been reported. It could be carried out either in Cd-rich solutions or in Se-rich solutions in the 1000-1100°C temperature range, in thick walled silica ampoule. It appears nevertheless to be a borderline case of application of this growth technique.

4.1.c. $Cd_{1-x}Mn_xS$

The CdS-MnS phase diagram, determined by Cook (1968) and Wiedemeier and Khan (1968) displays a general shape similar to the two previous (Cd,Mn) chalcogenides diagrams. A solid solution range with wurtzite structure extending from $x_{Mn} = 0$ to ~ 0.5, followed by a miscibility gap, the solidus and liquidus practically indistinguishible on the CdS-rich side of the diagram, with the melting point decreasing very slightly as x_{Mn} increases (Fig. 9). The essential difference are the high melting points (1400°C for CdS) higher than the softening point of silica, which prevents from using it. The growth of (Cd,Mn)S crystals by the Bridgman method under high inert gas pressure (40 - 100 kg/cm^2) in carbon crucibles has been reported (Ikeda et al, 1968) ; Komura and Kando, 1975). Compressed pellets of finely powdered MnS and CdS samples were used as starting charges. CdS and MnS can be prepared from stoichiometric mixtures by heating the ampoule to about 1000°C for both cases with the cooler end at ~ 450°C. The Bridgman crystals of (Cd,Mn)S, of wurzite structure, are quite homogeneous along the growth axis as a consequence of the near coincidence of liquidus and solidus. $Cd_{1-x}Mn_xS$ crystals of 1 cm^3 in size have also been grown from the melt by vertical Bridgman using induction heating and self-sealing graphite crucibles as reported by Debska et al, 1984. Chemical transport methods are particularly appropriate in a so high temperature range. Iodine can be used as the transport agent (~ 10 mg/cm^3) from mixed pure binary compounds in the desired ratio as source material. After one week at 950°C, small crystals of good quality are reported to be obtained, (Giriat and Furdyna, 1988) of composition close to that of the starting material.

Pajaczkowska and Rabenau, 1978, obtained (Cd,Mn)S crystals by High Pressure Solution (HPS) from the binary compounds as starting material and 1 M HCl saturated with H_2S as a solvent. The autoclave was filled up to 30 % and the temperature was 450°C. After three days, crystals smaller than 0.1 mm with 26 mole % MnS and wurtzite type structure were obtained.

Small size and poor purity of the crystals show definitely the limits inherent to these techniques of "chemical" growth.

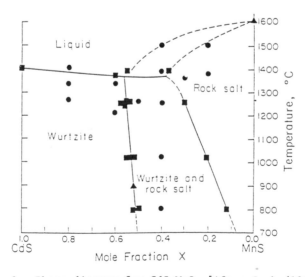

Fig. 9. Phase diagram for CdS–MnS. [After Cook (1968).]

4.1.d. $Zn_{1-x}Mn_xTe$

The phase diagram of the ZnTe-MnTe system has not been established so far. It is clear that the high temperature required for the growth of (Zn,Mn)Te crystals explain why this system has been less studied than the (Cd,Mn)Te one. A single phase solid solution of ZB structure over a wide range of Mn concentration ($\sim$ 0.85) has been reported (Pajaczkowska, 1978).

The general approach used in the growth of (Cd,Mn)Te crystals, as described above, can be applied to the (Zn,Mn)Te alloys. $Zn_{1-x}Mn_xTe$ crystals have been prepared by the vertical Bridgman technique, using tellurium solution with the aim to decrease the liquidus temperature (Desjardins-Deruelle et al, 1986).

The starting $Zn_{1-x}Mn_xTe$ mixtures, loaded in vitreous carbon crucibles contained in silica ampoules, were of general formula $(Zn_{1-x}Mn_x)_{0.4}Te_{0.6}$, with x = 0.05, 0.1, 0.2, 0.4, 0.7. The composition profiles of the ingots associated with these starting materials are presented in Fig. 10. The segregation, which can be important for the high Mn concentrations expresses that, contrarily to the (Cd,Mn)Te case, the liquidus and solidus are not in near coincidence in the (Zn,Mn)Te system, but that the width of the two phases field increases, as usual, when increasing the Mn content.

The segregation can be calculated, in such a normal freezing process in solution, according to the approach proposed in annex D. In contrast to (Cd,Mn)Te, the (Zn,Mn)Te crystals do not present any lamellar twinned structure, as a result of the weaker ionicity of the Zn-Te bond beside the Cd-Te one.

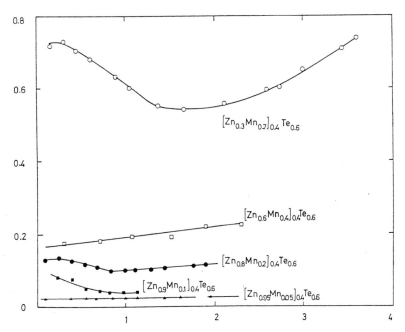

Fig. 10. Composition profile of (Zn,Mn)Te Bridgman ingots, along their growth axis, for different concentrations of the starting materials.

The Travelling heater Method has been nevertheless applied to some ingots in order to solve the problem of the Mn purity. The composition profile of an ingot grown by THM from a Bridgman grown source material (with $x_{Mn} = 0.4$) is displayed in Fig. 11.

The THM zone refining process emphasizes the segregation of Mn as expected, but is found to improve highly the purity of the crystals. The THM grown (Zn,Mn)Te crystals have been found more generally of distinctly better quality than their Bridgman counterparts.

4.1.e. $Zn_{1-x}Mn_xSe$

The maximum solubility of MnSe in ZnSe is about 0.57. According to Juza et al, 1956, the solid solution exhibits the ZB structure up to 35 mole % MnSe and then the wurtzite form in the $0.35 < x < 0.57$ composition range. As predictable (cf (Cd,Mn)Te) considerable twinning is observed in the ZB phase, increasing in density as x approaches the transition value of ~ 0.30. Inside the miscibility gap, for $x > 0.57$, two phases, with wurtzite and rocksalt structures (α MnSe), are observed in the mixture. No solubility of ZnSe in MnSe was detected.

Pre-reacted binary ZnSe and MnSe are commonly used as starting materials. ZnSe and MnSe can be prepared by several different methods. The synthesis of stoichiometric mixture loaded in evacuated sealed silica ampoules can be initiated manually by heating with a gas torch, with visual control of the reaction. The reaction is then completed by homogeneous heating at 1000°C during several days. Alternatively, the stoichiometric mixture of the elements can be slowly brought up to 1000°C, with the cooler end of the ampoule at $\sim$ 700°C, to avoid excessive pressures. Several days can also be necessary to achieve a complete reaction.

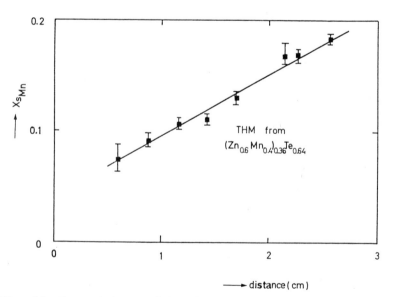

Fig. 11. Composition profile along the growth axis of a THM ingot grown using a Bridgman ingot as source material.

Pressed pellets of mixed powder of the two binaries, in desired proportions, are then reacted for several days in the same temperature range. The resulting sintered material can be used as starting source for crystallization.

Given the high melting points of the alloys (over 1500°C throughout the entire solid solution range), chemical transport turns out to be the simplest method for single crystals growth. Iodine can be used, as usual, as the transport agent ($\sim$ 10 mg/cm^3). The charge is brought up to 1000°C with a critical temperature difference $\Delta T_{crit.}$ of $\sim$ 20°C. The resulting crystals are of rather small size, as usual with such a method.

As in the case of ZnSe, the Bridgman technique under an inert gas pressure allowing to reduce the sublimation of the charge can be achieved. But it necessitates the use of a rather complex and expansive equipment. Crystals of $Zn_{1-x}Mn_xSe$ with x up to 0.10 have been grown by Twardowski et al, 1983.

An elegant method of "self sealing graphite crucible" has also been proposed by Debska et al, 1984, which eliminates completely the need of silica ampoules. According to this method, schematized in Fig. 12, the crucible assembly consists of a closed graphite tube into which is fitted a long threaded plug. This graphite crucible is inverted, after loading the powdered charge, so that the plug comes at the bottom.

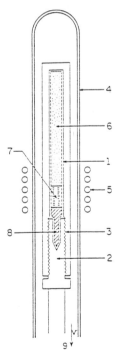

Fig. 12. Self-sealing graphite crucible used the gowth of $Zn_{1-x}Mn_xSe$: (1) graphite crucible; (2) threaded plug; (3) thread; (4) outside silica tube; (5) induction heating coils; (6) powdered material; (7) molten zone; (8) crystallized material; (9) exit to dynamic vacuum. [After Debska et al. (1984).]

An induction coil, coupled with the graphite crucible which acts as the susceptor, allows to increase slowly the temperature. Beyond the "melting point" of the charge, some of the material flows down, solidifies in the space between crucible wall and plug, thereby sealing the crucible.

A molten zone is then achieved by RF heating and moved up along the charge, at a rate of about 10 mm/h. The process is conducted under dynamic vacuum in the 10^{-4} - 10^{-6} Torr range. Single crystals of $Zn_{1-x}Mn_xSe$ have been produced over the whole solid solution range. Ingots are typically 9 mm in diamter, 70 to 90 mm in length, with crystals about 5 to 10 mm in length, exceptionaly up to 5 to 8 cm.

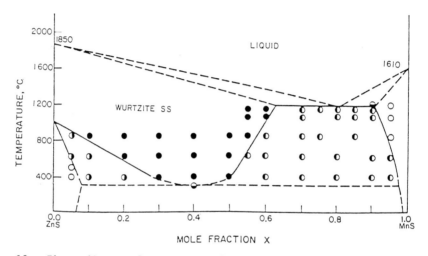

Fig. 13.   Phase diagram for ZnS-MnS. [After Sombuthawee et al. (1978).]

4.1.f. $Zn_{1-x}Mn_xS$

The pseudo-binary ZnS-MnS phase diagram shown in Fig. 13 has been investigated and proposed by Sombuthawee et al, 1978. Solid solutions of ZB structure exist for x < 0.1 and of wurtzite structure for 0.1 < x < 0.6, above the transition points of ZnS and its solid solutions with Mn.

ZnS and MnS have respectively the highest melting points of $A^{II}B^{VI}$ compounds and of the Mn chalcogenides (1722°C and 1530°C). Chemical transport is thus quite appropriate in such a case for the growth of $Zn_{1-x}Mn_xS$ crystals. The conditions are similar to those described for $Zn_{1-x}Mn_xSe$. Monocrystals of $\sim$ 0.5 cm³ can be obtained by this process, with x < 0.15. The crystals tend to be smaller for higher x values (Nitsche et al, 1961).

The hydrothermal growth (HPS) can also be used, starting from the binary compounds and 1M HCl saturated with H_2S as a solvent as described by Pajaczkowska and Rabenau, unpublished.

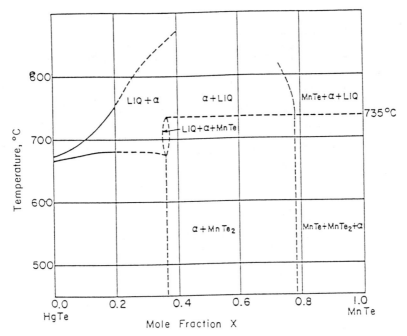

Fig. 14. Phase diagram for HgTe—MnTe. [After Delves and Lewis (1963).]

Finally, in spite of the very high melting temperatures of this system, Bridgman growth under inert gas pressure is not out of the performances of the furnaces used in the ZnSe case and could nevertheless be applied to $Zn_{1-x}Mn_xS$.

4.1.g. $Hg_{1-x}Mn_xTe$

The phase diagram of this system, investigated and proposed by Delves and Lewis, (1963), is shown in Fig. 14.

Solid solutions of zinc blende structure exist over a very wide range of composition, up to about 80 % MnTe. Nevertheless the presence of $MnTe_2$ is found in alloys containing more than 35 % mole MnTe.

In such a low temperature range (670 - 800°C) the growth of crystals of $Hg_{1-x}Mn_xTe$ by vertical Bridgman is quite appropriate. These alloys present several similarities with the $Hg_{1-x}Cd_xTe$ ones :

(i) a high segregation occurs in their Bridgman growth as expected from the shape of the HgTe-MnTe phase diagram.

(ii) the use of alternative growth techniques as THM (Dziuba, 1979) or quench anneal solid state recrystallization allows also to reduce the segregation found after the use of the classical normal freezing techniques.

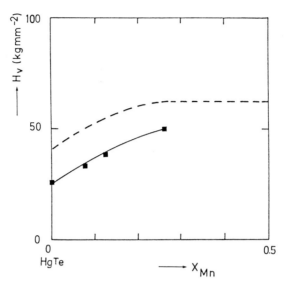

Fig. 15. Vickers microhardness in the HgTe-MnTe system, as a function of composition for the Hg-rich solutions. The dashed curve shows by comparison the correspondng variation in the CdTe-MnTe system.

(iii) the $Hg_{1-x}Mn_xTe$ crystals tend to crystallize with a slight Hg deficiency and then to be p-type. Classical stoichiometric annealing (200°C under saturated mercury vapour pressure for very long times) allow to restore stoichiometry. As in the case of the (Cd,Mn)Te aloys, the variation of the microhardness of the (Hg,Mn)Te alloys as a function of composition as shown in Fig. 15 (Triboulet, unpublished work), although given only for a reduced composition range, expresses a very weak solution hardening probably related to a destabilization of the Hg--Te bond when adding Mn, as suggested from the study of the thermal stability of (Hg,Mn)Te/CdTe superlattices (Staudenmann et al, 1987). On the same graph is shown by comparison the variation of the microhardness of the $Cd_{1-x}Mn_xTe$ alloys as a function of composition (dashed line). Both hehaviors are quite similar.

4.1.h. $Hg_{1-x}Mn_xSe$

The phase equilibrium in solid and liquid state of the HgSe-MnSe system has been investigated by Pajaczkowska and Rabenau, 1977 from lattice constants measurements on pressed samples, heated at 700°C, of stoichiometric mixtures of HgSe and MnSe.

The solubility limit of solid phase is 38.5 mole % MnSe at 700°C. These authors determined the solidus-liquidus equilibrium from DTA measurements, and proposed the phase equilibrium diagram shown in Fig. 16.

Given the temperatures brought into play, the vertical Bridgman technique can still be used as in the case of the (Hg,Mn)Te alloys.

4.1.i. $Hg_{1-x}Mn_xS$

Pajaczkowska and Rabenau, 1977, investigated this system and determined the range of solubility from lattice constants measurements. They found that $Hg_{1-x}Mn_xS$ forms single-phase solid solutions of zinc-blende structure with a limit of solubility of 37.5 % MnS at 600°C. They determined the conditions of synthesis and crystal growth in hydrothermal process (HPS) using aqueous solutions of HCl/H₂S, NH₄Cl and NH₄SCN and grew crystals up to 2 mm of platelet edge.

The stable phase of HgS being cinnabar below 300°C, Mn is found to stabilize the zinc blende structure and thus to favour the tetrahedral coordination.

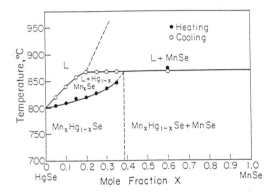

Fig. 16. Phase diagram for HgSe-MnSe (after Pajaczkowska and Rabeneau (1977a).

4.1.j. $Hg_{1-x-y}Cd_xMn_yTe$

A supplementary degree of freedom exists in these quaternary alloys beside the ternary ones : different values of the band gap can be obtained for a same amount of Mn, depending on the Hg/Cd ratio, while alloys with different Mn concentrations can present the same band gap. The interest for these alloys is thus stimulated by the ability to independently control band parameters and the exchange contribution.

Technological difficulties in the preparation of homogeneous crystals have so far limited progress in the research of these quaternary II-VI alloys. Crystals of good quality but of very weak Mn concentration ($y < 0.022$) have been grown by means of a modified zone melting method (Kim et al, 1982). Epilayers with a composition gradient perpendicular to the surface have been prepared by the close spaced isothermal method (Debska et al, 1981).

Homogeneous crystals have been obtained by a modified evaporation-condensation-diffusion method with Cd in the range $x = 0.14 - 0.27$ and Mn in the range $y = 0.016 - 0.115$ (Piotrowski, 1985). Compositional variations along the growth direction do not exceed the values $\Delta x = 0.025$ and $\Delta y = 0.015$ in these crystals.

The possibility of using the Traveling Heater Method, according to a technological arrangment described by Triboulet et al, 1985, for the growth of homogeneous (Hg,Cd)Te crystals and by Triboulet et al, 1986, for (Hg,Zn)Te, has been demonstrated for the growth of (Hg,Cd,Mn)Te crystals (Triboulet, unpublished work).

The source material is a cylinder consisting of two cylindrical segments - one HgTe, the other (Cd,Mn)Te - whose cross sections are in the ratio corresponding to the desired composition. The use of (Cd,Mn)Te cylindrical segments of uniform composition is made possible by the absence of segregation in this system, as shown in 4.1.a.

$Hg_{1-x-y}Cd_xMn_yTe$ crystals with x in the range 0.4 - 0.25 and y in the range 0.3 - 0.25 have been grown. If the Mn segregation has been found very weak in these crystals (the segregation of Mn is very weak in CdTe but higher in HgTe), the presence of Mn is found to enhance the segregation of Cd. The longitudinal composition profile (from electron microprobe analysis), as shown in Fig. 17, presents a plateau, corresponding to a steady state where the composition of the ingot matches the composition of the source material, only in the last centimeter over a whole length of five centimeters.

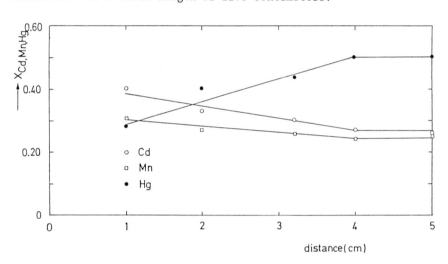

Fig. 17. Composition profile along the growth axis of a quaternary (Hg,Cd,Mn)Te ingot grown by THM.

4.2. 2D DMS, films and superlattices

2D DMS microstructures appear as an emerging field, attracting great interest. Numerous studies are devoted to the preparation and study of DMS films, heterostructures, and superlattices. Particularly studied are the CdMnTe/CdTe multiquantum well structures and SL's (Bicknell et al, 1967), (Gunshor et al, 1985), and to a lesser extent, the HgMnTe/CdTe (Chu et al, 1987) and ZnMnSe/ZnSe ((Gunshor et al, 1986) SL's, all grown by MBE.

DMS films have also been grown by ALE, an equilibrium growth method, in contrast to MBE (Herman et al, 1984, Tammenmaa et al, 1985).

The MBE growth of such binary compounds of NiAs structure as MnTe or MnSe on substrates of zinc blende structure has also allowed to stabilize their "hypothetical" zinc blende phase (Gunshor et al, to be published). The epitaxy opens thus the way to entirely new systems that cannot be obtained in the bulk.

Multilayers structures of $Cd_{1-x}Mn_xTe$, $Hg_{1-y}Mn_yTe$, $Hg_{1-x-y}Mn_yCd_xYe$ and $Cd_{1-x}Mn_xTe/CdTe$ have been grown recently at low temperature by MOVPE using MeMn $(CO)_5$ (Pain et al, 1990).

It is clear that, as in the case of the growth of non magnetic semiconductors, these techniques are destined for considerable development.

V. MISCELLANEOUS DMS COMPOUNDS

5.1. $A^{II}_{1-x}Fe_xB^{VI}$ alloys

While the $A^{II}_{1-x}Mn_xB^{VI}$ DMS have attracted great attention for their magnetic properties since about ten years, the DMS including iron atoms in selenium and tellurium II-VI compounds have only recently begun to emerge. The presence of Fe^{++} ions in the lattice not only leads to magnetic properties quite different that with Mn^{++} but directly influences the electrical properties of the related alloys.

For the reasons discussed in 2.a, (prohibitive energy to provide to add an electron with opposite spin), the solubility of Fe in II-VI compounds is considerably less than the Mn one. It appears to be considerably greater in the selenium based II-VI compounds than in the tellurium ones, as illustrated in Table 2.

Table 2

MATERIALS	SOLUBILITY %	GROWTH TECHNIQUE	REFERENCES
$Hg_{1-x}Fe_xTe$	0.06	bulk	Guldner et al, 1980
$Cd_{1-x}Fe_xTe$	0.02	"	Myczielski
$Zn_{1-x}Fe_xTe$	0.05	"	
$Hg_{1-x}Fe_xSe$	0.12	"	Vaziri et al, 1985
$Cd_{1-x}Fe_xSe$	0.15	"	Lewicki et al, 1986
$Zn_{1-x}Fe_xSe$	0.20	"	
"	1	MBE	Jonker et al 1988

Such alloy as $Zn_{1-x}Fe_xSe_x$, and possibly other systems, has been shown to present an exceptional predisposition to be grown by MBE : layers with x ranging from 0 to 1 have been grown by this technique (Jonker et al, 1988). This property could give new insights in the study of these alloys and into the MBE growth dynamics.

In order to make use of the most appropriate techniques of crystal growth for the $A^{II}_{1-x}Fe_xB^{VI}$, the phase diagrams of the $A^{II}B^{VI}$ - FeB^{VI} systems, and more generally of the ternary A^{II} - Fe - B^{VI} systems, are still to be completely investigated. They will allow to give accurately

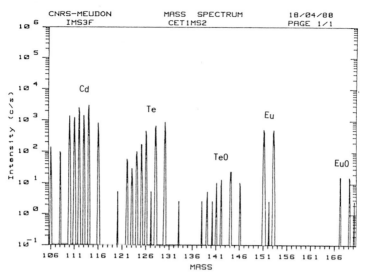

Fig.18. SIM's spectrum of an Eu-doped CdTe crystal

the limits of solubility. Up to now, in the absence of a perfect knowledge of the phase equilibria, the $A^{II}_{1-x}Fe_xB^{VI}$ alloys are generally submitted to bulk growth methods similar to those of their Mn counterparts (Mizera et al, 1980).

5.2. Rare earths in II-VI's

Diluted magnetic semiconductors of general formula $A^{II}_{1-x}RE_xB^{VI}$, with RE for rare earth, could be of great interest. The incomplete 3d orbitals of the transition metals, as Mn, are here replaced by the incomplete 4f ones.

According to the approach followed with Mn, among the RE gadolinium presents an exactly half filled 4f orbital, with all seven spins parallel in accordance to Hund's rule, which can in this sense be considered as complete. It most resembles group II elements and should thus present the highest solubility in $A^{II}B^{VI}$ compounds. However the factor limiting solubility of rare earths in II-VI's is their large atomic radius.

Europium has been successfully introduced in CdTe, either under its atomic form or using EuTe by Bridgman technique from a Te-rich melt of formula $Cd_{0.75}Eu_{0.05}Te_{1.2}$ in vitreous carbide crucibles contained in silica ampoules, with a maximum furnace temperature of 975°C and a growth rate of 250 /h.

The solubility of Eu in CdTe has been estimated from SIM's measurements (Fig. 18) and electron microprobe analysis to be < 0.1 % (Triboulet, unpublished).

The CdTe -Gd_2Te_3 equilibrium phase diagram has been studied by Agaev et al, 1983, by physicochemical methods, Fig. 19. A solubility of Gd_2Te_3 about 1.5 % in CdTe at room temperature has been stated from X diffraction measurements.

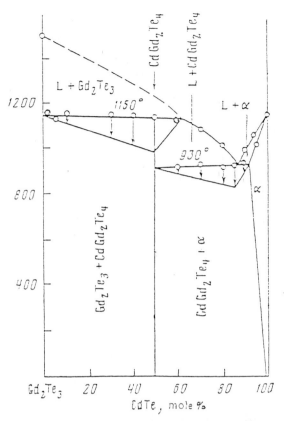

Fig. 19. Equilibrium diagram of the CdTe-Gd$_2$Te$_3$ system.

Amounts of Gd < 0.5 % have been introduced from Gd$_2$Te$_3$ in CdTe according to the same process used for Eu (Triboulet, unpublished). The HgTe-EuTe phase diagram has been studied by Gavaleshko et al, 1985 (Fig. 20). Although very incomplete, it could let expect a certain solubility of Eu in HgTe. Bridgman growth has been performed from (HgTe)$_{0.95}$(EuTe)$_{0.05}$ liquids. The resulting crystals are inhomogeneous and contain many phases : a HgTe matrix with irregular EuTe features and a quite regular dotted network of spherical Hg$_{1-x}$Eu$_x$Te droplets 1 μm of less diameter with various x, ranging in a wide composition range (0.07 to $\sim$ 0.56) (Fig. 21).

The dynamics of EuTe dissolution in liquid HgTe are probably very low (El Hanany and Triboulet, unpublished).

5.3. Mn and rare earths in IV-VI

Literature reveals an increasing interest for Pb$_{1-x}$Mn$_x$B^{VI} alloys. The miscibility of Mn in PbB^{VI} compounds appears to be smaller than in A^{II}B^{VI} materials.

The crystal growth techniques commonly used to prepare these alloys are Bridgman growth for bulk crystals and hot wall epitaxy for thin epitaxial layers with growth parameters similar to those successfully used for the parent lead salts binaries.

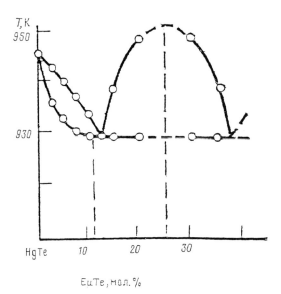

Fig. 20 HgTe-EuTe pseudo binary phase diagram

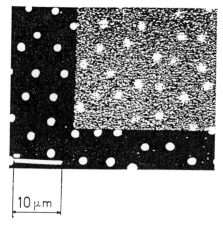

Fig. 21 Electron microprobe image of a Eu doped HgTe crystal

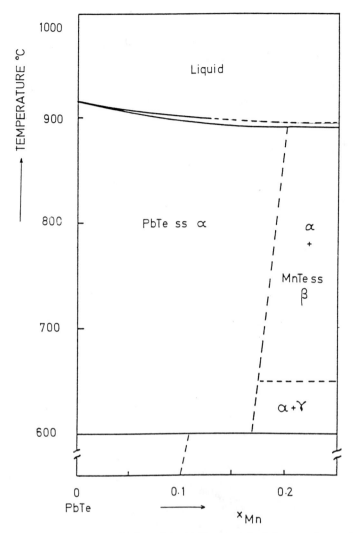

Fig. 22 Pb-rich side of the PbTe-MnTe pseudo binary phase diagram

The PbTe-MnTe phase diagram has been partially studied by DTA, X diffraction and electron microprobe measurements (Triboulet and Guergouri, unpublished) to about 25 % Mn. This diagram displays a solid solution of cubic structure extending from 0 to about 0.1 as shown in Fig. 22.

These results are confirmed by Bridgman crystal growth experiments and X diffraction studies. The compositional profiles of $Pb_{1-x}Mn_xTe$ ingots grown with x = 0.03 and x = 0.1 can be fitted by the classical Pfann expression (Annexe D) $C_s = kCo\ (1-g)^{k-1}$ accounting for segregation in normal freezing, leading to segregation coefficients respectively of 0.8 and 0.85, as expressed from the narrow two phases field (Fig. 23). On the contrary, the compositional profiles with x = 0.15 and x = 0.2 cannot be fitted by Pfann's expression indicating that the corresponding alloys do not present a single phase.

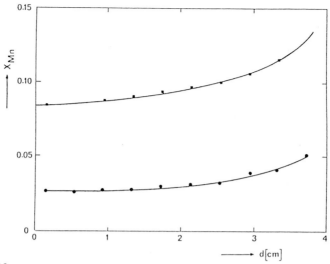

Fig. 23 Mn concentration profile along growth axis of a (Pb,Mn)Te
Bridgman ingot. Black dots experimental points ; curve : fit of the
experimental points using Pfann's expression

The variation of the lattice parameter as a function of composition
is displayed in Fig. 24. The Vegard's law is only followed between 0 and
0.10.

Excellent $Pb_{1-x}Mn_xTe$ crystals, as revealed by all their studied
physical properties, have been grown by the Travelling Heater Method,
using Te as the solvent, at a growth temperature of 600°C (Triboulet,
1985, unpublished). In such tellurium rich conditions, the maximum
solubility of Mn has been found to be 0.054 at this growth temperature.

As for the $A^{II}_{1-x}Fe_xB^{VI}$ alloys, the lead salt alloys including rare
earths show extended miscibility range when grown by MBE, as x $\sim$ 0.40
for $Pb_{1-x}Yb_xTe$ (Partin, 1983) compared to non-MBE growth.

As in the case or Mn-based lead salts, Bridgman growth is
classically used for bulk crystals preparation, and hot wall epitaxy for
thin films ($Pb_{1-x}Eu_xTe$, $Pb_{1-x}Cd_xTe$).

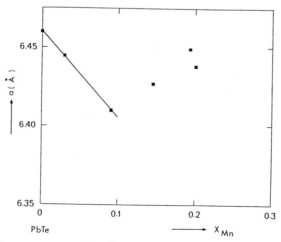

Fig. 24 Lattice parameter in the Pb rich side of the PbTe-MnTe system.

REFERENCES

AGAEV A.B., ZUL'FUGARLY DZh.I, ALIEV O.M. and AZADALIEV R.A.
Russian Journ. Inorg. Chem. 28(1) (1983) 142.

BALZAROTTI A., CZYZYK M., KISIEL M., MOTTA A.
Phys. Rev. B30 (1984) 2295.

BAUDOUR J.L., GRANGER M.M., TOUPET L., GRANGER R. and TRIBOULET R.
J. Phys. Chem. Solids 50, 2 (1989) 309.

BICKNELL R.N., GILES N.C., SCHETZINA J.F.
Appl. Phys. Lett. 50 (1987) 691.

COOK W.R.
J. American Ceramic Soc. 51 (1968) 518.

DEBSKA U., DIETL M., GRABECKI G., JANIK E., KIEREK-PECOLD E. and
KLIMKIEWICZ M.
Phys. Stat. Solidi (a) 64 (1981) 707.

DEBSKA U., GIRIAT W. HARRISON H.R. and YODER-SHORT D.R.
J. Cryst. Growth 70 (1984) 339.

DELVES R.T. and LEWIS B
J. Phys. Chem. Solids 24 (1963) 549.

DESJARDINS-DERUELLE M.C., LASCARAY J.P., COQUILLAT D. and TRIBOULET R.
Phys. Stat. Sol. (b), 135 (1986) 227.

DZIUBA Z.E.
J. elektrochemical Society 116 (1969) 104.

GARY SIGAI A. and WIEDEMEIER H.
J. Chem. Themodynamics 6 (1974) 983.

GAVALESHKO N.P., KAV'YUK P.V., LOTOTSKII V.B., SOLONCHUK L.S.,
KHOMAYAK V.V.
Izv. Akad. Nauk SSSR, Neorg. Mater. 21 (1985) 1965.

GILLE P.
Cryst. Res. Technol. 28 (1988) 481.

GIRIAT W. and FURDYNA J.K.
Semicond. and Semimetal, Vol. 25, Acad. Press Inc. (1988).

GULDNER Y. RIGAUX C., MENANT M., MULLIN D.P. and FURDYNA J.K.
Solid State Comm., 33 (1980) 133.

GUNSHOR R.L., OTSUKA N., YAMANISHI M., KOLODZIEJSKI L.A.
BONSETT T.C., BYLSMA R.B., DATTA S., BECKER W.M. and FURDYNA J.K.
J. Cryst. Growth, 72 (1985) 294.

GUNSHOR R.L., KOBAYASHI M. KOLODZIEJSKI L.A., NURMIKKO A.V.
J. Cryst. Growth, to be published.

GUNSHOR R.L., KOLODZIEJSKI L.A., OTSUKA N., CHANG S.K., NURMIKKO   A.V.
J. Vac. Sci. Technol. A (1986) 4, 2117.

HERMAN A., JYLHA O.J. and PESSA M.
J. Crystal Growth 66 (1984) 480.

HOBGOOD H.M., SWANSON B.W. AND THOMAS R.N.
J. Crystal Growth 85 (1987) 510.

IKEDA M. ITOH H. and SATO H.
J. Phys. Soc. Japan 25 (1968) 455.

JONKER B.T., KREBS J.J., QADRI S.B. and PRINZ G.A.
Appl. Phys. Lett. 50 (1987) 848.

JUZA R., RABENEAU A. and PASCHER G.
Z. Anorg. allg. Chem. 19 (1961) 229.

KANIEWSKI J., WITKOWSKA B., GIRIAT W.
J. Crystal Growth 60 (1982) 179.

KIM R.S., MUITA Y., TAKEYAMA S. and NARITA S.
in Proc.4th Intern. Conf. on the Physics of Narrow Gap Semicond.

LINZ (1981) Lecture Notes in Physics, Vol. 152 (Springer, Berlin
1982) 316.

KOMURA H. and KANDO Y.
J. Appl. Phys. 46 (1975) 5294.

LEWICKI A., MYCIELSKI A. and SPALEK J.
Acta Phys. Polon. A 69 (1986) 1043.

MAHESWARANATHAN P., SLADEK R.J. and DEBSKA U.
Phys. Rev. B, 31, 8 (1985) 5212.

MIZERA E., KLIMKIEWICZ M., PAJACZKOWSKA A. and GODWOOD K.
Phys. Stat. Sol.(a) 58 (1980) 361.

NITSCHE R., BOLSTERLI H.U. and LICHTENSTEIGER M.
J. Phys. Chem. Solids 21(1961) 199.

PAIN G.N., BHARATULA N., CHRISTIANSZ G.I., KIBEL M.H., SANDFORD C.,
DICKSON R.S., ROWE R.S., Mc GREGOR K. DEACON G.B., WEST B.O.,
GLANVILL S.R., HAY D.G., ROSSOW C.J. and STEVENSON A.W.
Proc. 5th Int. Conf. II-VI Compounds, J. Cryst. Growth, (1990) to
be published.

PAJACZKOWSKA A.
Prog. Cryst. Growth Charact. Vol. 1, (1978) 289.

PARTIN D.L.
J. Vac. Sci. Technol., B1 (1983) 174.

PERKOWITZ S., SUDHARSANAN R. WHROBEL J.M., CLAYMAN B.P. and
BECLA P.
Phys. Rev. B, 38, 8 (1988) 5565.

PIOTROWSKI T.
J. Cryst. Growth, 72 (1985) 117.

PODGORNY N., and ZIMMAL STARNAWSKA M.
Phys. Rev. B30, (1984) 2295.

STAUDENMANN J.L., KNOX R.D., FAURIE J.P.
J. Vac. Sci. Technol. A 5(5) (1987) 3161.

SOMBUTHAWEE C., BONSALL S.B. and HUMMEL F.A.
J. Solid State Chemistry 25 (1978) 391.

TAMMENMAA M., KOSKINEN T., HILTUNEN L., NIINISTO L. and LESKELA M.
Thin Solid Films 124 (1985) 125.

TRIBOULET R., HEURTEL A and RIOUX J.
J. Crystal Growth, (1990) to be published.

TRIBOULET R.T., NGUYEN-DUY T. and DURAND A.
J. Vac. Sci. Technol. A(3) (1985) 95.

TRIBOULET R. LASBLEY A. TOULOUSE B., GRANGER R.
J. Cryst. Growth 79 (1986) 695.

TWARDOWSKI A., DIETL T. and DEMIANIUK M.
Solid State Comm. 48 (1983) 845.

VARIZI M., DEBSKA V. and REIFENBERGER R.
Appl. Phys. Lett. 47 (1985) 407.

WEBB C., KAMINSKA M., LICHETENSTEIGER M., and LAGOWSKI J.
Solid State Comm. 40, (1981) 609.

WIEDEMEIER H. and KHAN A.
Trans. Metall. Sc. AIME, 242 (1968) 1969.

WU A.Y. and SLADEK R.J.
J. Appl. Phys. 53 (1982) 8589.

YODER-SHORT D.R., DEBSKA U. and FURDYNA J.K.
J. Appl. Phys. 58, (1985) 4056.

Given the similarity between blende and wurtzite structures, both tetrahedrally coordinated, each structure being based on a closest packing of spheres with a stacking order ABCABC... along the [111] direction for blende and ABABAB... along the c axis for wurtzite, it is meaningful to consider the mean cation - cation distance dc of an alloy without distinction of structure, according to the approach of Yoder-Short et al, 1985.

For zinc blende structure, $dc = a_o/\sqrt{2}$, for wurtzite, $d_c = a_o$ and $d_c = \sqrt{3/8}\ c_o$.

Yoder-Short et al proposed the universal plot of Fig. 25, giving the mean cation - cation distance d_c versus manganese concentration x of all $A^{II}_{1-x}Mn_xB^{VI}$ alloys;

d_b, the tetrahedral covalent bond length, is found to be $d_b = \sqrt{3/8}\ d_c$ for both structures by geometrical considerations. The d_b values for the Mn chalcogenides can thus be extrapolated from the ternary alloys. They are more reliable than those determined from X ray experiments, because the Mn chalcogenides present frequently some mixed structure (NaCl and NiAs) at room temperature. From these values and the tetrahedral covalent radii of the elements determined by van Vechten and Philips, 1970, Yoder-Short et al obtained a reliable value for the tetrahedral radius of Mn :

$$r_{Mn} = (1.326 - 0.018)\ \overset{\circ}{A}$$

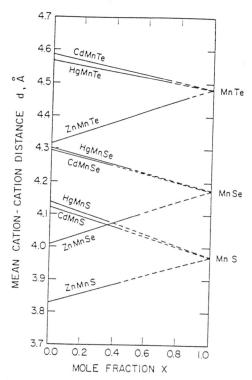

Fig. 25 Mean cation-cation distances d as a function of Mn mole fraction x for $A^{II}_{1-x}Mn_xB^{VI}$ alloys. Analytic expressions corresponding to the figures are given in Table III. [After Yoder-Short et al. (1985).]

ANNEXE B : THERMODYNAMICS OF THE CdSe - MnSe SYSTEM

after : G. Sigai and H. Wiedemeier, 1974

The Knudsen - effusion technique has been used to derive the activities of CdSe and MnSe in the (Cd,Mn)Se solid solutions as a function of composition and temperature.

It has been shown that only CdSe vaporizes from $Mn_{1-x}Cd_xSe$ solid solutions. Since CdSe vaporizes to monoatomic metal and diatomic selenium species, under the conditions used (850 - 980 K), the vaporization reaction can be described by the equation :

$$Mn_{1-x}Cd_xSe \text{ (s)} \rightleftharpoons (Mn_{1-x}Cd_{x-y})Se_{1-y}(s) + y\ Cd(g) + (y/2)\ Se_2(g) \quad (1)$$

Reaction (1) can be represented by :

$$CdSe \text{ (ss)} \rightleftharpoons Cd \text{ (g)} + 1/2\ Se_2 \text{ (g)} \quad (2)$$

where ss refers to CdSe in solid solutions.

The activity a (CdSe, ss) of CdSe in solid solution is related to the equilibrium constant by the equation :

$$K_p = P_{Cd}\ P^{1/2}_{Se_2}\ /a\ (CdSe\ ss) \quad (3)$$

Likewise, the equilibrium constant associated with the reaction

$$CdSe(s) \rightleftharpoons Cd(g) + 1/2\ Se_2(g) \quad (4)$$

is expressed by $K_p = P_{Cd}P^{1/2}_{Se2}$

in this case, the total pressure P can be writen as :

$$P' = P_{CdSe} + P_{Cd} + P_{Se_2} \text{ or, in first approximation, as } P' = P_{Cd} +$$

P_{Se_2}, P_{CdSe} being negligible beside P_{Cd} and P_{Se_2}.

P is minimum when its total differential is nil :

$$\delta P'/\delta P_{Cd} = \frac{\delta P'}{\delta P_{Se_2}} = 0 \quad (6)$$

$$\frac{\delta P'}{\delta P_{Cd}} = 1 - \frac{2K^2}{P^3_{Cd}} \implies P_{Cd} = 2^{1/3}\ K^{2/3}$$

$$\frac{\delta P'}{\delta P_{Se_2}} = 1 - \frac{K}{2P_{Se_2}^{3/2}} \qquad P_{Se_2} = \left(\frac{K}{2}\right)^{2/3}$$

and $P_{Cd} = 2P_{Se_2}$

$$P' = P_{Cd} + P_{Se_2} = K^{2/3}[2^{1/3} + (1/4^{1/3})]$$

$$\implies K' = \frac{2\sqrt{3}}{9} P'_T{}^{3/2} \tag{7}$$

Following the same approach for eq. (3), for which

$$P_T = P_{Se_2} + P_{Cd}, \text{ one finds that } Ka = \frac{2\sqrt{3}}{9} P_T{}^{3/2}$$

and thus

$$a(CdSess) = (P_T/P'_T)^{3/2} \tag{8}$$

The total pressure for reaction (2) is obtained from Knudsen measurements on solid solutions, according to the classical expression

$$P_T = \frac{3m(2\pi RT)^{1/2}}{at(M_{Se_2}{}^{1/2} + 2M_{Cd}{}^{1/2})} \text{ atm.}$$

where p_T is the total equilibrium pressure in atmosphere, m is the weight loss in grams, T the absolute temperature, a is the effective orifice area in square centimeters, t the time in seconds and M the molecular and atomic weight, respectively, of the effusing species (A.S. Pashinkin, 1964) and the total pressure for the sublimation of pure CdSe is calculated from previous thermochemical data (Sigai and Wiedemeier, 1982) by means of standard thermodynamic equations.

The activity of CdSe in solid solution is then calculated using equation (8).

The activity of MnSe in solid solution is calculated by use of the Gibbs-Duhem equation in the form :

$$\int d\ln [f(MnSe)] = \int [x/(1-x))] \, d\ln [f(CdSe)]$$

where x is the mole fraction of CdSe.

The activities of CdSe and MnSe are plotted as a function of temperature for various compositions of the solid solution in fig. 26.

For compositions in the CdSe rich-phase (x > 0.50) the activity coefficients of CdSe, very close to unity, express ideal behavior. For x = 0.6 and 0.8, the activity of CdSe is practically independent of temperature. In the MnSe rich phase, the activity of CdSe is higher than expected for ideal behavior, but approaches ideal behavior with increasing temperature.

The activities of CdSe and MnSe as a function of composition at 900 K are shown in fig. 27. The dotted lines represent the curves expected if Raoult's law were followed for complete solid solution. The activities for x-0.5 were obtained by extrapolating the experimentally determined values in the CdSe-rich phase to the phase limit (x=0.5). In the CdSe rich hexagonal phase, the activities of CdSe and MnSe follow Raoult's law. In the MnSe-rich cubic phase, the activity of CdSe shows positive deviations from ideality. The MnSe activity appears initially ideal but shows negative deviations from raoult's law with increasing CdSe content of the cubic phase.

The variation of the activity of CdSe and of MnSe as a function of temperature and composition can be explained in terms of the electronic structures of the respective cations in the cubic and hexagonal phases.

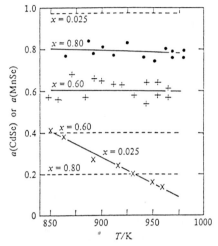

Fig. 26. The activity of CdSe (solid lines) and MnSe (dashed lines) as a function of temperature for fixed compositions in $Mn_{1-2}Cd_xSe$.

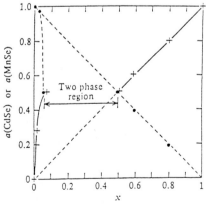

Fig. 27. The activity of CdSe (solid lines) and of MnSe (dashed lines) as a function of composition of the solid solution for a fixed temperature of 900 K in $Mn_{2-2}Cd_xSE$. The straight dashed lines represent the curves expected if Raoult's law were followed for complete solid solution.

ANNEXE C

Thermodynamics of the chemical vapor transport in a simple system : ZnSe.

Any equilibrium between a solid phase and a gazeous phase can be writen as :

$$aA(s) + bB(g) \rightleftharpoons cC(g) + dD(g) \tag{1}$$

where a, b, c, d are the stoichiometry coefficients.

The equilibrium constant of the reaction is given by :

$$Kp = \frac{P_C^c \cdot P_D^d}{P_A^a \cdot P_B^b} \tag{2}$$

related to the Gibbs free energy by the equation :

$$\Delta G° = -RT \, Ln \, Kp \tag{3}$$

The Gibbs free energy is given by :

$$\Delta G° = \Delta H°(T) - T\Delta S°(T) \tag{4}$$

The variations of standard enthalpies and entropies are given by :

$$\Delta H°(T) = H°_{298} + \int_{298}^{T} Cp \cdot dT$$

$$\Delta S°(T) = S°_{298} + \int_{298}^{T} Cp \, \frac{dT}{T}$$

The knowledge of $H°_{298}$, $S°_{298}$ and Cp turns out to be essential.

In a first approximation, the following equations can be proposed in the case of ZnSe.

$$ZnSe + I_2 \rightleftharpoons ZnI_2 + 1/2 \, Se_2 \tag{5}$$

$$ZnSe + 2I \rightleftharpoons ZnI_2 + 1/2 \, Se_2 \tag{6}$$

$$ZnSe + I \rightleftharpoons ZnI + 1/2 \, Se_2 \tag{7}$$

$$ZnSe(s) \rightleftharpoons Zn(g) + 1/2 \, Se_2 \tag{8}$$

$$ZnI_2(g) \rightleftharpoons Zn(g) + I_2 \, (g) \tag{9}$$

$$I_2(g) \rightleftharpoons 2I(g) \tag{10}$$

From (3) and (4), it is possible to express the equilibrium constants as a function of entropies and enthalpies, for a growth temperature of 800°C. They are given in the following table

Reactions	Kp
5	0.5
6	6.2×10^{-14}
7	1.3×10^{-3}
8	2.45×10^{-20}
9	3.3×10^{-8}
10	5.98×10^{-14}

The reaction controlling the system is thus essentially

$$ZnSe + I_2 \rightleftharpoons ZnI_2 + 1/2 \; Se_2$$

for a growth temperature of 800°C.

SEGREGATION

I - NORMAL FREEZING

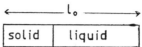

If $C_s = kC_L$, g the fraction of the solid which is frozen at t, Vo the initial volume of liquid, V its volume at t is expressed by :

$$g = (Vo - V)/Vo \qquad V = Vo(1-g)$$

The concentration of impurities in the liquid is $C_L = m_L/V$ with m_L being the mass of these impurities.

To an increase dg of the frozen fraction will correspond a variation $dV = -Vodg$ and a variation dm_L such as

$$-dm_L = C_s (-dV) = k C_L (-dV) = k \frac{m_L}{Vo(1-g)} Vodg$$

$$\frac{dm_L}{m_L} = - k \frac{dg}{1 - g}$$

If, at the beginning, the mass is m_0 $\qquad m_L = m_0 (1-g)^k$

$$C_L = \frac{m_L}{V} = Co (1-g)^{k-1}$$

$$C_s = kC_L = kCo (1-g)^{k-1}$$

II - Zone melting

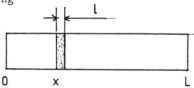

Let us calculate the concentration of the frozen solid.
s   quantityof solute in the liquid zone
l   width of the liquid zone
x   volume of the frozen solid

Constant section equal to 1 cm^2

For a displacement dx of the liquid zone, the variation ds of "solute" in this zone is expressed by :

$$ds = Codx - kC_Ldx$$

$$ds = (Co - k \frac{s}{l})dx$$

$$ds = (- Co-s) \frac{1}{k} \frac{kdx}{l} \quad or \quad \frac{ds}{1/k\, Co - s} = - \frac{k}{l} dx$$

$$1/k\, Co-s = (1/k\, Co - s_0)e^{-kx/l}$$

V_i : molar volume of component i

x_L, x_S atomic fraction $N_B/(N_A + N_B)$ of the solid and liquid phase respectively

y_L atomic fraction $N_C/(N_A + N_B + N_C)$ of the liquid phase.

Fig. 1 and 2 show respectively the axial distribution x_S measured by electron probe analysis in a $Pb_{1-x}Sn_xTe$ crystal grown by the Travelling Heater Method (THM) from a Te-rich solution zone, and the logarithmic representation of the axial distribution x_S in a $Pb_{1-x}Sn_xTe$ crystal grown by THM from a Te-rich solution zone revealing the segregation coefficient k and the effective zone length l_{eff}.

IV - SEGREGATION IN BRIDGMAN GROWTH IN SOLUTION

The example which will be dealt with is the determination of the Mn concentration in the liquid phase of a $Zn_{1-x}Mn_xTe$ ingot grown in Te excess, the concentration x of the solid being known. The general formula for the liquid phase is :

$$(Zn_{1-z}Mn_z)_y\ Te_{1-y}$$

and for the solid :

$$(Zn_{1-x}Mn_x)_{0.5}Te_{0.5}$$

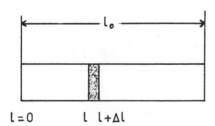

$$l = 0 \qquad l \quad l + \Delta l$$

<u>For Te</u> :

$$(1 - y\ (1))\ (l_0 - 1) = (1 - y\ (1 + \Delta 1)\ (l_0 - 1 - \Delta 1) + 0.5\ \Delta 1$$

Neglecting $(\Delta 1)^2$ terms, we obtain :

$$\frac{dy}{dl}(l_0 - 1) = y - 0.5$$

$$\frac{l_0 - 1}{dl} = - \frac{1}{dln(l_0 - 1)}$$

$$- \frac{dy}{dln(l_0 - 1)} = y - 0.5 \qquad ln(l_0 - 1) = x$$

$$- \frac{dy}{dx} = y - 0.5 \qquad\qquad y - 0.5 \rightarrow Y$$

$$\frac{dY}{Y} = - dx$$

$$LnY = -x + C$$

$$Y = Ae^{-x}$$

$$y = Ae^{-x} + 0.5$$

$$= Ae^{-ln(l_o-l)} + 0.5 = \frac{A}{l_o - 1} + 0.5$$

For $1 = 0$, $y = y_o$

$$A = l_o(y_o - 0.5)$$

$$y = \frac{l_o}{l_o - 1} (y_o - 0.5) + 0.5$$

__For Mn :__

$$z(l)y(l)(l_o-1) = z(1+\Delta 1)y(1+\Delta 1)(l_o-1-\Delta 1) + 0.5\ x(1).\Delta 1$$

neglecting $(\Delta 1)^2$ terms

$$- z(1).y(1)+(l_o - 1)\frac{d}{dz}[z(1).y(1)] + 0.5\ x(1) = 0$$

$$\frac{d}{dl}[z(1).y(1).(l_o-1)] = -0.5\ x(1)$$

$$z(1).y(1).(l_o-1) = - 0.5 \int_0^l x(1').dl' + cst$$

For $1=0$, $z(o)=Zo$, $y(0) = y_o$ and $= 0$

$$z(1)y(1)(l_o-1) = z_oy_ol_o - 0.5 \quad x(1')dl'$$

$$z(1) = \frac{z_oy_ol_o - 0.5 \int_o^l x(1')dl'}{l_oy_o - 0.5\ 1}$$

$y(1)$ and $z(1)$ can thus be calculated, and the segregation estimated, from the experimental results $x=f(1)$.

BAND STRUCTURE AND THEORY OF MAGNETIC INTERACTIONS

IN DILUTED MAGNETIC SEMICONDUCTORS

K. C. Hass

Research Staff, Ford Motor Company
Dearborn, Michigan 48121-2053
U. S. A.

INTRODUCTION

These lecture notes attempt to summarize the current state of understanding of the electronic structure and magnetic interactions in diluted magnetic semiconductors (DMS) as derived within an "effective" one-electron framework. The major portion of these notes focuses on the prototypical case of $A_{1-x}^{II}Mn_xB^{VI}$ DMS in which the magnetic element Mn is substituted for the group II element in a tetrahedral II-VI semiconductor. Several brief reviews of the microscopic theory of these materials have already been published.[1-4] The present survey is intended to be more pedagogical and comprehensive. Emphasis is placed on the development of a consistent picture based on first-principles spin-polarized band calculations, semi-empirical tight-binding calculations and experiment. Unresolved issues related to $A_{1-x}^{II}Mn_xB^{VI}$ DMS are also addressed. The notes conclude with a few comments on other DMS materials for which additional theoretical and experimental challenges remain.

MICROSCOPIC THEORY OF $A_{1-x}^{II}Mn_xB^{VI}$ DMS

Electronic Structure

One of the attractive features of $A_{1-x}^{II}Mn_xB^{VI}$ DMS from a theoretical perspective is the fact that the band structures of the II-VI host compounds are well known.[5] The valence (conduction) bands in these compounds are derived primarily from bonding (antibonding) combinations of sp^3 hybrid orbitals on neighboring atoms. The bonding is more polar (ionic) than in the case of group IV or III-V compounds but tetrahedral zinc-blende (zb) and wurtzite structures are still favored. All $A^{II}B^{VI}$ compounds with A^{II} = Zn, Cd, or Hg and B^{VI} = S, Se, or Te exhibit direct band gaps at $k=0$. This, along with the wide energy range of these gaps (0-3.8 eV), make II-VI's particularly promising for optical applications.

Semimagnetic Semiconductors and Diluted Magnetic Semiconductors
Edited by M. Averous and M. Balkanski, Plenum Press, New York, 1991

59

In addition to having two valence s electrons, each group II cation has a fully occupied d shell that lies relatively shallow in energy (> -1 Ryd). Recent studies[6,7] have shown that these d electrons are not completely core-like and do contribute weakly to the bonding and other properties of II-VI's. For the most part we will ignore this complication, although it is useful to contrast the $(n-1)d^{10}ns^2$ configuration of a conventional group II element with the $3d^5 4s^2$ configuration of Mn. Useful analogies also exist between $A^{II}_{1-x}Mn_x B^{VI}$ DMS and I-VII tetrahedral compounds (e.g. CuBr) in which the d^{10} levels of the noble metal are strongly mixed with the upper sp valence bands.[8]

In the absence of a magnetic field, the semiconducting properties of $A^{II}_{1-x}Mn_x B^{VI}$ DMS are remarkably similar to those of conventional II-VI alloys.[9] This is the first clue that the valence s electrons of Mn behave very much like those of Zn, Cd or Hg and that the half-filled Mn d shell is relatively inert. In atomic Mn, the five d electrons combine, according to Hund's first rule, to give a spin S of $5/2$ in the ground state. Local moments consistent with $S=5/2$ are also observed in Mn-containing DMS. To a first approximation, therefore, the Mn d electrons in these materials may be assumed to be localized and independent of the semiconducting sp bands.

Even with the Mn d levels neglected, the problem of calculating the electronic structure of $A^{II}_{1-x}Mn_x B^{VI}$ DMS is non-trivial, as is the case for any alloy. When work on this problem first began,[10] a further complication existed in that the maximum Mn concentration that had been achieved in a single phase tetrahedral sample was approximately 70%. (Non-diluted MnB^{VI} compounds are six-fold coordinated.) Usually in alloy theory one begins with a knowledge of the limiting crystal properties and then interpolates using an approach such as the virtual crystal approximation (VCA), if alloy

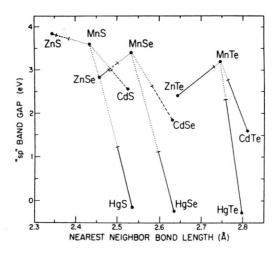

Fig. 1.   Variation (assumed linear) of low temperature sp band gaps with average nearest neighbor bond lengths. Solid, dashed and dotted lines represent zinc-blende, wurtzite and extrapolated regions, respectively. (After Ref. 2.)

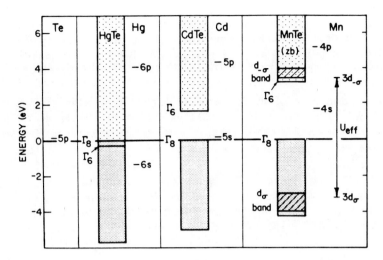

Fig. 2.   Limits of *sp* upper valence and lower conduction
          bands (lighter and darker boxes, respectively)
          relative to *sp* atomic energy levels. Also shown
          are spin-split Mn *d* levels and associated *d* "bands"
          (cross-hatched). (After Ref. 1.)

scattering can be neglected, or the coherent potential approximation (CPA),
if it cannot.[11]   In view of the experimental inaccessibility of the
tetrahedral $x=1$ limits, the first alloy calculations[10] for the *sp* "bands" in
$A_{1-x}^{II}Mn_xB^{VI}$ relied heavily on Harrison's universal tight-binding scheme[5] and on
extrapolated experimental information. The validity of such extrapolations
is illustrated in Fig. 1, which shows the observed low temperature *sp* band
gaps in $A_{1-x}^{II}Mn_xB^{VI}$ DMS as a function of the average nearest neighbor bond
lengths, which are themselves linear in x. The dotted lines correspond to
compositions for which tetrahedral bulk samples are unavailable. For each
B^{VI} anion, the experimental *sp* gaps at low Mn concentrations extrapolate to
a unique value for $x=1$. Recently, tetrahedral $A_{1-x}^{II}Mn_xB^{VI}$ DMS with up to 100%
Mn have been grown using epitaxial techniques.[12] The measured band gaps in
these cases are in excellent agreement with the extrapolations in Fig. 1.

A key feature of the Harrison tight-binding approach is the use of
atomic energy levels to ensure a reasonable description of chemical trends.[5]
Using this approach, Hass and Ehrenreich[10] attributed the rapid increase in
sp band gaps with x in $Hg_{1-x}Mn_xTe$ and $Cd_{1-x}Mn_xTe$ to the higher energy of Mn *4s*
states compared to Hg *6s* or Cd *5s*. Figure 2 shows the limiting crystal *sp*
empirical-tight-binding bands assumed in this work on the same absolute
energy scale as the atomic *s* and *p* levels used in the parametrizations. The
zero of energy is the CdTe valence band maximum. This Γ_8 level is derived
primarily from Te *5p* electrons and was therefore assumed to lie at the same
energy in HgTe and zb-MnTe. (It is now believed that the HgTe Γ_8 level
actually lies about 0.35 eV higher.[13]) The Γ_6 level, which represents the
conduction band minimum in CdTe and zb-MnTe, is instead derived primarily
from the valence *s* levels of the cation. The higher energy of the Cd *5s*

level compared to the Hg $6s$ level is responsible for the change from the inverted gap, semimetallic nature of HgTe to the 1.6 eV open gap in CdTe. The further increase to a 3.2 eV sp gap in zb-MnTe is a consequence of the still higher Mn $4s$ level.

The cation s level differences in Fig. 2 also give rise to relatively strong scattering effects in the alloys (which are believed to be random). Comparisons of VCA and CPA results for these systems indicate that the most pronounced disorder effects occur in regions where the cation s projected density of states is largest: in the lower part of the upper valence bands (near -4 eV) and about 2 eV above the conduction band minimum.[10] In accordance with the atomic level differences, the magnitude of the disorder effects is comparable in $Hg_{1-x}Cd_xTe$ and $Cd_{1-x}Mn_xTe$ and is largest in $Hg_{1-x}Mn_xTe$. Near the band edges, the sp bands are reasonably described by the VCA. A more complete CPA analysis that includes the effects of Mn d electrons is discussed below.

The presence of local magnetic moments in $A_{1-x}^{II}Mn_xB^{VI}$ DMS suggests that the Mn $3d^5$ electrons are strongly interacting. As is the case for many other transition metal compounds, conventional band theory, by which is meant the restricted Hartree-Fock approximation or some variant thereof, is completely inadequate for treating this problem.[14] A conventional band calculation for a tetrahedral MnB^{VI} compound would predict metallic behavior with the Fermi level at the midpoint of the d band complex.[15] The inconsistency between this prediction and the known semiconducting behavior of $A_{1-x}^{II}Mn_xB^{VI}$ DMS is similar to the much discussed case of NiO. As first pointed out by Mott,[16] NiO is non-metallic because the Coulomb energy U_{eff} associated with d shell charge fluctuations, $d^n + d^n \rightarrow d^{n-1} + d^{n+1}$, is much larger than the kinetic energy gained by delocalization of the d charge. Despite the enormous effort that has gone into this problem, a fully adequate theory of such "Mott-Hubbard" insulators is still lacking.[17] Fortunately, in the special case of Mn, because the d shell is exactly half full, a meaningful one-electron description may still be constructed, as we will see below. The more complicated case of Fe and other transition elements will be discussed briefly in a later section.

The traditional methods for treating the behavior of a partially filled d shell in a solid are those of crystal- and ligand-field theory.[18] The former includes only electrostatic interactions with the neighboring environment while the latter also includes some covalency (hybridization) effects. These theories are concerned with the internal degrees of freedom of individual magnetic ions in their solid state environments. The transition elements of interest here fall into the so-called weak field coupling scheme in which the environment is only a weak perturbation of the multiplet structure associated with the Coulomb interactions between d electrons; spin-orbit effects are weaker still. In the case of Mn, the non-degenerate ^6S ground state in the free atom, which has a spin of $5/2$ and zero angular momentum ($L=0$), becomes the $^6A_1$ ground state in the solid. The next highest free atom multiplet, ^4G ($S=3/2$, $L=4$), splits in a tetrahedral environment into a three-fold degenerate $^4T_1$ state, a three-fold degenerate $^4T_2$ state, a two-fold degenerate ^4E state and a non-degenerate $^4A_1$ state. Optical transitions between these states have been observed[9] in some $A_{1-x}^{II}Mn_xB^{VI}$

DMS, with the lowest energy excitation, $^6A_1 \rightarrow {}^4T_1$, occurring at approximately 2.2 eV. These transitions involve the flipping of a single electron spin within the d^5 manifold at a single Mn site. Because of their neutral, localized nature, we will refer to them as Frenkel excitons.

Of greater concern to the present discussion are charged quasiparticle excitations involving d electrons. In the case of $A_{1-x}^{II}Mn_xB^{VI}$ DMS, this problem is meaningfully addressed within the context of spin-polarized band theory.[19] The underlying assumption in this approach is that the effective one-electron potential in the solid is spin dependent. Such a potential is particularly useful in the case of Mn since it provides a natural separation between occupied and unoccupied d states, which correspond to opposite spin polarizations at each site.

Spin-polarized band theory has become almost synonymous in recent years with the local spin density approximation (LSDA) to density functional theory.[20] First-principles studies based on this approach yield good agreement between different computational schemes. While it is tempting to associate the resulting eigenvalues with quasiparticle excitation energies, it should be kept in mind that *this interpretation is not rigorous*[21] (i.e., Koopmans' theorem does not apply). Nevertheless, LSDA calculations for $A_{1-x}^{II}Mn_xB^{VI}$ DMS have been extremely valuable in providing qualitative insight, quantitative estimates and input to more empirical calculations. For computational reasons, use of the LSDA is limited at present to structures with long range order, both compositional and magnetic.

Figure 3 shows the LSDA density of states calculated by Larson, et al.[22] for zb-MnTe with an assumed "type-I" antiferromagnetic (AF) ordering. Similar results would be obtained for an AF type-III ordering, which is believed to be the preferred ground state.[9] Part (a) of the figure shows the total density of states for the upper valence and lower conduction bands with the zero of energy defined as the valence band maximum, E_v. Part (b) shows the corresponding projections of the density of states on the majority and minority spin d orbitals at a given Mn site. (The direction of the "majority" spin is opposite, of course, on the different AF sublattices). The majority spin component is degenerate with the sp valence bands and is centered ($\epsilon_d^\uparrow$) at -2.2 eV. The minority spin component is mostly unoccupied and lies within the sp band gap. Both d level contributions are broadened significantly by hybridization with the sp bands. (The nearest neighbor Mn-Mn separation of 4.5 Å is too large to allow direct d-d interactions.) The net spin-polarization at a given Mn site is found to be 4.2 electrons. Qualitatively similar results have also been obtained for tetrahedral MnSe and MnS[22] and by other groups.[23]

The LSDA calculations successfully reproduce the fact that the sp band gap is larger in tetrahedral MnB^{VI} compounds than in most conventional II-VI's. The prediction that unoccupied Mn d states lie within the sp gap is now believed to be incorrect, however.[24] These states have the character of single particle excitations and should represent the addition of an extra d electron to the system, primarily on a Mn site that already has five d electrons. Because of an overscreening of Coulomb interactions in the LSDA,

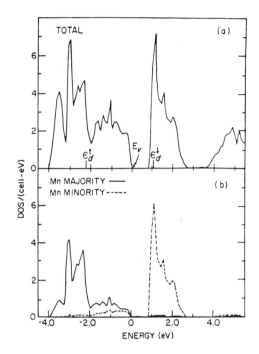

Fig. 3.   LSDA (a) total and (b) projected Mn d
densities of states for zb-MnTe with
AF-I ordering. Centroids of majority
and minority spin d states are labeled
$\epsilon_d^\uparrow$ and $\epsilon_d^\downarrow$, respectively. E_v is the
valence band maximum. (After Ref. 22.)

the predicted energy of such a charged excitation (not to be confused with
the neutral Frenkel excitons discussed above) is significantly
underestimated.[21,25]   This overscreening is more serious for relatively
localized d or f states than for extended sp bands. In addition to placing
unoccupied d states too low in energy, it also generally places occupied d
states too high. Occupied d states represent hole excitations that involve
the removal of a d electron from the system. The spin splitting in the case
of Mn is thus determined primarily by the energy to transfer a d electron
from one site to another: i.e., by the quantity U_{eff} defined earlier. As will
be seen below, the U_{eff} value in Fig. 3 is too small by ~3-4 eV compared to
experiment. As a consequence, the ground state spin polarization in the LSDA
is also underestimated.

To go beyond the LSDA, most semi-empirical studies of $A_{1-x}^{II}Mn_xB^{VI}$ DMS have
been based on a *diluted* Anderson lattice (DAL) Hamiltonian:[2,15,22,26]

$$H = H_{sp} + H_d + H_{sp\text{-}d} .\tag{1}$$

Here H_{sp} describes unperturbed, semiconducting sp bands, H_d describes strongly
interacting, atomic-like Mn d states and $H_{sp\text{-}d}$ describes sp-d hybridization.
These quantities are often expressed in terms of the tight-binding creation
operators $b_{j\alpha\sigma}^\dagger$, which creates an sp electron with spin σ in orbital α at site

j, and $d_{i\mu\sigma}^+$, which creates a Mn d electron with spin σ in orbital μ at Mn site i. The sp term,

$$H_{sp} = \sum_j \sum_{\alpha\sigma} \epsilon_{j\alpha} b_{j\alpha\sigma}^+ b_{j\alpha\sigma} + \sum_{jj'} \sum_{\alpha\alpha'\sigma} t_{jj'}^{\alpha\alpha'} b_{j\alpha\sigma}^+ b_{j'\alpha'\sigma} , \qquad (2)$$

is then characterized by the on-site energies $\epsilon_{j\alpha}$ and hopping integrals $t_{jj'}^{\alpha\alpha'}$, which may be either ordered or random. H_d is conveniently described by the U, U', J model of Kanamori,[14,27]

$$H_d = {\sum_i}' \sum_{\mu\sigma} n_{i\mu\sigma} \left[\epsilon_\mu + \frac{U}{2} n_{i\mu,-\sigma} + \frac{U'}{2} \sum_{\mu' \neq \mu, \sigma'} n_{i\mu'\sigma'} - \frac{J}{2} \sum_{\mu' \neq \mu} n_{i\mu'\sigma} \right] , \qquad (3)$$

where $n_{i\mu\sigma} = d_{i\mu\sigma}^+ d_{i\mu\sigma}$ is a number operator, ϵ_μ is an on-site energy for orbital μ, U (U') is an intra-atomic Coulomb integral between two electrons in the same (different) orbital(s), and J is an intra-atomic exchange integral. Usually $U>U'>J$. The final term in Eq. (1) is characterized by the additional hopping integrals $V_{ij}^{\mu\alpha}$:

$$H_{sp-d} = {\sum_i}' \sum_{j \neq i} \sum_{\alpha\mu\sigma} V_{ij}^{\mu\sigma} (d_{i\mu\sigma}^+ b_{j\alpha\sigma} + h.c.) . \qquad (4)$$

The primed sums in Eqs. (3) and (4) are over only those sites occupied by Mn.

The many-body nature of Eq. (1) complicates matters even in the special cases $x \to 0$ (Anderson impurity) and $x=1$ (Anderson lattice). It is useful, however, to consider a spin-polarized one-electron treatment based on the *unrestricted* Hartree-Fock approximation (UHFA).[2,14,15] Equation (3) reduces in the UHFA to

$$H_d = {\sum_i}' \sum_{\mu\sigma} n_{i\mu\sigma} \left[\epsilon_\mu + U \langle n_{i\mu,-\sigma} \rangle + U' \sum_{\mu' \neq \mu, \sigma'} \langle n_{i\mu'\sigma'} \rangle - J \sum_{\mu' \neq \mu} \langle n_{i\mu'\sigma} \rangle \right] , \qquad (5)$$

where $\langle \; \rangle$ denotes a ground state expectation value. For $|V_{ij}^{\mu\alpha}| \ll U$, self-consistency is unimportant and $\langle n_{i\mu\sigma} \rangle$ may be taken to be one or zero for Mn, depending on whether σ is parallel or anti-parallel to the net spin on site i, respectively. Equation (5) then reduces further to

$$H_d = {\sum_i}' \sum_{\mu\sigma} n_{i\mu\sigma} (\tilde{\epsilon}_\mu + U_{eff} \langle n_{i\mu,-\sigma} \rangle) , \qquad (6)$$

where $\tilde{\epsilon}_\mu = \epsilon_\mu + 4U' - 4J$ and $U_{eff} = U + 4J$. Since Koopmans' theorem remains valid in the UHFA, one can check that this value for U_{eff} is the same as that obtained by computing the energy required to transfer a Mn d electron from one site to another (with net spin anti-parallel) using Eq. (3). By contrast, the energy required to flip a spin on a single Mn site according to Eq. (3) is only $4J$. The U, U', J model thus distinguishes clearly between neutral and charged excitations.

Parameters for Eqs. (2), (4) and (6) have generally been obtained from both theory and experiment. Aspects of the tight-binding parametrization of the sp bands have already been discussed. To correct for the above-mentioned errors in the LSDA, the DAL calculations discussed below are based on the modified d level positions shown on the right hand side of Fig. 2. These are obtained by shifting the majority spin d states (d_σ) down to approximately 3.4 eV below the valence band maximum, as suggested by photoemission experiments,[28] and the minority spin d states $(d_{-\sigma})$ up above the sp conduction band minimum. The actual location of unoccupied d states in $A^{II}_{1-x}Mn_xB^{VI}$ DMS is still not very well known. Inverse photoemission experiments, in principle, provide the most direct probe. The only such data reported to date[29] suggests that the minority spin d levels in $Cd_{0.8}Mn_{0.2}Te$ are centered about 1 eV higher than the position assumed in Fig. 2.

To complete the parametrization, it is necessary to distinguish between Mn d orbitals with e_g symmetry $(x^2-y^2, 3z^2-r^2)$ and those with t_{2g} symmetry (xy, yz, zx). The latter have slightly higher orbital energies $(\tilde\epsilon_\mu)$ in a tetrahedral environment and also hybridize more strongly with neighboring anion states. The strongest mixing occurs with anion p orbitals, which lie closest in energy (cf. Fig. 2). This mixing is conveniently described by the Slater-Koster[30] two-center integrals $(pd\sigma)$ and $(pd\pi)$. Most DMS workers have reduced this to a single hybridization parameter by assuming the "universal" ratio $(pd\sigma)/(pd\pi)=-2.18$ [with $(pd\pi)<0$] given by Harrison.[5] Using the d level positions in Fig. 2 and fitting to the valence band edge exchange constants $N_0\beta$ discussed in the next section gives values for

$$V_{pd} \equiv E_{x,yz}(111) = \left[\frac{(pd\sigma)}{3} - \frac{2(pd\pi)}{3\sqrt{3}} \right] \tag{7}$$

on the order of -0.5 eV in all Mn-containing II-VI's.[22] Similar values are obtained by fitting to the LSDA bands directly, i.e., without shifting the d level positions.

One final approximation that is often made to assume that the Mn spins in $A^{II}_{1-x}Mn_xB^{VI}$ alloys are Ising-like with up ($\uparrow$) and down ($\downarrow$) spin sites distributed at random.[1-3,15,26] This is, of course, a gross oversimplification of the true *vector* spin glass ground state expected for intermediate x,[9] but it does simulate the dominant effects of *magnetic* disorder on the electronic structure.

Figures 4 and 5 show CPA results[2,3] based on this approximation for the three-component alloy $Cd_{1-x}(Mn\uparrow)_{x/2}(Mn\downarrow)_{x/2}Te$. The E vs. k curves in Fig. 4 represent VCA sp bands for $x=0.6$ along with the dispersionless Mn d "bands" obtained in the absence of sp-d hybridization. (Mn d levels with t_{2g} symmetry are assumed here to lie 0.5 eV higher than those with e_g symmetry.) Figure 4 also shows normalized CPA spectral functions at various k points, which represent the probabilities of finding electrons in the alloy with particular energies and wavevectors. These curves include sp-d hybridization and differ from the δ-functions found in crystalline solids because of the presence of both compositional and magnetic disorder. It should be noted that the use of a sophisticated alloy theory such as the CPA is essential in this case because of the absence of d orbitals on Cd sites and the large splitting (U_{eff}) between d levels of a given spin on Mn$\uparrow$ and Mn$\downarrow$ sites.

The spectral features associated with Mn e_g states in Fig. 4 are slightly broadened but unshifted relative to their unhybridized locations.

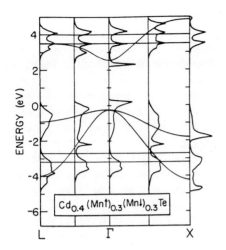

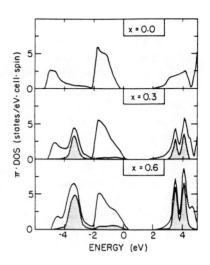

Fig. 4. Normalized CPA spectral densities as function of energy at various *k* points. Unhybridized VCA *sp* bands and Mn *d* levels shown for comparison. (After Ref. 2.)

Fig. 5. CPA total and projected Mn *d* (shaded) densities of states for $Cd_{1-x}Mn_xTe$ with various *x* values. The CdTe valence band maximum is at -0.3 eV. (After Ref. 2.)

By contrast, the occupied and unoccupied Mn t_{2g} states are repelled by the Te-*p*-like upper valence band region to lower and higher energy, respectively. This effect is also seen in the total densities of states in Fig. 5 in which the Mn *d* contributions are shaded. The e_g-t_{2g} splitting for unoccupied *d* states increases from the unhybridized value in Fig. 4 while the corresponding splitting for occupied *d* states disappears. The shapes of the Mn *d* contributions are relatively insensitive to *x* since they are determined primarily by hybridization with Te nearest neighbors. The shape of the occupied *d* state contribution, including the significant *d* admixture within 2 eV of the valence band maximum, is consistent with photoemission and resonant photoemission experiments[28] on $Cd_{1-x}Mn_xTe$.

The *k*-resolved spectra in Fig. 4 also agree qualitatively with angle-resolved photoemission experiments.[26] In addition, the broadened structure throughout the upper valence band region is consistent with the rapid smearing of reflectivity[31] and ellipsometry[32] data for $A_{1-x}^{II}Mn_xB^{VI}$ DMS with increasing *x*. This broadening is a consequence of the magnetic disorder in the system and is much more pronounced than that which occurs in alloys where this effect is absent (e.g. $Hg_{1-x}Cd_xTe$). The magnetic disorder results from the fact that an *sp* electron of a given spin hybridizes only with *d* orbitals of the same spin; such orbitals correspond to occupied states on some Mn sites and unoccupied states on others. For reasons that will become clear in the next section, the effects of magnetic disorder on states near the valence band maximum are essentially equivalent to those of spin-disorder scattering.[33] Near the conduction band minimum, on the other hand, magnetic disorder plays a much smaller role due to the absence of a significant d admixture. The slight broadening of the conduction band edge in Fig. 4 is due to the cation s level disorder discussed earlier. The additional effects of sp compositional disorder near -4 eV are obscured somewhat in Figs. 4 and 5 by the Mn d contributions.

Further analysis including relativistic effects[34] shows that the admixture of Mn d states into the valence band maximum in $A_{1-x}^{II}Mn_xB^{VI}$ tends to reduce the zone-center spin-orbit splitting, Δ_0. This is analogous to what occurs in tetrahedral I-VII compounds[8] and is believed to be the dominant cause of the net decrease of Δ_0 with x in $Cd_{1-x}Mn_xTe$.[35] The d admixture may also produce a net shift of the valence band maximum, as can be seen in Figs. 4 and 5. This shift is given approximately by

$$\Delta E_v = \frac{x}{2} (4V_{pd})^2 \left[\frac{1}{E_v - \epsilon_d} - \frac{1}{\epsilon_d + U_{off} - E_v} \right] , \qquad (8)$$

where E_v and ϵ_d are the unhybridized positions of the valence band maximum and the Mn t_{2g} levels, respectively.[34] The quantity ΔE_v is generally quite small since the two terms in Eq. (8) tend to cancel. Nevertheless, even a small shift may be important in the context of band offsets in DMS heterostructures. Most theories of such offsets neglect $sp\text{-}d$ hybridization entirely.[36] Unfortunately, the parameters required in Eq. (8) to estimate this effect are too uncertain at present to have much predictive value. The shift in Fig. 4, for example, is most likely exaggerated, since most experimental evidence[37,38] for $Cd_{1-x}Mn_xTe$ is consistent with less than a 100 meV variation in the valence band maximum as a function of x.

The above picture of the electronic structure is believed to apply to all $A_{1-x}^{II}Mn_xB^{VI}$ DMS. Photoemission measurements indicate that the position of the occupied d state emission is remarkably constant across this series.[28] (It should be kept in mind, however, that the observed separations of ~3.5 eV from the valence band maximum include the effects of hybridization.) Since U_{eff} is largely an atomic property, it too should be relatively constant, although a slight increase with ionicity is expected because of a reduction in screening.[22]

The most significant chemical trend expected in $A_{1-x}^{II}Mn_xB^{VI}$ alloys is an increase in the degree of $sp\text{-}d$ hybridization as the anion is changed from Te to Se to S. Harrison's universal tight-binding scheme predicts that $V_{pd} \propto d^{-7/2}$, where d is the Mn-B^{VI} bond length.[5] The 10% shorter bond lengths in sulfides compared to tellurides (cf. Fig. 1) are thus associated in this scheme with a nearly 50% increase in hybridization. Although the magnitude of this estimate is highly uncertain, the direction of the increase is well supported by photoemission[28,39] and LSDA studies,[22] and by the exchange interactions discussed in the next two sections.

Mn d - sp Band Exchange

The largest magnetic exchange interaction in $A_{1-x}^{II}Mn_xB^{VI}$ DMS is that between the local Mn spins (S_i) and the spin densities $[s(r)]$ associated with "sp" valence and conduction states.[9] The dominant (isotropic) interaction of this kind may be described by the phenomenological Kondo-like Hamiltonian,

$$H_K = -\sum_i{}' J_{sp\text{-}d}(r-R_i)\, S_i \cdot s(r) \quad , \tag{9}$$

with an exchange function $J_{sp\text{-}d}$ that falls off rapidly with the distance from a Mn site. This interaction is responsible for the enhanced magneto-optical and magneto-transport properties of DMS that have attracted both scientific and technological interest.[9] Of greatest importance are the effects of this interaction on the conduction and valence band edges. Within a k-space, mean-field treatment of Eq. (9), the relevant exchange constants (proportional to matrix elements of H_K) are generally denoted $N_0\alpha$ and $N_0\beta$, respectively. Experimental values of these quantities are on the order of 0.2 and -1.0 eV, respectively, throughout the entire class of $A_{1-x}^{II}Mn_xB^{VI}$ DMS.

The differences in sign and magnitude of $N_0\alpha$ and $N_0\beta$ were first explained by Bhattacharjee, Fishman and Coqblin.[40] These authors noted that, for a general sp band state, ψ_{nk}, there are two competing contributions to the associated $sp\text{-}d$ exchange constant. The first is an ordinary "potential" exchange term, of the form

$$J_{sp\text{-}d}^{pot}(nk) = \sum_\mu \int\!\!\int d^3r\, d^3r'\, \phi_{d\mu}^*(r)\, \psi_{nk}^*(r')\, v_{sc}(|r-r'|)\, \phi_{d\mu}(r')\, \psi_{nk}(r) \quad , \tag{10}$$

where $\phi_{d\mu}$ is a Mn d function with symmetry μ and v_{sc} is a screened Coulomb interaction. For $\phi_{d\mu}$ and ψ_{nk} suitably orthogonalized, this interaction is always ferromagnetic (F) (i.e., $J_{sp\text{-}d}^{pot}>0$). The second contribution is an AF "kinetic" exchange term due to hybridization of the ψ_{nk} state with the d orbitals.[41] As discussed in Ref. 40 and in the preceding section, $sp\text{-}d$ hybridization is negligible in $A_{1-x}^{II}Mn_xB^{VI}$ DMS near the conduction band minimum; in the absence of disorder, it is strictly forbidden for the Γ_1 (non-relativistic) or Γ_6 edge, by symmetry. The conduction band exchange $N_0\alpha$ is thus entirely determined by the F interaction in Eq. (10). By contrast, the valence band edge does hybridize strongly with the d states, as we have already seen. Usually when such hybridization occurs for $3d$ transition elements, kinetic exchange dominates. It follows that the valence band exchange $N_0\beta$ should be AF and larger in magnitude than $N_0\alpha$.

Further support for this explanation and quantitative estimates of $N_0\alpha$ and $N_0\beta$ have been obtained by first-principles LSDA calculations. Hass, et al.[42] demonstrated that, by examining the spin-polarized band structures of hypothetical F-DMS, it is possible to extract the desired $sp\text{-}d$ exchange constants from the observed band edge spin splittings, in analogy with the experimental determination of $N_0\alpha$ and $N_0\beta$ in an external magnetic field. To illustrate this approach, Fig. 6 shows spin-polarized LSDA bands for F-zb-MnTe. Because the net Mn spin $<S_z>$ in this case is no longer zero (as it was in Figs. 3-5), the spin-up and spin-down bands are no longer degenerate. In particular, the spin-up d states, which now correspond to the majority spin direction on every Mn site, are completely occupied and the spin-down d states are completely empty. As a result, the spin-up (spin-down) valence band maximum Γ_{15} is repelled upwards (downwards) in energy by hybridization with the lower (higher) lying d states. The associated spin splitting, $-\Delta E_v$,

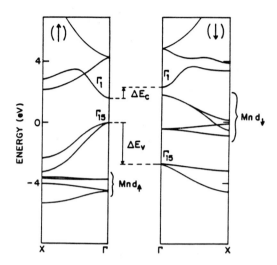

Fig. 6. LSDA energy bands along Γ - X for
majority (left) and minority (right)
spin directions in F-zb-MnTe. ΔE_c and
ΔE_v denote the conduction and valence
band edge spin splittings, respectively.
(After Ref. 22.)

is equal to $N_0\beta$ times $<S_z>$. Similarly, the spin-splitting of the Γ_1 conduction band edge, ΔE_c, is equal to $N_0\alpha$ times $<S_z>$. Because this latter state does not hybridize with the d levels, ΔE_c is due entirely to the different spin-up and spin-down *potentials*. The predicted values of $N_0\alpha$ and $N_0\beta$ from Fig. 6 are given in Table 1 (first line) along with additional LSDA results[22,23] obtained in a similar manner. The agreement between these estimates and experimental exchange constants[9] (e.g., $N_0\alpha$=0.22 eV and $N_0\beta$=-0.88 eV in $Cd_{1-x}Mn_xTe$ with small x) is surprisingly good in view of the errors in the LSDA d level positions discussed earlier.

Using the well-known Schrieffer-Wolff transformation[41] and neglecting the small potential exchange contribution, one can derive an analytic expression for $N_0\beta$ starting from the DAL Hamiltonian of Eq. (1). The result,[22,40]

$$N_0\beta \approx -\frac{2}{5}(4V_{pd})^2 \left[\frac{1}{E_v - \epsilon_d} + \frac{1}{\epsilon_d + U_{eff} - E_v} \right] ,\qquad (11)$$

is qualitatively different from Eq. (8) in that the two terms in brackets are now separated by a plus sign. Equation (11) is useful both for constraining the electronic structure parameters which appear and for exploring chemical trends. For example, both experiments[9] and the LSDA results in Table 1 show a clear increase in $|N_0\beta|$ from tellurides to selenides to sulfides; Eq. (11) explains this effect in terms of the previously discussed increase in $|V_{pd}|$ in this series.

One minor complication that arises due to spin-orbit splitting is that the valence band exchange constants associated with the Γ_7 (lower) and Γ_8

Table 1. LSDA estimates of sp-d and d-d exchange constants.

	$N_0\alpha$ (eV)	$N_0\beta$ (eV)	J_1 (K)	Ref.
MnTe	0.33	-1.05	-17	22
	0.48	-1.16	-16	23
$Cd_{0.5}Mn_{0.5}Te$	0.32	-1.12	—	22
	0.54	-1.30	—	23
MnSe	0.33	-1.35	-24	22
MnS	0.36	-1.50	-26	22

(higher) states (both derived from Γ_{15}) may differ.[22,34] Equation (11) still applies in this case although the quantity E_v must be re-interpreted as the corresponding unhybridized Γ_7 or Γ_8 level energy. For typical Γ_8-Γ_7 splittings of < 1 eV, the resulting difference in exchange constants is less than 10%, which is consistent with the limited magneto-optical data available for the Γ_7 state.[43]

Mn d - Mn d Exchange

Also of great interest are the magnetic exchange interactions among the Mn spins themselves.[9] The dominant interaction of this kind is again isotropic and is usually described by the phenomenolgical, spin-5/2, diluted Heisenberg Hamiltonian,

$$H_H = -\sum_{i \neq j}{}' J_{ij} S_i \cdot S_j \quad . \tag{12}$$

The exchange constants J_{ij} in Eq. (12) fall off rapidly with distance and are predominantly AF. The nearest neighbor exchange, which will be denoted J_1, is on the order of -10 K (-1 meV) in all $A_{1-x}^{II}Mn_xB^{VI}$ DMS.[9,44] (Note that the total interaction between two spins according to Eq. (12) is $-2J_{ij}S_i \cdot S_j$.)

A simple estimate of J_1 may be obtained from the LSDA total energy difference between AF and F ordered structures. By relating this difference to that given by Eq. (12) with nearest neighbor interactions only, Larson, et al.[22] obtained a value of $J_1 = -17$ K in zb-MnTe. This, and the additional LSDA results[22,23] in Table 1, are too large by about a factor of three. Such overestimates are typical of the LSDA[45] and are not surprising in view of the errors in the LSDA d level positions and the assumption of a nearest neighbor only Hamiltonian.

Considerable effort has also gone into understanding the mechanism responsible for the Mn-Mn exchange. To avoid semantic arguments, it should be kept in mind that there is often considerable overlap between different mechanisms, as emphasized by Herring.[46] In the case of $A_{1-x}^{II}Mn_xB^{VI}$ DMS, one can confidently neglect both direct exchange (because of the large Mn-Mn separation) and itinerant exchange (which occurs in transition *metals*). Using Herring's nomenclature, one is left with only indirect exchange[47,48] and superexchange.[49,50] The former results from the polarization of the

intervening medium and the latter is due to the covalent mixing of magnetic
and non-magnetic orbitals.

The clearest case of indirect exchange occurs for completely localized
spins (e.g., nuclear spins) interacting with the intervening medium through
a contact interaction $[J_{sp-d}(\mathbf{r}-\mathbf{R}_i) \propto \delta(\mathbf{r}-\mathbf{R}_i)$ in Eq. (9)]. Two separated spins
then feel each other's influence because of their surrounding polarization
clouds. This effect is easily calculated in second-order perturbation theory
and leads to the well-known Ruderman-Kittel-Kasuya-Yosida (RKKY) interaction[47]
in metals or the analogous Bloembergen-Rowland (BR) interaction[48] in materials
with a band gap. The RKKY interaction falls off asymptotically as R_{ij}^{-3}, where
$R_{ij}=|\mathbf{R}_i - \mathbf{R}_j|$, while the BR interaction falls off exponentially. Many early
theoretical studies[51,52] of Mn-Mn exchange in $A_{1-x}^{II}Mn_xB^{VI}$ DMS focussed
exclusively on indirect mechanisms starting from experimental values of $N_0\alpha$
and $N_0\beta$. Of particular interest was the zero-gap case,[51] which yields an
algebraic decay faster than R_{ij}^{-3}. None of these studies of indirect exchange,
however, succeeded in explaining the observed magnitudes of the dominant near
neighbor exchange constants in $A_{1-x}^{II}Mn_xB^{VI}$ DMS. The main problem with indirect
mechanisms is that they do not allow for the possibility that the electrons
contributing to a given spin may themselves transfer to a different Mn site
because of sp-d hybridization.[22]

The exchange interaction resulting from hybridization of the d levels
with *occupied sp* orbitals is generally referred to as superexchange.[50] This
interaction is most easily visualized in real space as a spreading out of the
magnetic orbitals, with a corresponding decrease in energy, for AF aligned
spins.[4] A theory of superexchange was first proposed by Kramers[49] in 1934 and
was later developed in more detail by Anderson.[50] Superexchange is now
believed to be the dominant mechanism in most magnetic insulators. Its role
in $A_{1-x}^{II}Mn_xB^{VI}$ DMS was first examined theoretically by Larson, et al.[22] using
a perturbative, k-space formalism developed by Falicov and co-workers.[53]
Although somewhat unusual for a superexchange calculation, a k-space approach
is almost essential for these materials because of their relatively broad (~5
eV) upper sp valence bands; alternative real-space descriptions[54] may still
be useful heuristically but are poorly suited for quantitative calculations.

The calculations of Larson, et al.[22] begin with the DAL Hamiltonian of
Eq. (1). The effects of the H_{sp-d} term on the spin degrees of freedom are
assumed to be reasonably described by Eq. (12). Use is made of the fact that
the exchange integral J_{ij} is proportional to the matrix element $<1|H_H|2>$
between the states $|1>=|M_i=5/2,M_j=3/2>$ and $|2>=|M_i=3/2,M_j=5/2>$ which differ
only in the magnetic quantum numbers (M) associated with the spin operators
at sites i and j. Evaluation of this matrix element to lowest non-vanishing
order in perturbation theory[53] yields an expression for J_{ij} that is fourth
order in H_{sp-d}. The contributions of various exchange mechanisms may then be
classified by intermediate states. Larson, et al.[22] define the superexchange
contribution to be that consisting of all processes that involve the creation
of two holes in the sp valence bands. A typical term of this type is shown
pictorially in Fig. 7a. The remaining processes involve either the creation
of only one sp hole and one sp electron (in the conduction band), or two sp
electrons. The former of these is associated in Ref. 22 with BR exchange

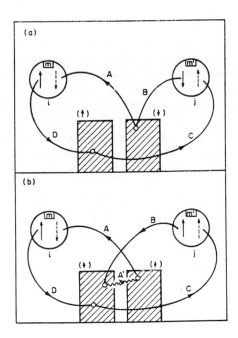

Fig. 7.   Schematic representation of processes
contributing to (a) J_{ij} and (b) D_{ij}.
Circles represent Mn d orbitals with
symmetry m at site i and m' at site j.
Spin of electron in orbital changes
from solid to dashed arrows. A, B, C
and D represent spin-conserving hoppings
and A' a spin-orbit induced spin flip.
(After Ref. 55.)

since the creation of an electron-hole pair is a polarization process.   It
should be noted that this method of calculating the BR contribution is
formally different from that of Refs. 51 and 52.   Neither approach includes
all possible polarization effects, but the approximation made by Larson, et
al. (neglect of sp-d potential exchange) appears to provide a more reasonable
description of the *total* Mn-Mn exchange at short distances.

The most recent numerical results[55] obtained using this approach and
electronic structure parameters similar to those discussed earlier yield a
value for J_1 of approximately -5.6 K in $Cd_{1-x}Mn_xTe$ for small x.   This is in
excellent agreement with the experimental value of -6.3 K, which is known
with a high degree of confidence.[44]   The ratio J_2/J_1, where J_2 is the Mn-Mn
next nearest neighbor exchange, was similarly calculated[22] to be about 0.1,
which is also in good agreement with experiment.[44]

More important than the quantitative results themselves is the fact that
these calculations[22] clearly establish superexchange as the dominant
contribution to J_1 and J_2.   Numerically, superexchange accounts for about 95%

of the total exchange at these distances in $Cd_{1-x}Mn_xTe$, with the BR mechanism making up most of the remainder. Roughly the same percentages hold in all $A_{1-x}^{II}Mn_xB^{VI}$ DMS. The dominance of superexchange results from the strong hybridization between the Mn d levels and the anion p-like upper valence bands. The much weaker hybridization with the sp conduction bands causes the remaining processes (including BR) to be negligible by comparison.

Because of the similarity of the electronic structure throughout the entire class of $A_{1-x}^{II}Mn_xB^{VI}$ DMS, Larson, et al.[22] were able to extract an approximately universal form for the superexchange contribution:

$$J_{ij} \approx -b^2 \left[\frac{1}{U_{eff}} + \frac{1}{\epsilon_d + U_{eff} - E_v} \right] f(R_{ij}/a) \quad , \tag{13}$$

where

$$b = V_{pd}^2 / (\epsilon_d + U_{eff} - E_v) \tag{14}$$

and a is the lattice constant. With the present definition of V_{pd}, the function $f(R/a)$ is given numerically by

$$f(R/a) \approx 4.1 \, e^{-5(R/a)^2} \quad . \tag{15}$$

The first term in parentheses in Eq. (13) gives the usual Anderson form of superexchange,[50] $-b^2/U$, while the second term represents an additional contribution that was first recognized by Sawatzky and co-workers in a more general context.[56]

Like Eq. (11) for $N_0\beta$, Eq. (13) provides an explicit connection between the electronic structure parameters for $A_{1-x}^{II}Mn_xB^{VI}$ DMS and the exchange constants. The fact that V_{pd} enters to the fourth power suggests a strong dependence on the Mn-B^{VI} bond length d. This is consistent with the near doubling of the magnitude of experimental J_1 values[44] as the anion is changed from Te to Se to S. Larson, et al.[22] have argued that a similar decrease in d is the primary cause of the observed[44] increase in $|J_1|$ from Cd- to Zn-based DMS. Equation (13) does not include bond angle distortions, however, which may also play a role in this case[54] (see below). The equation is further limited in that it cannot account for the observed difference in J_1 for inequivalent pairs of Mn nearest neighbors in wurtzite DMS[44] and for the long range behavior of J_{ij} that affects spin-glass properties.[57] The former is about a 10% effect that Larson[58] has subsequently explained in terms of sp-d hybridization beyond nearest neighbors. The asymptotic behavior of J_{ij} is poorly described by Eq. (13) because of the neglect of both BR contributions and sp-d potential exchange. Although a more reliable calculation in this regime is still lacking, it is unlikely that such a calculation would yield anything other than an exponential decay in an open gap system; the algebraic decays extracted in the experimental analyses of Twardowski and co-workers[57] are thus somewhat puzzling.

Besides the isotropic exchange, a number of smaller *anisotropic* spin interactions[4] are also present in $A_{1-x}^{II}Mn_xB^{VI}$ DMS. The largest of these is the Dzyaloshinsky-Moriya (DM) interaction[59,60]

$$H_{DM} = -\sum_{i \neq j}{}' D_{ij} \cdot S_i \times S_j \quad . \tag{16}$$

This interaction was first shown by Dzyaloshinsky to be allowed by group theory in crystals that lack inversion symmetry.[59] Moriya later demonstrated that it arises naturally in the theory of superexchange when spin-orbit interactions are included.[60] As is clear from Eq. (16), the DM interaction prefers spins to lie canted, away from a collinear AF alignment, in the plane perpendicular to D_{ij}. Experimental evidence for this interaction in $A_{1-x}^{II}Mn_xB^{VI}$ DMS is provided by spin resonance linewidths,[55,61] neutron diffraction studies of noncollinear short range order at large x,[62] and the widths of magnetization steps.[44] Anisotropic superexchange undoubtedly also plays a role in determining the spin glass behavior and spin dynamics in these materials.[63]

Early studies of anisotropic superexchange in nonmetals associated the DM interaction with the spin-orbit splitting of the magnetic ion itself.[64] As shown theoretically by Larson and Ehrenreich,[55] and experimentally by Samarth and Furdyna,[61] however, the principal source of anisotropy in $A_{1-x}^{II}Mn_xB^{VI}$ DMS is instead the spin-orbit splitting λ of the B^{VI} anion. The calculations of Ref. 55 are very similar to those of Ref. 22 with the additional feature that an sp band electron is allowed to flip its spin with a probability proportional to λ. A perturbation expansion for D_{ij} is constructed by considering the matrix element between the state |1> defined previously and the new state $|2>=|M_i=5/2,M_j=5/2>$. As in the calculation of J_{ij}, the dominant terms in this expansion involve only the sp valence bands. A typical term is shown in Fig. 7b. The only difference between this term and that in Fig. 7a is the additional spin-flip process A'. Numerical calculations[55] indicate that the nearest neighbor DM interaction, D_1, has a magnitude that ranges from 5% of $|J_1|$ in Te-based DMS to more than an order of magnitude smaller in S-based systems. These results are in good agreement with experiment[55] and are consistent with the fact that $\lambda \propto Z^2$, where Z is the anion atomic number. The ratio $|D_{ij}/J_{ij}|$ is given approximately by the universal expression $\lambda/(\epsilon_d+U_{eff}-E_v)$. Because of the rapid decay of J_{ij} with distance, interactions beyond D_1 are unlikely to be important. The direction of D_1 itself (which may differ for different pairs of Mn neighbors) is perpendicular to the plane containing the two Mn sites of interest and the intervening anion.[55] In this sense, the DM exchange in $A_{1-x}^{II}Mn_xB^{VI}$ DMS is similar to that in metallic spin systems containing spin-orbit impurities.[63]

Other anisotropic interactions in $A_{1-x}^{II}Mn_xB^{VI}$ DMS are generally much smaller.[55] Pseudodipolar terms are proportional to λ^2 and are safely neglected. Single-ion anisotropies are also small (~40 mK or less) because of the nearly spherical Mn d^5 ground state. Ordinary dipolar interactions are less than 20 mK at near neighbor distances but become dominant in the very dilute regime because of their R_{ij}^{-3} falloff.

Unresolved Issues

Despite the above successes, the microscopic theory of $A_{1-x}^{II}Mn_xB^{VI}$ DMS is by no means complete. Some of the remaining challenges and opportunities include:

Effects of d levels on bonding: This topic has yet to be explored in detail although it may have important implications for the ultimate materials quality that can be achieved. Both LSDA[23] and tight-binding[3] calculations indicate that _sp-d_ hybridization lowers the energy of occupied states and thereby increases the cohesive energy, as it does in other systems containing partially occupied _d_ shells.[65] The significance of this result is unclear at present since an alloy is only as strong as its weakest link. A more relevant concern may be whether the presence of Mn has a stabilizing or a de-stabilizing effect on the A^{II}-B^{VI} bond; this is of particular interest in Hg-alloys where the Hg-B^{VI} bond is notoriously weak.

On a related note, ultrasonic measurements have revealed that the elastic moduli of $A^{II}_{1-x}Mn_xB^{VI}$ DMS are anomalously small compared to conventional II-VI alloys.[66] This has again been attributed to _sp-d_ hybridization in view of the analogous softening that occurs in tetrahedral I-VII compounds.[5] The combination of this effect with the preference of Mn-B^{VI} compounds to be six-fold coordinated raises questions concerning the validity of simple valence force models[67] for describing the local structure of DMS.

k-dependence of sp-d exchange: Perhaps the most glaring deficiency in existing theoretical models of $A^{II}_{1-x}Mn_xB^{VI}$ DMS is their failure to account for the observed weakness of the effects of _sp-d_ exchange on the E_1 optical transition.[68] This transition is usually associated with the L point of the zb Brillouin zone. Typical E_1 splittings in a magnetic field are on the order of 1/16 times the corresponding splittings at Γ. Simple analyses based on the DAL tight-binding model[69] described above yield reduction factors closer to 1/4. LSDA predictions, obtained by extending the approach in Fig. 6 to include the L point,[3,70] are in even worse agreement with experiment. These discrepancies are disturbing because they call into question the general understanding of _sp-d_ exchange. The most likely explanation for the failure of the tight-binding model is an improper description of the "_sp_" conduction band at L due to the limited basis set.[71] Potential sources of error in the LSDA result have been alluded to previously. A better understanding of the _k_-dependence of _sp-d_ exchange in $A^{II}_{1-x}Mn_xB^{VI}$ DMS is essential for modeling such technologically important properties as the Faraday rotation[72] and the temperature dependence of the fundamental band gap.[73] Resolving the E_1 problem might also lead to new insight into the magneto-optical properties of other DMS materials, particularly Mn-substituted IV-VI compounds[74] (see below), whose fundamental gaps are at L.

x-dependence of sp-d exchange: Even for the _sp-d_ exchange at Γ, there is still some question as to how $N_0\alpha$ and $N_0\beta$ vary with Mn concentration. That there must be some x dependence is obvious from the fact that these quantities differ slightly (< 20%) in $Zn_{1-x}Mn_xB^{VI}$ and $Cd_{1-x}Mn_xB^{VI}$ at small x,[9] yet must converge as $x\rightarrow1$. The theoretical considerations discussed earlier suggest that the x dependence of $N_0\alpha$ and $N_0\beta$ is weak and monotonic. It is thus unlikely that the anomalously small spin-splittings observed in magneto-optic experiments[75] on large x samples of $Cd_{1-x}Mn_xTe$ are due to significant reductions in _sp-d_ exchange constants. It would also be surprising to see a dramatic enhancement of $N_0\alpha$ and $N_0\beta$ in very dilute samples, although published data for $Cd_{1-x}Mn_xS$ suggests that this is the case.[76]

Angular dependence of superexchange: A possible relationship between angular distortions and superexchange in DMS was first suggested by Spalek, et al.[54] in an attempt to explain the 20-30% increase in $|J_1|$ from Cd to Zn-based alloys. These authors recognized that, according to extended X-ray absorption fine structure measurements,[67] the local structure in $A^{II}_{1-x}Mn_xB^{VI}$ DMS, as in other semiconductor alloys, is distorted from that of a perfect lattice due to the much stronger bond-stretching forces compared to bond-bending forces in these systems; individual A^{II}-B^{VI} and Mn-B^{VI} bond lengths vary only slightly from their limiting crystal values even though the average lattice constant obeys Vegard's law. Based on this, Spalek, et al. argued that the most significant change upon replacement of Cd by Zn is a decrease in the Mn-B^{VI}-Mn bond angle due to the shorter Zn-B^{VI} bond. This argument was later extended[77] to relate the observed x dependence[78] of J_1 to bond angle changes as well. Strong evidence for an angular dependence of the form

$$J_1 = J_{\pi/2} + (J_\pi - J_{\pi/2}) \cos^2\theta \quad , \tag{17}$$

where θ is the angle which the anion makes with the neighboring magnetic ions, is provided by previous theoretical and experimental studies of superexchange[79]. To be consistent with the above trends, $J_{\pi/2}$ must be larger in magnitude than J_π in $A^{II}_{1-x}Mn_xB^{VI}$ DMS. A similar inequality is needed to explain the results of a recent neutron diffraction study of magnetic ordering in strained MnSe/ZnSe heterostructures.[80] This latter study is particularly convincing since it does not involve any bond length changes. In contrast, the trends considered by Spalek, et al. are somewhat ambiguous since even small changes in the Mn-B^{VI} bond lengths may dominate over any bond-angle effect.[22]

Given that superexchange in $A^{II}_{1-x}Mn_xB^{VI}$ DMS does vary as in Eq. (17), it is not at all clear why $|J_{\pi/2}|$ should be greater than $|J_\pi|$. The opposite relationship holds in Fe^{3+} compounds[79] and is what one generally expects for d^5 systems based on the Goodenough-Kanamori rules.[50,81] With potential exchange neglected, these rules give $|J_{\pi/2}|<|J_\pi|$ because of the dominance of ($pd\sigma$) hopping over ($pd\pi$) hopping. Further work is clearly required to resolve this dilemma. One should note in this context that the original argument by Spalek, et al.[54] based on experimental exchange constants for rocksalt structure Mn compounds is not convincing since nearest neighbor Mn pairs in this structure have *two* equivalent intervening anions with $\theta=\pi/2$.

Many-Body Effects: Although the one-electron picture of $A^{II}_{1-x}Mn_xB^{VI}$ DMS reviewed here has been remarkably successful and is attractive because of its conceptual simplicity, there is still a need for a more complete many-body theory of these and other strongly correlated materials. The most striking limitation of the one-electron picture is the failure to account for satellite peaks ~6-8 eV below the valence band maximum in photoemission spectra.[39] These peaks do appear naturally in various configuration interaction cluster calculations[39] and in a more recent many-body treatment of the Anderson impurity Hamiltonian.[71] The main valence peak and its satellite in these theories are associated primarily with $d^5\underline{L}$ and d^4 final states, respectively, where $\underline{L}$ represents a ligand hole. Correlation and hybridization effects are included in a manner similar to the Heitler-London theory of molecules. Although even these theories have their limitations,[17] they have already provided important new insights into the electronic structure and magnetic interactions in a wide range of transition metal compounds,[56,82] including high temperature superconductors.[83] Further applications to $A_{1-x}^{II}Mn_xB^{VI}$ DMS would also be useful, particularly for clarifying the relationship between neutral and charged excitations.

COMMENTS ON OTHER DMS SYSTEMS

Mn-Substituted IV-VI Compounds

Closely related to the $A_{1-x}^{II}Mn_xB^{VI}$ system considered up until are DMS alloys of the form $C_{1-x}^{IV}Mn_xB^{VI}$ where C^{IV} = Pb, Sn.[74] The main difference in this case is that the bonding in IV-VI's is almost entirely ionic, with each C^{IV} cation transferring two valence p electrons to a B^{VI} anion. The resulting rocksalt structures differ from those of zb and wurtzite in two important respects. First, the Mn-B^{VI} bond lengths tend to be about 20% longer than in $A_{1-x}^{II}Mn_xB^{VI}$. Second, both first and second nearest neighbor pairs of Mn atoms have only one intervening anion forming angles of $\pi/2$ and π, respectively. Both of these differences have important implications for the magnetic properties of $C_{1-x}^{IV}Mn_xB^{VI}$ DMS. All other things being equal, the longer bond lengths, combined with Harrison's scaling law, suggest an order of magnitude smaller superexchange interaction compared to II-VI's [cf. Eqs. (13) and (14)]. Second nearest neighbor Mn-Mn interactions, however, should be more important than in II-VI's and comparable in magnitude to the nearest neighbor interactions. Both of these predictions are consistent with available magnetic susceptibility data[84] for $Pb_{1-x}Mn_xB^{VI}$.

Other microscopic properties of $C_{1-x}^{IV}Mn_xB^{VI}$ DMS remain to be explored. LSDA calculations for MnB^{VI} compounds with the rocksalt structure suggest that the locations and effects of Mn d levels are similar to those in tetrahedral systems.[70] Both these calculations and a recent tight-binding study[15] find the "sp" band gap in rocksalt MnTe to be indirect, with the valence band maximum at L and the conduction band minimum at X. Since the gap in PbTe is direct with both extrema at L, a direct to indirect gap transition must take place in $Pb_{1-x}Mn_xTe$ as a function of x; unfortunately, this effect may occur outside of the accessible range of Mn concentrations (0-11%).[74]

Non-Mn-Based DMS

A more recent trend in DMS research has been the study of systems containing magnetic elements other than Mn.[9] The most widely considered alternatives are the transition elements Fe and Co and the rare earth elements Eu and Gd. Rare earth DMS are qualitatively much different than Mn-containing systems because of the more localized nature of $4f$ electrons and the more prominent role of spin-orbit interactions. Hybridization-induced superexchange is certainly less important in these systems. The dominant f-f exchange mechanism is most likely indirect BR exchange due to polarization effects associated with an sp-f contact interaction.

In principle, most of the theory in this paper should be extendable to $A_{1-x}^{II}T_xB^{VI}$ DMS with T = Fe, Co, ... Unfortunately, LSDA calculations for the $x=1$ limits in this case are not even qualitatively correct,[70] for reasons similar to the failure of the LSDA for transition metal oxides.[14,83] As mentioned earlier, the LSDA includes the difference in potential between up and down spin states, but not between occupied and unoccupied states. For the half-filled d shell of Mn, these conditions fortuitously coincide. In Fe and Co systems, however, some minority spin d states are also occupied and the LSDA predicts metallic behavior[70], which is contrary to experiment.

It is precisely this complication that shows the full power of the U, U', J model[14] of Eq. (3). Within the UHFA, this model predicts that the singly occupied minority spin state in an Fe d^6 configuration lies approximately $4J$ higher in energy than the majority spin states, while the unoccupied d states lie higher still by an amount on the order of $U-J$. The separation between majority and minority spin occupied d states is consistent with photoemission experiments on $A_{1-x}^{II}Fe_xB^{VI}$ alloys, which show the principal Fe d emission again about 3-4 eV below the valence band maximum with a much smaller d signal right near the top of the band.[85] The spltting between majority spin and unoccupied d states has not been measured for Fe but should be smaller than in the case of Mn ($U+3J$ instead of $U+4J$). As a result, both energy denominators in Eq. (13) should also be smaller and one might expect a larger superexchange interaction in Fe-based DMS. This is in fact what is observed.[44,86] The still larger exchange interactions in Co-substituted[44] II-VI's are also consistent with this argument since the unoccupied d states in Co should lie even lower in energy. A significant increase in superexchange from Mn to Fe to Co also occurs in transition metal oxides.[56] Inverse photoemission experiments for $A_{1-x}^{II}T_xB^{VI}$ DMS would clearly be useful to test whether the unoccupied d state location is indeed the determining factor. It is also possible, of course, that the strength of T d - B^{VI} p hybridization increases from Mn to Co, but this seems less likely according to tabulated d orbital radii.[5]

One final issue that is worth considering is whether there are any other transition element substitutions that might lead to a *ferromagnetic* DMS. This is unlikely in the case of II-VI compounds since the tetrahedral angle always allows some AF superexchange channel for any d level occupancy.[3] The most promising situation for an F interaction, according to the Goodenough-Kanamori rules,[81] are two d^3 ions at an angle of $\pi/2$ relative to the anion. This occurs, for example, in the compound $CdCr_2Se_4$. A possible DMS system with this characteristic is $C_{1-x}^{IV}V_xB^{VI}$. For this to work, the vanadium must enter the rocksalt structure substitutionally and maintain a high spin d^3 configuration.

ACKNOWLEDGEMENTS

The author has benefited greatly from a long-term collaboration with A. E. Carlsson, H. Ehrenreich, and B. E. Larson. He is also grateful to P. M. Hui, N. F. Johnson, R. J. Lempert, and P. M. Young for their contributions to this work and to L. C. Davis for helpful discussions.

REFERENCES

1. H. Ehrenreich, K. C. Hass, B. E. Larson, and N. F. Johnson, in "Proc. MRS Symposium on Semimagnetic (Diluted Magnetic) Semiconductors," Vol. 89, edited by R. L. Aggarwal, J. K. Furdyna and S. von Molnar (MRS, Boston, 1987) p. 187.

2. K. C. Hass and H. Ehrenreich, Acta Physica Polonica A 73, 933 (1988).

3. K. C. Hass and H. Ehrenreich, J. Cryst. Growth 86, 8 (1988).

4. B. E. Larson and H. Ehrenreich, J. Appl. Phys. 67, 5084 (1990).

5. W. A. Harrison, "Electronic Structure and the Properties of Solids," (Freeman, San Francisco, 1980).

6. K. C. Hass and D. Vanderbilt, J. Vac. Sci. Technol. A 5, 3019 (1987).

7. S.-H. Wei and A. Zunger, Phys. Rev. Lett. 59, 2831 (1987); Phys. Rev. B 37, 8958 (1988).

8. A. Goldman, Phys. Stat. Sol. (b) 81, 9 (1977).

9. For recent reviews, see J. K. Furdyna, J. Appl. Phys. 64, R29 (1988) and other contributions to the present volume.

10. K. C. Hass and H. Ehrenreich, J. Vac. Sci. Technol. A 1, 1678 (1983).

11. H. Ehrenreich and L. M. Schwartz, in "Solid State Physics," Vol. 31, edited by H. Ehrenreich, F. Seitz and D. Turnbull (Academic, New York, 1976) p. 149.

12. L. A. Kolodziejski, R. L. Gunshor, N. Otsuka, B. P. Gu, Y. Hefetz and A. V. Nurmikko, Appl. Phys. Lett. 48, 1482 (1986).

13. N. F. Johnson, P. M. Hui and H. Ehrenreich, Phys. Rev. Lett. 61, 1993 (1988) and references therein.

14. B. H. Brandow, Adv. in Phys. 26, 651 (1977).

15. J. Mašek, B. Velický and V. Janiš, J. Phys. C 20, 59 (1987).

16. N. F. Mott, Proc. Roy. Soc. London, Ser. A 62, 416 (1949).

17. Z.-X. Shen, C. K. Shih, O. Jepsen, W. E. Spicer, I. Lindau and J. W. Allen, Phys. Rev. Lett. 64, 2442 (1990).

18. S. Sugano, Y. Tanabe and H. Kamimura, "Multiplets of Transition-Metal Ions in Crystals," (Academic, New York, 1970); J. S. Griffith, "The Theory of Transition Metal Ions," (Cambridge, London, 1961).

19. J. C. Slater, Phys. Rev. 82, 538 (1951).

20. A. R. Williams and U. von Barth, in "The Inhomogeneous Electron Gas," edited by S. Lundqvist and N. H. March (Plenum, New York, 1983).

21. J. P. Perdew and M. Levy, Phys. Rev. Lett. 51, 1884 (1983); L. J. Sham and M. Schlüter, Phys. Rev. Lett. 51, 1888 (1983).

22. B. E. Larson, K. C. Hass, H. Ehrenreich and A. E. Carlsson, Solid State Commun. 56, 347 (1985); Phys. Rev. B 37, 4137 (1988).

23. S.-H. Wei and A. Zunger, Phys. Rev. Lett. 56, 2391 (1986); Phys. Rev. B 35, 2340 (1987).

24. Y. R. Lee, A. K. Ramdas and R. L. Aggarwal, Phys. Rev. B 33, 7383 (1986).

25. A. E. Carlsson, Phys. Rev. B 31, 5178 (1985).

26. B. Velický, J. Mašek, G. Paolucci, V. Chab, M. Shurman and K. C. Price, Festkörperprobleme 25, 247 (1985).

27. J. Kanamori, Prog. Theor. Phys. 30, 275 (1963).

28. A. Wall, S. Chang, R. Philip, C. Caprile, A. Franciosi, R. Reifenberger and F. Pool, J. Vac. Sci. Technol. A 5, 2051 (1987) and references therein.

29. A. Wall, A. Franciosi, Y. Gao, J. H. Weaver, M. H. Tsai, J. D. Dow and R. V. Kasowski, J. Vac. Sci. Technol. A 7, 656 (1989).

30. J. C. Slater and G. F. Koster, Phys. Rev. 94, 1498 (1954).

31. T. Kendelewicz, J. Phys. C 14, L407 (1981).

32. P. Lautenschlager, S. Logothetidis, L. Viña and M. Cardona, Phys. REv. B 32, 3811 (1985).

33. T. Dietl, this volume.

34. P. M. Hui, H. Ehrenreich and K. C. Hass, Phys. Rev. B 40, 12346 (1989).

35. T. Kendelewicz, Solid State Commun. 36, 127 (1980).

36. W. A. Harrison and J. Tersoff, J. Vac. Sci. Technol. B 4, 1068 (1986); J. Tersoff, Phys. Rev. Lett. 56, 2755 (1986).

37. S.-K. Chang, A. V. Nurmikko, J.-W. Wu, L. A. Kolodziejski and R. L. Gunshor, Phys. Rev. B 37, 1191 (1988).

38. M. Taniguchi, L. Ley, R. L. Johnson, J. Ghijsen and M. Cardona, Phys. Rev. B 33, 1206 (1986).

39. M. Taniguchi, M. Fujimori, M. Fujisawa, T. Mori, I. Souma and Y. Oka, Solid State Commun. 62, 431 (1987); L. Ley, M. Taniguchi, R. L. Johnson and A. Fujimori, Phys. Rev. B 35, 2839 (1987).

40. A. K. Bhattacharjee, G. Fishman and B. Coqblin, Physica B&C 117-118, 449 (1983).

41. J. R. Schrieffer and P. A. Wolff, Phys. Rev. 149, 491 (1966).

42. K. C. Hass, B. E. Larson, H. Ehrenreich and A. E. Carlsson, J. Mag. Magn. Mater. 54-57, 1283 (1986).

43. A. Twardowski, E. Rokita and J. A. Gaj, Solid State Commun. 36, 927 (1980).

44. Y. Shapira, J. Appl. Phys. 67, 5090 (1990), and present volume.

45. T. Oguchi, K. Terakura and A. R. Williams, Phys. Rev. B 28, 6443 (1983).

46. C. Herring, in "Magnetism," Vol. IV, edited by G. T. Rado and H. Suhl, (Academic, New York, 1963) p.99.

47. M. A. Ruderman and C. Kittel, Phys. Rev. 96, 99 (1954).

48. N. Bloembergen and T. J. Rowland, Phys. Rev. 97, 1697 (1955).

49. H. A. Kramers, Physica 1, 182 (1934).

50. P. W. Anderson, in "Solid State Physics," Vol. 14, edited by F. Seitz and D. Turnbull, (Academic, New York, 1963) p. 99.

51. G. Bastard and C. Lewiner, Phys. Rev. B 20, 4256 (1979).

52. V.-C. Lee and L. Liu, Phys. Rev. B 29, 2125 (1984).

53. C. E. T. Goncalves da Silva and L. M. Falicov, J. Phys. C 5, 63 (1972); B. Koiller and L. M. Falicov, J. Phys. C 8, 695 (1975).

54. J. Spalek, A. Lewicki, Z. Tarnawski, J. K. Furdyna, R. R. Galazka and Z. Obuszko, Phys. Rev. B 33, 3407 (1986).

55. B. E. Larson and H. Ehrenreich, Phys. Rev. B 39, 1747 (1989).

56. J. Zaanen and G. A. Sawatzky, Can. J. Phys. 65, 1262 (1987).

57. A. Twardowski, H. J. M. Swagten, W. J. M. de Jonge and M. Demianiuk, Phys. Rev. B 36, 7013 (1987).

58. B. E. Larson, J. Appl. Phys. 67, 5240 (1990).

59. I. Dzyaloshinsky, Phys. Chem. Solids 4, 241 (1958).

60. T. Moriya, Phys. Rev. 120, 91 (1960).

61. N. Samarth and J. K. Furdyna, Solid State Commun. 65, 801 (1988).

62. U. Steigenberger and L. Lindley, J. Phys. C 21, 1703 (1988).

63. K. Binder and A. P. Young, Rev. Mod. Phys. 58, 801 (1986) and references therein.

64. J. J. Pearson, Phys. Rev. 126, 901 (1962).

65. H. Ehrenreich, Science 235, 1029 (1987) and references therein.

66. P. Maheswaranathan, R. J. Sladek and U. Debska, Phys. Rev. B 31, 5212 (1985).

67. A. Balzarotti, M. Czyzyk, A. Kisiel, N. Motta, M. Podgorny and M. Zimnal-Starnawska, Phys. Rev. B 30, 2295 (1984).

68. D. Coquillat, J. P. Lascaray, M. C. D. Deruelle, J. A. Gaj and R. Triboulet, Solid State Commun. 59, 25 (1986) and references therein.

69. A. K. Bhattacharjee, Phys. Rev. B 41, 5696 (1990).

70. A. E. Carlsson, private communication.

71. O. Gunnarson, O. K. Anderson, O. Jepsen and J. Zaanen, Phys. Rev. B 39, 1708 (1989).

72. P. A. Wolff, private commmunication.

73. A. K. Bhattacharjee, Solid State Commun. 65, 275 (1988).

74. G. Bauer, this volume.

75. J. P. Lascaray, D. Coquillat, J. Deportes and A. K. Bhattacharjee, Phys. Rev. B 38, 7602 (1988).

76. V. G. Abramishvili, S. I. Gubarev, A. V. Komarov and S. M. Ryabchenko, Sov. Phys.: Solid State 26, 666 (1984).

77. A. Lewicki, J. Spalek, J. K. Furdyna and R. R. Galazka, Phys. Rev. B 37, 1860 (1988).

78. T. M. Giebultowicz, J. J. Rhyne, J. K. Furdyna and P. Klosowski, J. Appl. Phys. 67, 5096 (1990).

79. C. Boekema, F. Van der Woude and G. A. Sawatzky, Int. J. Magnetism 3, 341 (1972).

80. N. Samarth, P. Klosowski, H. Luo, T. M. Giebultowicz, J. J. Rhyne, J. K. Furdyna, B. E. Larson and N. Otsuka, unpublished.

81. J. B. Goodenough, Phys. Rev. 100, 564 (1955); J. Kanamori, Phys. Chem. Solids 10, 87 (1959).

82. L. C. Davis, J. Appl. Phys. 59, R25 (1986) and references therein.

83. K. C. Hass, in "Solid State Physics," Vol. 42, edited by H. Ehrenreich and D. Turnbull, (Academic, New York, 1989) p. 213.

84. M. Gorska and J. R. Anderson, Phys. Rev. B 38, 9120 (1988).

85. M. Taniguchi, Y. Ueda, I. Morisada, Y. Murashita, T. Ohta, I. Souma and Y. Oka, Phys. Rev. B 41, 3069 (1990) and references therein.

86. A. Twardowski, J. Appl. Phys. 67, 5108 (1990).

TRANSPORT PHENOMENA IN SEMIMAGNETIC SEMICONDUCTORS

Tomasz Dietl

Institute of Physics, Polish Academy of Sciences
Al. Lotników 32/46, PL 02-668 Warszawa, Poland

I. INTRODUCTORY SECTION

1. Scope of the chapter

In retrospection, transport properties of semimagnetic semiconductors or diluted magnetic semiconductors (DMS) have attracted considerable attention for two interrelated reasons. Firstly, the studies of these compounds help us to understand how the presence of open magnetic shells may affect transport phenomena – a problem widely discussed in the context of magnetic semiconductors,[1,2] and still not satisfactorily solved. Secondly, it turns out that DMS offer novel opportunities to verify a number of general ideas about electron transport in solids.

The purpose of this chapter is to supplement the previous review articles[3-5] concerning transport properties of DMS with some recent experimental and theoretical developments. We limit ourselves primarily to II–VI DMS since this group of compounds is the most familiar to the author and, at the same time, it has been the most thoroughly studied one.

We begin our contribution by recalling various models of electronic states near to the metal–insulator transition in semiconductors doped with hydrogenic–like impurities and its driving mechanisms. We also present basic information about the location of energy levels of transition metals in II–VI semiconductors. Next, we discuss in detail the effects the sp–d exchange interaction may have on electron transport in semiconductors containing shallow impurities. This question is best addressed to Mn–based compounds because Mn in II–VI's acts neither as a donor nor as an acceptor and does not directly perturb the density–of–states near the Fermi energy. We consider the ranges of impurity concentrations that correspond to metallic and insulating phases, respectively, paying particular attention to the transitory region between the above two extreme cases. It is near to the metal–insulator transition, where the presence of the exchange interaction manifests itself in a rather non–standard way, and – which is perhaps more important – constitutes a valuable tool to elucidate the driving mechanisms of the transition. The subsequent part of the article concerns such DMS, in which transition metals act as resonant impurities. We discuss, in particular, the remarkable properties of $Hg_{1-x}Fe_xSe$ in the mixed–valence region of Fe donors. We also touch upon the interesting but

Semimagnetic Semiconductors and Diluted Magnetic Semiconductors
Edited by M. Averous and M. Balkanski, Plenum Press, New York, 1991

83

unexplored topic of electron localization in the case when the potential of the donating impurity has no bound states. Since DMS exist almost exclusively in a paramagnetic or spin–glass phase, effects of long–range magnetic ordering or magnons on the transport phenomena is out of the scope of our article.

When analyzing experimental results it is tempting to disregard the imperfect crystallographic structure of real samples. In order to illustrate how strongly the low–temperature transport properties can be altered by such imperfections, in the final part of the chapter the conduction along grain boundaries in p–$Hg_{1-x}Mn_x$Te as well as to apparent superconductivity in HgSe:Fe will be discussed.

2. Doped semiconductors and metal–insulator transition

It is known that electronic properties of semiconductors doped with hydrogenic–like impurities depend crucially on the ratio of the mean distance between impurities $\bar{r} = (3/4\pi n)^{1/3}$ to their effective Bohr radius $a_B = \hbar^2 \kappa_0/e^2 m^*$. In the dilute case, $\bar{r} \gg a_B$, electrons are bound to individual impurities, and low–temperature conduction proceeds by means of phonon–assisted tunneling between occupied and empty states (which are always present in real materials as a result of compensation). Since there are no phonons at zero temperature, the hopping conductivity vanishes in this limit, and the material can be regarded as an insulator. In this strongly localized regime, correlation effects are known to be of primary importance. For instance, it is the Coulomb repulsion between electrons localized on the same site that precludes the existence of doubly occupied states at low temperatures.

In the opposite limit, $\bar{r} \ll a_B$, electrons reside in the conduction band, and low–temperature mobility is determined by ionized impurity scattering. Here, the correlation effects are insignificant as, according to the Landau theory of Fermi liquids, they lead to the qualitatively and, for $\bar{r} \ll a_B$, quantitatively unimportant renormalization of the effective mass. Besides, provided that the temperature is smaller than the Debye and Fermi temperatures, no temperature dependence of conductivity is observed, because under these circumstances inelastic processes are inoperative and the electron gas is degenerate.

It has been experimentally established that the transition from the metallic phase [where $\sigma(T) > 0$ for $T \to 0$] to the insulating phase [where $\sigma(T) \to 0$ for $T \to 0$], the Anderson–Mott transition, occurs for roughly $\bar{r} = 2.5a_B$. This means that n–InSb remains metallic down to $n \simeq 10^{14} \text{cm}^{-3}$ but n–CdSe, for instance, changes from being a metal to being an insulator at $n \simeq 3 \times 10^{17} \text{cm}^{-3}$. In terms of the wave function of electrons at the Fermi level, the metal–insulator transition (MIT) marks a transformation from extended states into localized states. Thus, as the kinetic energy involves space derivatives of the wave function, the MIT is obviously associated with an increase of the kinetic energy of electrons. Therefore, quite generally, the MIT will occur provided that the localization results in such a drop of the potential energy which overcompensates the above mentioned increase in the kinetic energy.

A considerable experimental and theoretical effort has been devoted over the recent years to establish the dominant driving mechanisms of the MIT in bulk doped semiconductors [6,7]. In this section some of the proposed models are briefly described. Various effects of magnetic ions upon n_c is discussed in Sec. II. 3.

2.1 <u>Mott–Hubbard transition</u>: Numerical studies of electronic states in a crystal built up of hydrogen atoms suggest that the MIT occurs for $n_c^{1/3}a_B = 0.22$.[8] The driving force of this transition is the energy of the Coulomb interaction between electrons on the same site, U. For an appropriately large lattice constant the increase of the kinetic energy due to localization becomes smaller than the decrease of the potential energy related to the fact that electrons can no longer appear on the same site. In the insulating phase there is one electron localized around each atom, and thus the model explains how local magnetic moments can be formed. The MIT is, presumably, discontinuous in the sense that at n_c extended states undergo transformation to localized states with the localization radius $\sim a_B$. In contrast, in doped semiconductors, because of a random distribution of impurities and the presence of minority impurity charges, the localization could occur gradually. Within this scenario, the MIT marks the end–point of the transformation of the extended states with energies $\epsilon \leq \epsilon_F$ into the singly occupied local states characterized by the localization radius $\sim a_B$.

2.2 <u>Wigner crystallization</u>: In this model the MIT is also driven by the Coulomb interactions among electrons, but the impurities are assumed to act only as a uniformly charged neutralizing background. To some extend this situation is accomplished in the recently designed modulation doped heterostructures.[9] It has been also argued that such a model should apply if the conducting electrons originate from resonant donors with bound states above the Fermi energy ϵ_F.[10]. Standard arguments demonstrate that the critical concentration of the Wigner condensation should scale with a_B^{-3}.[10,11] The proportionality factor is anticipated to be, however, at least one order of magnitude smaller than that observed in doped semiconductors.

2.3 <u>Percolation transition</u>: Imagine a degenerate liquid of electrons to be placed in a randomly fluctuating potential produced by impurities. If the rms amplitude of the potential fluctuations increases, the volume occupied by electrons shrinks. Once the electron droplets turn to occupy less than $\sim 16\%$ of the sample volume, they cease to form a percolation cluster, and the materials turns into an insulator.[12] Such an MIT is continuous. Experimental studies[13] of a mixture of conducting and nonconducting materials imply that the conductivity vanishes with the critical exponent $t = 1.8 \pm 0.2$. When approaching the MIT from the insulating phase, the dielectric susceptibility diverges with the critical exponent $s = 0.7 \pm 0.1$. Obviously, the percolation model disregards effects of electron–electron interactions as well as quantum phenomena, such as tunneling and interference, which tend to delocalize and localize electrons, respectively.

2.4 <u>Anderson localization</u>: Anderson's MIT is the quantum localization of the Fermi liquid induced by scattering. It occurs due to a collective action of many scattering centers and would take place even if the potential of a single impurity, or defect, had no bound states. In macroscopically homogeneous systems, the Anderson localization is thought to set–in before the percolation transition. An interesting exception is probably the localization in 2d systems under the quantum Hall effect conditions.[14]

It appears, by now, that there are two main mechanisms by which scattering may give rise to the localization.[7,15,16] One of them is the quantum interference of scattered waves. It is known that, when the positions of scattering centers are spatially correlated (as, e.g. , atoms in a crystal) the interference enhances or diminishes the transition probability between two points in space depending on the electron wavelength. If, on the other hand, the scattering centers are distributed randomly, the interference terms will vanish after averaging over all possible configurations of scattering centers. In other words, the mean transition probability between two points is well approximated by the sum of probabilities of all possible electron paths, in agreement with the classical Boltzmann description of diffusion. This is no longer true, however, if we consider the probability of returning to the starting point. This is because the "optical paths", and hence the phases of the transition probability amplitudes, are identical for the clockwise and counterclockwise trajectories of an electron wave along the same sequence of scattering centers. Since such interference is constructive, it increases the probability of return and thus, diminishes the conductivity.

Another consequence of scattering by impurities is a substantial modification of the interactions between electrons. A simplified picture behind it is that the diffusion motion allows two quasiparticles to meet and interact several times. If the two quasiparticles have similar energies they will acquire a small phase difference between the successive meetings. In such a case the successive interaction events cannot be regarded as independent. This invalidates several conclusions of the standard Fermi liquid theory and, in particular, leads to corrections to the Boltzmann conductivity.

The role played by the above quantum effects grows with the increase of $\hbar/\epsilon_F\tau$, where τ is the momentum relaxation time. Since, according to the Brooks–Herring formula (and its modification by Moore[17]) $\epsilon_F\tau$ for ionized impurity scattering depends only on $n^{1/3}a_B$, the model in question is compatible with the proportionality of n_c to a_B^{-3}. An interesting aspect of these quantum corrections is their unusual sensitivity to the temperature and to the so–called symmetry lowering perturbations. The corrections depend on the inelastic scattering processes (appearing at $T > 0$, $\tau_{in} \sim T^{-p}$, where $p \geq 1$) as well as on the magnetic field, spin–orbit– and spin–disorder scattering, because these change the phase of the wave functions. The contribution from the electron–electron interactions is affected additionally by thermal broadening of the Fermi–Dirac distribution function, and by the spin–splitting, as the latter influences the time–evolution of the phase difference if the two quasiparticles come from the different spin subbands. Depending on the symmetry lowering perturbations present (magnetic field, spin–splitting, spin–orbit– or spin–disorder scattering) various universality classes have been identified, each characterized by a different set of critical exponents controlling the behavior of relevant variables in the immediate vicinity of the MIT.[16]

The models of the MIT listed above by no means exhaust all the possibilities. In particular, an interesting question arises whether it is justified to treat, say, phonons or a subsystem of interacting localized spins only as a source of scattering. While for the inelastic processes the corresponding scattering rates vanishes at $T \to 0$, it seems plausible that in this limit the dynamic aspects of these couplings could show up, affecting the localization.

3. Transition metals in semiconductors: an overview

It has been known for a long time that the influence of magnetic atoms on transport properties is ultimately related to the energetic position of d or f–like open shells in respect to the Fermi level and band edges. This position as well as the degree of electron localization on the magnetic shells are known to reflect the competition between two effects. The first is the on-site Coulomb repulsion between electrons (which tends to conserve the local character of d or f shells), while the second is the delocalizing influence of mixing (hybridization) between extended states and magnetic orbitals. Additionally, the nature of the magnetic ground state is determined by local distortions and spin–orbit coupling.

In the case of interest here, i.e., for transition metals in II–VI compounds, these effects result in a rich spectrum of possible situations. For instance, the uppermost occupied one-electron level created by Mn in the 2+ charge state is superimposed on the valence-band continuum–of–states, whereas that of Sc^{2+} lies probably above the bottom of the conduction band in, e.g., CdSe. A remarkable empirical observation is that the

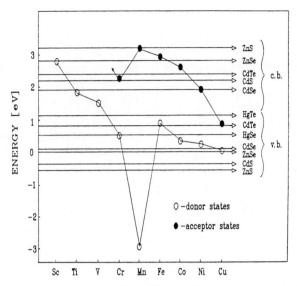

Figure 1. Approximate energy level positions of transition metals in the 2+ (open symbols) and 1+ (solid symbols) charge states relative to band edges of II–VI compounds (after Refs. 18–20).

distances between the energy levels of different transition metals remain, to a good approximation, the same in various hosts. These distances are schematically shown in Fig. 1 together with distances to the band edges of various II–VI semiconductors.[18–20] This plot enables an approximate determination of the level position when no direct experimental result is available. Moreover, it points to the existence of a common "reference energy", from which band offsets between various semiconductors can be obtained.[19–21]

In the Mn-based compounds,[22] the Pauli exclusion principle allows such a distribution of electrons between states created by Mn^{2+} that the total spin attains its maximum possible value, $S = 5/2$. This configuration reduces strongly the Coulomb-repulsion energy, and hence the uppermost level of Mn^{2+} resides deeply in the valence band. At the same time, because an extra electron must have an opposite spin orientation, the correlation energy U is very large. As sketched in Fig. 1, the level of Mn^{1+} is located above the bottom of the conduction band. Therefore, Mn in II–VI compounds has rather unique properties: it acts neither as a donor, nor as an acceptor, and does not perturb directly the density–of–states in the vicinity of the Fermi energy. However, since there is a localized spin on Mn ions, there exists a Kondo–like interaction, $-J\,S\cdot s$, between the spin of effective mass electrons and the localized spins. This interaction is brought about by the exchange part of the Coulomb interaction between the two subsystems as well as by quantum hopping (hybridization) between Mn and band states (kinetic exchange). The latter operates (and, in fact, dominates) for electronic states of a symmetries compatible with those of the impurity levels. This is the case for electrons (holes) from Γ_{15} band states, which are coupled to t states originating from the d–shell.

The exchange interaction causes spin–disorder scattering, spin splitting of electronic states, and formation of magnetic polarons. Because of the small wave vector of electrons in semiconductors, the spin–disorder scattering is usually weak, at least when compared to ionized impurity scattering. The spin splitting, in turn, being proportional to the macroscopic magnetization of d–spins, can become, at appropriately low temperatures, as large as $1-3$ meV in the external field of 1 kOe. Thus, in wide–gap compounds, it is not only greater than the Zeeman splitting, but also several times greater than the cyclotron energy. Therefore Mn–based compounds offer an interesting opportunity to examine the influence of the spin–splitting on transport phenomena. Still another effect of the exchange interaction is the existence of ferromagnetic clouds around the singly occupied local electronic states. The formation of such complexes – known as bound magnetic polarons – enhances the binding energy and shrinks the localization radius of localized electrons.

In many respects, Fe–based compounds[23] exhibit entirely different properties than the materials containing Mn. In particular, since there is an even number of d–electrons involved, the Kramers theorem does not apply, and the ground state of the Fe^{2+} ion is a magnetic singlet. Furthermore, according to Fig. 1, the uppermost occupied level of Fe^{2+} resides at relatively high energies, in most cases above the top of the valence band. This state is of e symmetry, and therefore, in the limit of weak spin–orbit interaction is not coupled to the band states from the vicinity of the center of the Brillouin zone. This weak coupling is indirectly confirmed by magnetooptical studies of the exchange interactions between holes and d–spins in $Zn_{1-x}Fe_xSe$.[24] These studies imply a negative sign of the exchange integral demonstrating that the kinetic exchange involves primarily t states, which are located below the top of the valence band. As shown by A. Mycielski et al.,[25] a particularly interesting situation occurs in the zero–gap $Hg_{1-x}Fe_xSe$, where the Fe–related level is superimposed on the conduction–band continuum–of–states. Thus, Fe acts as a resonant donor in HgSe. If the concentration of these donors is greater than the number of states available in the conduction band below the Fe states, the Fermi level is "pinned" by those states. Under such conditions various aspects of mixed valence systems can be studied.

Worth mentioning is also the case of CdSe:Sc and HgTe:Cu, where the transition metals appear to form resonant donors and acceptors, respectively. The problem of electron localization by impurities with no bound states can be studied in these systems.

II. EFFECTS OF THE sp–d EXCHANGE INTERACTION ON TRANSPORT PROPERTIES

1. Metallic region

1.1 Spin–splitting: Perhaps the most studied property of DMS is the exchange–induced spin–splitting of effective mass states. According to unnumerable magnetooptical works,[26] this spin–splitting – in the case of extended states – is proportional to macroscopic magnetization of magnetic ions, $M_0(T, H)$. Thus, for an s–type conduction band the total spin–splitting assumes the form

$$\hbar\omega_s = g^*\mu_B H + \frac{\alpha}{g\mu_B}M_0, \tag{1}$$

where the first term describes the Zeeman splitting of the conduction band; α is the s–d exchange integral, and $g \simeq 2.0$ is the Mn–spin Landé factor.

One of effects of spin–splitting on transport phenomena is a *redistribution of carriers between spin–subbands*. In degenerate semiconductors the redistribution starts when $\hbar\omega_s$ becomes comparable to the Fermi energy ε_F, and leads to an increase in the

Fermi wavevector of majority–spin carriers. Since the scattering rate depends usually on the Fermi wavevector, the redistribution may have an influence on transport phenomena. Such an approach is meaningful provided that neither the Landau quantization of the density of states, i. e, $\omega_c \tau \ll 1$, nor mixing of the spin states, i. e, $\omega_s \tau_s \ll 1$, are of significance, where τ and τ_s are the momentum and spin relaxation time, respectively. A detail quantitative discussion of the effect for various scattering mechanisms and degree of electron–gas degeneracy was given by Kautz and Shapira[27] in the context of their experimental results for EuS. The calculation was carried out within the first Born approximation and adopting the Thomas–Fermi dielectric function to describe the free–carrier screening of ionized impurity potentials. The above approximations should be valid deeply in the metallic phase, $n \gg n_c \approx (0.26/a_B)^3$. In this concentration range, Kautz and Shapira [27] predict up to twofold decrease of the low–temperature resistivity in the magnetic field, if mobility is limited by ionized impurity scattering. A similar calculation was performed for n-$Cd_{0.95}Mn_{0.05}Se$.[28] Since for this material in the metallic region $\hbar\omega_s < \varepsilon_F$, the field–induced redistribution of electrons between the spin–subbands is only partial, and thus its effect on the conductivity is rather small. In contrast, the exchange contribution to the spin–splitting modifies substantially the quantum (Shubnikov–de Haas) oscillations of the resistivity, visible best in narrow–and zero–gap DMS. This issue was reviewed in detail previously,[4,5] and will not be explored here.

As shown in the pioneering works of Bastard et al.,[29] Jaczyński et al.,[30] and Gaj et al.[31] particularly striking is the *influence of the p–d exchange interaction on the Γ_8 band*. According to Gaj et al.,[31] the exchange interaction splits the Γ_8 band into four components which become strongly non–parabolic and non–spherical. In the case of symmetry–induced zero–gap materials (such as $Hg_{1-x}Mn_xTe$ for $x \leq 0.07$) this splitting gives rise – for a certain range of x values and magnetic fields – to an overlap between the heavy hole and conduction bands.[29,30] Accordingly, the magnetoresistance of zero–gap $Hg_{1-x}Mn_xTe$ was found to contain a negative component,[32–35] in contrast to positive magnetoresistance of $Hg_{1-x}Cd_xTe$ with similar band structure parameters.[36] The negative magnetoresistance appears also in cases when no overlap seems to exist. It results presumably from delocalization of holes bound to resonant acceptor states.[34,35] In the case of open-gap DMS, the uppermost hole band was predicted to acquire the light–hole character in the direction perpendicular to the macroscopic magnetization (magnetic field), while to retain the heavy–hole character for the motion along the magnetic field.[31] Thus, in p–type metallic samples, a strong negative magnetoresistance is expected, as the redistribution of holes between the four subbands causes the majority of holes to change their transverse mass from the heavy mass to light mass. Experimental studies[37–42] of p–type open–gap DMS have indeed revealed the presence of a giant negative magnetoresistance associated with unusual anisotropy; the conductivity was larger in the transverse configuration. Except for one sample of Germanenko et al.,[40] the other studied samples were on the insulating side of the metal–insulator transition and, therefore, the magnetoresistance reflected the influence of the p–d exchange interaction on acceptor states or on the transition. We shall comeback to these results in further sections.

An interesting case represents magnetoresistance of metal–insulator–semiconductor structures containing p–type narrow–gap $Hg_{1-x}Mn_xTe$.[43] The a.c. conductivity of interfacial holes was probed by measuring absorption of far–infrared laser radiation in the accumulation regime. As shown in Fig. 2, a strong increase of conductivity was noted above a critical magnetic field, $B \simeq 2$ T. A quantitative calculation, depicted by a solid line in Fig. 2, has demonstrated that the critical field marks the onset of a spin–splitting–induced redistribution of holes between the heavy–hole and light–hole electric–subbands.

Another manifestation of the large spin–splitting could be the *extraordinary Hall effect*, known best from studies of ferromagnetic metals,[44] but also detected in non–magnetic semiconductors such as n–InSb and n–Ge.[45] The spin–dependent Hall effect appears under the presence of a spin–orbit interaction in the crystal, and its magnitude is proportional to the spin–polarization of the electron gas. Beside, the ratio of the extraordinary to ordinary Hall coefficient decreases with the electron mobility.[45] As far as DMS are concern, an anomaly of the Hall coefficient was noted by Brandt et al. and Gluzman et al.[46] near the spin–glass freezing temperature in $Hg_{1-x}Mn_xTe$. Though no detail microscopic mechanism has been proposed, it is probable that the effect is to be linked to a particularly large influence of the spin–orbit interaction, caused by the p–type character of the carrier wave functions in this narrow–gap semiconductor.

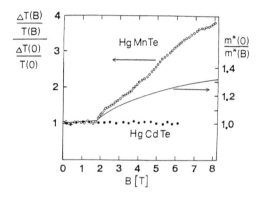

Figure 2. Transmission changes for the surface hole accumulation layer ($N_s = 10^{12} cm^{-2}$) vs. magnetic field, in p–$Hg_{0.87}Mn_{0.13}Te$ (open symbols) and p–$Hg_{0.74}Cd_{0.26}Te$ (solid symbols). The solid line presents the calculated inverse of the average effective mass at the Fermi level in an accumulation layer of p–$Hg_{0.87}Mn_{0.13}Te$ (after Chmielowski et al.[43]).

1.2 <u>Fluctuations of magnetization</u>: It is obvious that a replacement of non–magnetic cations by transition metals affects electron self–energy, as it changes crystalline potential, giving also rise to the appearance of a spin–dependent interaction. Within the first order perturbation theory, i.e., within the virtual–crystal and mean–field approximations, a modification of crystalline potential leads to a change of the energy gap, while the spin–dependent interaction produces a spin–splitting, which – according to Eq. 1 – is proportional to the macroscopic magnetization of d–spins. Proceeding further, the real part of the second–order term leads to an additional shift of electron energies,[47-52] while its imaginary part describes scattering between k–states.[27,53-57] Two distinct effects contribute to the second–order corrections to the electron self–energy in diluted magnetic systems: randomness of magnetic ion distribution (compositional fluctuations) and randomness of their spin orientations (fluctuations of magnetization). The latter depends on the spin–spin correlation function which, via the fluctuation–dissipation theorem and Kramers–Kronig relations, is determined by magnetic susceptibility of the system. The shift of electron energies induced by the fluctuations of magnetization has been shown to account for a non–standard dependence of the energy gap on the temperature T[49,58] and composition x[48] in DMS. It is also probable that the second order perturbation theory with the presence of the magnetic field taken into account [50,51] could explain some peculiarities in the dependence of $\hbar\omega_s$ on x[51] and T.[59,60]

We shall consider a contribution of scattering by magnetic ions to the momentum relaxation rate assuming $\hbar\omega_s \ll \varepsilon_F$, that is, disregarding the redistribution of electrons between the spin subbands. In such a case the scattering rate in question can be written as a sum of three independent components:[54,55]

$$\frac{1}{\tau} = \frac{1}{\tau_{al}} + \frac{2}{\tau_{sx}} + \frac{1}{\tau_{sz}}. \tag{2}$$

The first term describes the spin–independent part of chemical (alloy) scattering,

$$\frac{1}{\tau_{al}} = \frac{km^* N_0 x (1-x) V^2}{\pi \hbar^3}, \tag{3}$$

where k is the electron momentum, $N_0 x$ is the magnetic ion concentration, and $V N_0$ is the energy difference in the position of conduction band edges for the end–point materials, i.e., the heterostructure band offsets.[21] The remaining terms in Eq. 2 stands for electron scattering by fluctuations of magnetization. The corresponding scattering rate is determined by the time dependent spin–spin correlation function,

$$C\{S_i(t), S_j(0)\} = < (S_i(t) - x < S >), (S_j(0) - x < S >) >,$$

where indexes i and j run over all cation sites, and $-g\mu_B x < S >$ is the macroscopic magnetization. It is convenient to separate different mechanisms that contribute to the cross section. The first of them involves $< (S_i(t) - < S_i >), (S_j(0) - < S_j >) >$, and thus, via the fluctuation–dissipation theorem, the absorptive part of the dynamic susceptibility. This contribution corresponds to scattering by thermodynamic fluctuations of magnetization. Such scattering is inelastic and thus vanishes for $T \longrightarrow 0$. The remaining part of the correlation function, the static correlation function, $< (< S_i > -x < S >), (< S_j > -x < S >) >$, gives rise to elastic scattering. Two effects make the static correlation function to be non–zero in DMS: compositional fluctuations of magnetization specific to alloys, and spin–glass freezing.[55,56] Neglecting the latter we obtain

$$\frac{1}{\tau_{sx}} = \frac{km^* \alpha^2 k_B T}{4\pi \hbar^3 g^2 \mu_B^2} \tilde{\chi}_\perp(T, H); \tag{4}$$

$$\frac{1}{\tau_{sz}} = \frac{km^*}{4\pi\hbar^3} \left[\frac{\alpha^2 k_B T}{g^2 \mu_B^2} \tilde{\chi}_\parallel(T, H) + \frac{x(1-x)}{N_0} \left| \frac{d\hbar\omega_s(T, H)}{dx} \right|^2 \right], \tag{5}$$

where the second term in Eq. 5 arises from compositional fluctuations. The functions $\chi_{\perp,\parallel}(T, H)$ are given by

$$\tilde{\chi}_{\perp,\parallel} = \frac{\beta^2}{\pi} \int \frac{d\omega' \chi''_{xx,zz}(\omega')}{[\exp(\beta\omega') - 1][1 - \exp(-\beta\omega')]}, \tag{6}$$

where $\beta = \hbar/k_B T$; $\chi''_{xx}(\omega)$ and $\chi''_{zz}(\omega)$ denote, respectively, the transverse and longitudinal imaginary parts of the dynamic magnetic susceptibility of Mn spins for $q \leq 2k$.

If the energetic width of the Mn–spin excitation spectrum in the above q–range is smaller than $k_B T$, the denominator in Eq. 6 can be expanded to lowest order in $\beta\omega'$. In this high temperature case $\chi_{\perp,\parallel}(T, H)$ reduce, via the Kramers–Krönig relations, to the transverse and longitudinal static susceptibilities according to,

$$\tilde{\chi}_\perp(T, H) \simeq \chi_\perp(T, H) \equiv M(T, H)/H; \tag{7}$$

$$\tilde{\chi}_\parallel(T, H) \simeq \chi_\parallel(T, H) \equiv \partial M(T, H)/\partial H. \tag{8}$$

Inspection of Eq. 6 shows that the high temperature approximation (Eqs. 7,8) constitutes an upper limit estimate for actual values of $\chi_{\perp,\parallel}$. It is worth noting that while the efficiency of scattering by thermodynamic fluctuations decreases with the magnetic field, the efficiency of scattering by compositional fluctuations grows with the field. Thus, spin–disorder scattering in alloys may be responsible for both negative and positive magnetoresistance, depending on the field range, and character of the magnetization.

The above formalism can easily be adopted for *two–dimensional electronic systems*. If only one electric subbands is occupied, then k in Eqs. 3–5 is to be replaced by $\pi \int dz |f(z)|^4$, where $f(z)$ is the envelope function describing electron confinement, and the integration extends over the region where magnetic ions reside.[61,62]

In the *spin–glass phase*, in addition to inelastic scattering described by Eqs. 4 and 5, elastic scattering may operate. The corresponding scattering rate is proportional to the spin–glass order parameter $q = S^{-2} << S_i >^2>_i$.[55] Since at the spin freezing temperature T_f rather a slowing down of spin dynamics than a change in the spin configuration takes place, the total spin–disorder scattering rate should not exhibit any anomaly at T_f. This seems to be confirmed by experimental results for metallic spin glasses.[63] At the same time "after–effects" and a cusp in the Hall coefficient have been observed,[46,63] presumably because of the presence of remanent magnetization and the associated extraordinary Hall effect.

The formulae presented above describe the scattering rate within the lowest order Born approximation. After Kondo it is well known that higher order terms can modify the efficiency of scattering by the localized spin in a highly nontrivial way.[64] The sign of the *Kondo–effect corrections* depends on the sign of the s–d exchange integral α; for the ferromagnetic coupling the Born approximation overestimates the scattering rate. The corrections appears near and below the Kondo temperature, $T_k = \varepsilon_F \exp[-\rho(\varepsilon_F)|\alpha|/3]/k_B$. In semiconductors, T_k is very low because of the small density–of–states at ε_F, $\rho(\varepsilon_F) = m^* k_F/\pi^2\hbar^2$. Furthermore, the Kondo effect may be weakness even more by the fact that the concentration of itinerant electrons in DMS is usually much smaller than that of localized spins.[65] It seems, therefore, that the Kondo effect may not appear in DMS.

So far we have considered electrons with the s–type Bloch wave functions. The case of zinc–blende *zero–and narrow–gap semiconductors*, in which the conduction band wave function contains p–type components, was treated by Kossut.[54,66] The momentum relaxation time for alloy scattering is still given by Eq. 3 with V^2 replaced as follows:

$$V^2 \rightarrow V^2 a^4 - \frac{2}{3} V W a^2 (b^2 + c^2) + \frac{W^2}{3} \left[(b^2 + c^2)^2 + 2b^2 \left(\frac{1}{2} b - \sqrt{2} c \right)^2 \right], \tag{9}$$

where a represents a contribution of s–type component, while b and c p–type components in the Bloch function; W is the energy difference in the position of the Γ_8 valence–band edges for the end point materials, related to V and the energy gap $E_0 = E_{\Gamma_6} - E_{\Gamma_8}$ according to

$$\frac{dE_o}{dx} = N_o(V - W). \tag{10}$$

The treatment of spin–disorder scattering is more complex because the scattering rate depends on the direction of the electron momentum $\boldsymbol{k}$. Averaging over $\boldsymbol{k}$ orientations we obtain Eqs. 4 and 5 with

$$\alpha^2 \rightarrow \alpha^2 a^4 - \frac{2}{3}\alpha\beta a^2 \left(c^2 - \frac{1}{3}b^2\right) + \frac{\beta^2}{3}\left[c^4 + \frac{5}{6}b^4 + \frac{2b^2c^2}{3} + \frac{2\sqrt{2}}{3}b^3c\right], \tag{11}$$

where β is the p–d exchange integral.

The formulae above give the momentum relaxation rate which contributes to the resistivity. Another quantity of interest is the *electron lifetime* in a given state $\boldsymbol{k}$. The lifetime determines among other things the Dingle temperature,[67] that is, the damping of the quantum oscillations such as the Shubnikov – de Haas magnetoresistivity oscillations. For short range scattering potential and s–type electron wave functions, the momentum relaxation time and the lifetime coincide. If the electron wave function contains a p–type component, the lifetime is still determined by Eqs. 9 and 11, provided that the terms proportional to VW and $\alpha\beta$ are omitted.

Quantitative estimates show that spin–disorder scattering is usually much weaker than ionized impurity scattering in DMS. Its influence would therefore be hardly seen in standard resistivity or magnetoresistivity measurements. Nevertheless, the presence of spin–disorder scattering manifests itself in a number of situations. For instance, Wittlin et al.[68] have attributed a resistivity increase under EPR conditions of Mn spins to this scattering mode. As already noted, spin–dependent scattering play an important role near the metal–insulator transition, as it modifies quantum localization effects.[69] Furthermore, spin–disorder scattering, together with other spin relaxation mechanisms, determines the degree of luminescence polarization [70] and photomagnetization signal[71] under optical pumping conditions. As long as spin relaxation occurs in the conduction band the spin relaxation rate is $1/T_1 = 4/\tau_{sx}$, where τ_{sx} is given in Eqs. 4 and 7. The spin–disorder scattering contributes also to the width of the spin resonance line of conducting electrons. The corresponding transverse spin relaxation time assumes the form

$$\frac{1}{T_2} = \frac{2}{\tau_{sx}} + \frac{2}{\tau_{sz}}, \tag{12}$$

where τ_{sz} is given in Eqs. 5 and 8. Recently, a quantitative study of the linewidth of spin–flip Raman scattering in barely metallic $Cd_{0.95}Mn_{0.05}Se{:}In$ has been carried out.[57] As shown in Fig. 3, the analysis of experimental results demonstrates that both thermodynamic and compositional fluctuations of magnetization are of relevance. Besides, the linewidth was found to be enhanced by electron–electron interactions and interference phenomena, both of paramount importance because of the proximity to the metal–insulator transition.

Within the molecular field approximation the fluctuations of magnetization can be regarded as a source of an additional spin–splitting, inversely proportional to the square root of the volume occupied by the electron.[72,73] This effect would result in a gaussian broadening and a shift of the spin–resonance line unless an efficient averaging mechanism

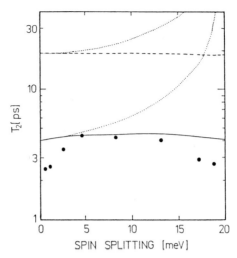

Figure 3. Spin–relaxation time versus spin–split energy for itinerant electrons, obtained by comparing linewidth of forward and backward spin–flip Raman scattering at 1.9 K in $Cd_{1-x}Mn_xSe$:In, $x = 0.05$, $n = 7 \times 10^{17} cm^{-3}$. The solid curve represents scattering by fluctuations of magnetization, including the enhancement due to electron–electron interactions and localization from nonmagnetic scattering. The dashed curve, Eqs. 4–8,12, neglects these enhancements. The dotted curves show the effect when compositional fluctuations are ignored (after Dietl et al.[57]).

exists. For delocalized electrons a transformation of the gaussian into the lorenzian line occurs due to the motional narrowing. Our estimates show that the resulting linewidth assumes the same form as that given above (Eqs. 4–8, 12). This indicates the equivalence of the two approaches to the linewidth. This conclusion may call into question the interpretation of magnetooptical spectra of n–$Hg_{0.9}Mn_{0.1}Te$ in terms of the fluctuation-enhanced spin–splitting.[73] Actually, similar spectra were recently detected in barely metallic non–magnetic n–$Hg_{0.8}Cd_{0.2}Te$.[74] They were interpreted as being due to the inter–impurity transitions, involving those donors which remained isolated.[74]

1.3 <u>RKKY interaction and free magnetic polaron</u>: So far we have neglected the influence of the electron liquid on the localized spins, and possible feedback effects on electron transport. It is well known that electrons mediate an exchange interaction between localized spins. Quite generally, the magnetization $M_0(T, H)$ which determines the spin-splitting (Eq. 1) and spin–disorder scattering rate (Eqs. 4,5) has to take the presence of this electron–mediated exchange interaction into account. In other words, disregarding a small diamagnetic correction as well as the paramagnetic term involving the band Landé factor g^*, $M_0(T, H)$ represents the true macroscopic magnetization of a given sample. For itinerant electrons in a narrow band, the carrier–mediated exchange interaction is to be described within the double exchange formalism.[75] In the case of a wide conduction band the celebrated RKKY model applies. According to the latter, if – as in DMS – the electron concentration is smaller than that of localized spins the resulting spin–spin interaction is ferromagnetic. Its effect on $\chi(T)$ can be estimated considering the Landau free energy of the system with the magnetization of d spins, M, and the net concentration of spin–down electrons, $\Delta n = n_\downarrow - n_\uparrow$, treated as order parameters,

$$\mathcal{F}[M, \Delta n] = \frac{M^2}{2\chi_0} - M \cdot H + \frac{(\Delta n)^2}{2\rho(\epsilon_F)\left(1 + \frac{F}{2}\right)} - \frac{1}{2}\left(g^*\mu_B H + \frac{\alpha M}{g\mu_B}\right)\Delta n. \qquad (13)$$

Here χ_0 is the magnetic susceptibility of localized spins in the absence of electrons, and F is the Fermi liquid parameter describing interactions between electrons. The individual terms in Eq. 13 represent respectively: (i) a change in the d–spin entropy and internal energy caused by the d–spin polarization; (ii) the Zeeman energy of d–spins; (iii) spin–polarization–induced change in the electron kinetic energy as well as in the potential energy of the electron–electron interaction; (iv) the Zeeman energy of electrons and the s–d energy in the molecular field approximation. By minimizing $\mathcal{F}[M, \Delta n]$ with respect to M and Δn we get their values corresponding to thermodynamic equilibrium, with the effects of electron–electron interactions taken into account. Inserting M and Δn to $\mathcal{F}$ we obtain the free energy F_{tot} of the system within the mean field approximation, which should be valid except for the proximity of the magnetic phase transition. The magnetic susceptibility $\chi_{tot} = -\partial^2 F_{tot}/\partial H^2$ becomes

$$\chi_{tot}(T) = \chi_0(T)\frac{\left[1 + \frac{1}{4}\left(1 + \frac{F}{2}\right)\rho(\epsilon_F)g^*\alpha/g\right]^2}{\left[1 - \frac{1}{4}\left(1 + \frac{1}{2}F\right)\rho(\epsilon_F)\frac{\alpha^2}{g^2\mu_B^2}\chi_0(T)\right]} + \frac{1}{4}\left(1 + \frac{F}{2}\right)\rho(\epsilon_F)g^{*2}\mu_B^2, \qquad (14)$$

and is seen to diverge for T given by

$$\frac{1}{4}\left(1 + \frac{1}{2}F\right)\rho(\epsilon_F)\frac{\alpha^2}{g^2\mu_B^2}\chi_0(T) = 1. \qquad (15)$$

Story et al.[76] discovered a ferromagnetic phase transition induced by free carriers in p–$Pb_{1-x-y}Sn_yMn_xTe$. The transition occurred once holes started to occupied a side band Σ, characterized by the large density of states $\rho(\epsilon_F)$.[77] The transition was accompanied by the appearance of a giant extraordinary Hall effect.[78]

A great deal of interest has attracted the question of magnetic polarons in DMS. At least three kinds of magnetic polarons can be distinguished: (i) bound magnetic polaron, i.e., electron localized by a defect or impurity potential, the s–d interaction supplying an additional binding energy; (ii) self–trapped magnetic polaron, i.e., electron in a localized, symmetry–broken state, the s–d coupling constituting the main binding force; (iii) free magnetic polaron, i.e., delocalized electron surrounded in its way through the crystal by a ferromagnetic cloud of d–spins.

The issue of bound magnetic polaron will be discussed in subsequent sections. As far as *self-trapped magnetic polaron* is concerned, its stability was calculated[79] to require a rather large sp–d coupling constant, effective mass, and magnetic susceptibility. Hence, such a complex can probably exist only in the form of the trapped heavy hole, particularly in systems of reduced dimensionality such as quantum wells.[80]

The case of the *free magnetic polaron* has proven to be very involved and, therefore, we limit ourselves to qualitative comments. Quantitative approaches[81] to this problem in DMS seem to suffer from improper treatment of the kinetic energy, and have turned out to be unsuccessful vis a vis experimental data.[57,60,82] The effects of the s–d interaction which we have discussed above were described within the adiabatic approximation, i.e., electronic energies were determined for frozen configurations of d–spins. From previous studies of the electron–phonon system we know that in order to describe the free polaron this approximation has to be relaxed. In the magnetic systems the adiabatic approximation is valid as long as electron–spin–density fluctuations decay faster than fluctuations of d–spin magnetization. This is not the case for quasiparticles with energies (calculated from the Fermi energy) smaller than the energetic width of the excitation spectrum of the d–spin subsystem. If, therefore, $k_B T$ is smaller than this width polaronic (dynamic) corrections to, say, transport coefficients may become important. It is worth noting in passing that just in this temperature range inelastic scattering by magnetic ions ceases to operate. Since there is no long range magnetic order in DMS, the magnetic excitations are, presumably, localized in space even at $T \to 0$. Thus, theoretical treatment of the problem cannot follow methods developed for magnetic semiconductors[83] or copper oxides.[84] Recently, Sadchev[85] has examined effects of the exchange coupling between the Fermi liquid electrons and a disordered system of localized spins. The analysis implies that the dynamic effects do affect the spin–spin correlation function of itinerant electrons and, in particular, the Fermi liquid parameter F. Whether these effects will perturb the conductivity is to be elucidated.

2. Insulating range

2.1 Spin–splitting: It is easy to show that if the spin–splitting of relevant donor levels and of the conduction band is the same, it does not affect the concentration of thermally activated electrons. Furthermore, within a standard Miller–Abrahams[86] model of the hopping conduction, no influence of the spin–splitting on the low–temperature conductivity is expected. A number of materials exhibits, however, a magnetoresistance in the hopping region. In order to account for the experimental findings, several modifications of the Miller–Abrahams model has been put forward. For instance, Goldman and Drew[87] in an attempt to explain the *negative magnetoresistance* of n-$Cd_{0.5}Mn_{0.1}Se$ have taken into consideration the influence of an exchange interaction between neighboring donor electrons upon the shape of the impurity band. As this inter–site exchange interaction is predominately antiferromagnetic, the spin–splitting may convert pairs of coupled electrons from singlets into triplets. According to Goldman and Drew,[87] this field–induced spin polarization diminishes the width of the impurity band, thereby increasing the density of states at ϵ_F, and thus the hopping conductivity.

A model for the spin–splitting–induced *positive magnetoresistance* in the hopping region has been proposed by Kurobe and Kamimura.[88] Their model assumes the width of the impurity band to be greater than the on–site correlation energy, so that singly and doubly occupied impurity states can coexist near ϵ_F. In such a case, in addition to hops between occupied and empty states, transitions between the two singly occupied neighboring impurities are also possible. Since the ground state of doubly occupied donors is singlet, these additional transitions are suppressed when electrons become spin–polarized. While the above models assumes the hopping process to be spin–conserving, Movaghar and Schweitzer[89] have argued that the hopping conduction is actually limited by the efficiency of transitions involving spin–flip of the electron. Although Movaghar and Schweitzer[89] have developed their model under assumption $\hbar \omega_s \ll k_B T$, which is usually not fulfilled in DMS, von Ortenberg et al.[90] have applied this model to describe the low–temperature magnetoresistance of n-$Zn_{1-x}Mn_xSe$.

Because of a complex influence of the exchange interaction on the fourfold degenerate *acceptor states*, the case of p–type materials requires a separate discussion. Delves[37]

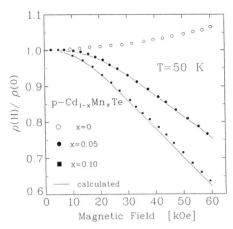

Figure 4. Magnetoresistance of p–CdTe and p–Cd$_{1-x}$Mn$_x$Te at $T = 50$ K, where conduction is due to free holes thermally activated to the valence band. Solid lines were calculated with no adjustable parameters taking orbital quenching of the acceptor spin-splitting and polaronic effects into account; the former accounts for about 90% of the effect (after Jaroszyński and Dietl[41]).

already in 1966 discovered a giant negative magnetoresistance of p–Hg$_{1-x}$Mn$_x$Te and interpreted it correctly as being caused by a different spin–splitting of the acceptor levels and the valence band. It has been theoretically demonstrated by Mycielski and Rigaux[91] that the p–d exchange spin–splitting of the acceptor state in the zinc–blende DMS is indeed about 20% smaller than that of the valence band at $k = 0$. Accordingly, the concentration of holes thermally activated to the valence band increases with the magnetic field. As shown in Fig. 4, this "boil off" effect, together with a destructive influence of the magnetic field on a polaronic contribution to the acceptor binding energy, explains quantitatively the negative magnetoresistance of p–Cd$_{1-x}$Mn$_x$Te at temperatures, where conduction is due to free holes in the valence band.[41] When the spin–splitting becomes comparable to the acceptor binding energy – a situation occurring in narrow–gap DMS at moderately strong magnetic fields – the spin–splitting is accompanied by a substantial increase in the transverse Bohr radius. This is because the uppermost valence subband acquires the light hole character. As noted by Mycielski and Mycielski,[92] and then elaborated by their successors,[93] this increasing trend turns into decreasing one when the magnetic length $l_H = (e\hbar/cH)^{1/2}$ becomes smaller than the Bohr radius. A number of groups[38–40,42,94] has undertaken magnetoresistivity studies of p–Hg$_{1-x}$Mn$_x$Te and p–Hg$_{1-x-y}$Mn$_x$Cd$_y$Te in the hopping region, where the conductivity is particularly sensitive to the acceptor wave function. These studies confirm the existence of both giant magnetoresistance and unusual anisotropy. As shown in Fig. 5, the magnetoresistance and its anisotropy change their sign in appropriately strong magnetic fields.[94] Such behavior supports theoretical expectations referred to above. A meaningful quantitative description of experimental results is, however, rather difficult since the change in the Bohr radius alters the hopping conductivity through at least three factors: (i) matrix element of transition probability;[93] (ii) width of the impurity band being determined by

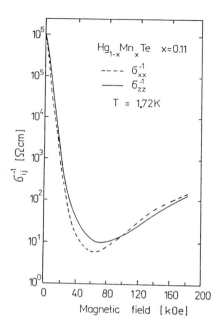

Figure 5. Inverse of the conductivity tensor components, σ_{xx}^{-1} and σ_{zz}^{-1}, versus magnetic field at 1.72 K for p–Hg$_{0.89}$Mn$_{0.11}$Te in the hopping region (after Wojtowicz et al.[94]).

fluctuations of composition;[95] (iii) energy of bound magnetic polarons.[38] Furthermore, it is unclear whether acceptor concentrations in the studied samples were low enough to neglect precursory effects of the insulator–to–metal transition.

2.2. Fluctuations of magnetization: It is clear that the hopping conductivity drops when the width of the impurity band increases, that is, the density–of–states near ϵ_F decreases. In ordinary semiconductors, the band width is primarily determined by electric fields of minority charges and defects. In DMS, an additional broadening of the impurity band arises from fluctuations of the composition and magnetization. For random distribution of magnetic ions this additional broadening of the donor band is gaussian with the field–dependent variance σ given by[96]

$$\sigma^2 = \left[\left(V - \frac{1}{2}\frac{d|\hbar\omega_s|}{dx}\right)^2 x(1-x) + \frac{\alpha^2}{2g^2\mu_B^2}k_BT\right]\frac{1}{8\pi a_B^3}. \tag{16}$$

If V and $d|\omega_s|/dx$ are of the same sign, the variance σ decreases with the magnetic field. A magnitude of the resulting negative magnetoresistance has been estimated for n–Cd$_{0.95}$Mn$_{0.05}$Se, and found to be small.[96] To the same conclusion leads a detail numerical study of the far–infrared light absorption in n–Cd$_{0.9}$Mn$_{0.1}$Se.[97]

It is worth mentioning that thermodynamic fluctuations may prompt electron hopping between impurities. Such hopping mechanism could dominate at very low temperatures, where the magnetic excitations appear to determine specific heat of the system. It seems that this interesting possibility has not yet been explicitly considered in the DMS literature.

2.3. Bound magnetic polarons. It has been realized by Kasuya and Yanase,[98] in the context of europium chalcogenides, that formation of bound magnetic polarons (BMP) may severely diminish the hopping conductivity. In principle, the DMS offer an appealing opportunity, from the one hand, to elucidate how the BMP affect transport phenomena, and on the other, to use the BMP as a tool for a better understanding of microscopic mechanisms underlying hopping conduction in doped semiconductors. Such favorable situation is due to detail spectroscopic measurements and their successful theoretical interpretation,[99] indicating that the basic properties of BMP in DMS are well understood. Those studies have demonstrated that the local magnetization which gives rise to zero–field spin–splitting is induced by a molecular field of the donor electron as well as by thermodynamic fluctuations of magnetization.[72]

The polaronic corrections to the impurity binding energy turned out to explain a strong increase of the acceptor ionization energy with the Mn concentration, estimated from conductivity measurements as a function of temperature in p–$Cd_{1-x}Mn_xTe$.[100] As already mentioned these corrections are also partly responsible for the negative magnetoresistance of p–$Cd_{1-x}Mn_xTe$, as depicted in Fig. 4.[41]

The influence of the BMP on the hopping conductivity in DMS has been considered by the present author and his co–workers[101] as well as by Ioselevich,[102] and Belyaev et al.[42] It has been suggested in the former work that the activation energy of the hopping conductivity in DMS is augmented because of a difference in the time–averaged spin–splitting of the occupied and empty impurity. If the thermodynamic fluctuations are taken into account, this model predicts a positive and negative magnetoresistance in small, $\hbar\omega_s \geq k_BT$, and strong, $g\mu_BH \geq k_BT$, magnetic fields, respectively. Such shape of the magnetoresistance is indeed observed experimentally.[90,101,103–106] Its calculated magnitude turns out, however, to be usually smaller than that observed.[101] This discrepancy was taken[101] as indicative of conduction proceeding through a simultaneous hopping of many electrons,[86] binding energy of each being altered by the polaronic corrections. It seems, however, by now that other precursory effects of the insulator–to–metal transition may also be involved.

Ioselevich[102] analyzed theoretically polaronic effects under assumption that resistances forming the Miller–Abrahams network are determined by the inverse of the time–averaged hopping probabilities. Because of thermodynamic fluctuations, the energy difference between the ground states of the two impurities fluctuates around its time–averaged value. It follows that the time–averaged probability, and thus the hopping conductivity, is even less affected by the BMP than that calculated from the previous model,[101] i.e., by using the time–averaged spin–splittings. Finally, Belyaev et al.[42] have detected two activation energies of the hopping conductivity in p–$Hg_{1-x}Mn_xTe$. The data have been analyzed supplementing Ioselevich's model by taking into account the influence of the exchange interaction on the localization radius. Obviously, the polaronic effects leads to a decrease of the ground–state localization radius, and to its increase if it happens that the carrier appears in the spin–reverse state. Thus, at high temperatures the hopping proceeds mainly through the spin excited states and is, therefore, characterized by a large activation energy. At low temperatures only the ground state is involved.

3. Vicinity of the metal–insulator transition

3.1. A look at experimental results: In order to visualize a strong and complex influence of the s–d exchange interaction on electron transport near the metal–insulator transition

(MIT) it is instructive to contrast dependencies of the resistivity on the magnetic field and temperature in non–magnetic and magnetic materials. The magnetoresistances of n–CdSe and n–Cd$_{0.95}$Mn$_{0.05}$Se at $T \simeq 2$ K are shown in Figs. 6 and 7.[96] The data span both sides of the MIT, as the Mott critical concentration n_c is about $3 \cdot 10^{17}$cm^{-3} in n–CdSe.[107] In the case of CdSe, the magnetoresistance contains both negative and positive components, the latter taking over at appropriately strong magnetic fields and low donor concentrations. The magnetoresistance of n–Cd$_{0.95}$Mn$_{0.05}$Se has a completely different character and much greater magnitude. As shown in Fig. 7, it is positive in low fields and negative in high fields. Qualitatively similar dependencies have also

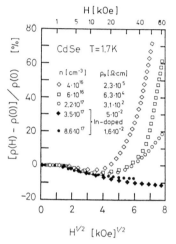

Figure 6. Magnetoresistance of n–CdSe as a function of the square root of the magnetic field at 1.7 K for selected electron concentrations n. The metal–insulator transition occurs at about 3×10^{17}cm^{-3} (after Dietl et al.[96]).

been observed for n–Cd$_{1-x}$Mn$_x$Se with other Mn concentrations,[103,104] as well as for n–Cd$_{0.9765}$Mn$_{0.0235}$S,[105] n–Zn$_{1-x}$Mn$_x$Se,[90] and n–Cd$_{0.95}$Mn$_{0.05}$Te.[106] An interesting aspect of the data is that the magnetoresistance exhibits a similar character on the both sides of the MIT, its magnitude decreasing away from the MIT. This suggests strongly that the magnetoresistance is primarily caused by the influence of the exchange interaction on the MIT. In light of this conclusion those descriptions of experimental results,[90,101] which have ignored the proximity of the MIT may require a reconsideration. Furthermore, since – according to Fig. 7 – n–type DMS exhibit a sizable negative magnetoresistance, its occurrence in p–type DMS may not entirely result from the peculiar influence of the p–d exchange interaction on the Γ_8 effective mass states.

DMS show also an unusual behavior in the absence of an external magnetic field. Figure 8 depicts a temperature dependence of the resistivity ρ in Cd$_{0.95}$Mn$_{0.05}$Se:In with the donor concentration corresponding to the close proximity to the MIT.[96] As shown, ρ increases with decreasing temperature T according to $\rho(T) \sim \exp(T_0/T)^\alpha$, where

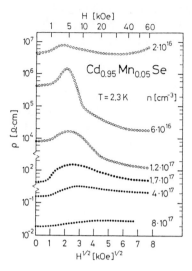

Figure 7. Resistivity of n–$Cd_{0.95}Mn_{0.05}Se$ as a function of the square root of the magnetic field at 2.3 K for selected electron concentrations n (after Dietl et al.[96]).

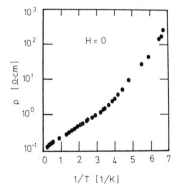

Figure 8. Resistivity of n–$Cd_{0.95}Mn_{0.05}Se$ as a function of the inverse temperature for $n = 4 \times 10^{17}cm^{-3}$ (after Dietl et al.[96]).

$\alpha > 1$. Such a dependence points out that this sample is on the insulating side of the transition. In similar samples of non–magnetic materials, particularly of n–CdSe,[107] the temperature dependencies of the resistivity are, however, much weaker: $1/4 \leq \alpha \leq 1/2$. The strong temperature dependence of ρ observed in magnetic materials is referred to as temperature dependent localization. A precursory of the effect was discovered by Sawicki et al.[108] for n–$Cd_{0.95}Mn_{0.05}Se$ on the metallic side of the MIT in the form of a sudden and unexpected increase of ρ below 0.5 K. The enhanced value of ρ disappeared under the presence of the magnetic field.[108] Similar anomalies were also detected by Germanenko at al.[35] in p–$Hg_{1-x}Mn_xTe$.

Thus, theoretical description of the experimental results ought to provide explanation for the low–field positive and high–field negative magnetoresistances as well as for the temperature dependent localization and the associated negative magnetoresistance. At the same time, the data on DMS may constitute a valuable tool to tell the dominant driving mechanisms of the MIT in doped semiconductors.

3.2. Influence of the s–d exchange interaction on the metal–insulator transition: Early experiments on magnetic semiconductors already demonstrated that the exchange interaction exerted a strong influence on the MIT.[1,2,109] Model descriptions of experimental findings have emphasized the role played by the redistribution of electrons between spin subbands,[110] spin–disorder scattering,[111] and formation of bound magnetic polarons.[110,112] Obviously, in what direction the perturbations in question will shift the critical carrier concentration n_c or how they will make n_c to vary with the magnetic field and temperature depends on the adopted model of the MIT.

As already mentioned, *the redistribution of electrons induced by the spin–splitting* increases the Fermi wavevector k_F of majority–spin electrons. From the previous discussion devoted to the models of the MIT we know that the increase of k_F will shift n_c towards lower electron concentrations in the case of both percolation transition and Anderson localization.[88,113] Making a natural assumption that the smaller n_c, the greater conductivity we see that within the mentioned models the spin–splitting leads to a negative magnetoresistance, appearing in relatively high magnetic fields $\hbar\omega_s \gtrsim \epsilon_F$. The spin–polarization of the electron–gas diminishes n_c also for the Wigner crystallization.[114] In contrast, the spin–polarization will increase n_c, and thus the resistivity if the MIT would be driven by the Hubbard–Mott mechanism. Indeed, since the redistribution of electrons between the spin subbands increases the screening length, it enforces the binding ability of the screened Coulomb potential.[110] A similar conclusion emerges when considering the insulator side of the MIT; the spin–polarization precludes quantum hopping between occupied states, increasing in this way an effective on–site correlation energy U, and thus n_c.

An interesting influence of the spin–splitting and spin–disorder scattering upon n_c shows up in the case of Anderson's localization. If electron–electron interactions are taken into consideration, these *symmetry lowering perturbations* shift n_c towards higher electron concentrations.[16] Accordingly, once $\hbar\omega_s$ is greater than k_BT and $\hbar/\tau_{si}$ – the spin–disorder scattering rate – the resulting magnetoresistance is positive. A contribution of these quantum effects can be analyzed quantitatively on the metallic side of the MIT, i.e., in the so–called weakly localized regime, where the perturbation approach should be valid.[15]

In principle, the mean free path, and thus n_c for the Anderson transition is directly altered by *alloy and spin–disorder scattering*. In most cases, however, the effect is of minor importance as these scattering modes are usually of a small efficiency in comparison to the ionized impurity scattering. Actually, the action of spin–disorder scattering upon n_c trough the quantum effects refereed to above appears to be stronger that through its direct influence on the mean free path. It is worth noting that near the MIT there may exist an additional source of spin–disorder scattering, that induced by those fluctuations of d–spin magnetization which are produced, via s–d coupling, by the inhomogeneous distribution of electron spin density, inherent to disordered systems.[96,110,115,116] If all electrons would be delocalized, a non–zero time–averaged local inhomogeneity in the s–spin density, and thus the fluctuation of d–spin magnetization may appear only under the presence of an external magnetic field.[110,116] The effect will operate as long as the magnetic field will not saturate all Mn spins in the crystal. A quantitative study[57] of the spin–flip Raman scattering spectra of conducting electrons in $Cd_{0.95}Mn_{0.05}Se$:In near the MIT does not point out to the significance of this scattering mechanism. Another source of additional fluctuations of magnetization could be the molecular field produced by localized electrons, i.e., *bound magnetic polarons*.[96,110,115] Obviously, localization associated with the polaron formation will lead to an increase of n_c independently of the model of the MIT. Since, in DMS, the polaron binding energy and the magnitude of the local magnetization drop with the temperature and magnetic field, the magnetic polarons may account for the temperature dependent localization, and the related negative magnetoresistance. A quantitative approach to the problem turns out, however, to be difficult because of our poor knowledge concerning mechanisms of the formation of the local electron moments near the MIT. We shall return to this topic when discussing experimental results.

As already mentioned, the s–d interaction, when treated going *beyond the adiabatic approximation*, mediates an interaction between the Fermi liquid electrons. Because of the crucial role played by electron–electron interactions in the localization phenomena, a contribution of the s–d coupling to the interaction amplitude may be of particular importance near the MIT. This interesting problem deserves further experimental and theoretical studies.

3.3 <u>Discussion of selected experimental results</u>: Results of one of the experimental attempts to detect the presence of quantum corrections to the conductivity of disordered metals are shown in Fig. 9, where the conductivity of the barely metallic CdSe:In and $Cd_{0.95}Mn_{0.05}Se$:In is plotted as a function of the magnetic field.[108] Though there is a qualitative difference between the results for CdSe and CdMnSe, in both materials the magnetoresistance (MR) appears at very weak magnetic fields, at which the cyclotron and spin splittings are much smaller than both the inverse momentum relaxation time and the Fermi energy. The Boltzmann–type theories of magnetoconductivity do not predict any MR in such a field range. In contrast, the theory[15] which takes the quantum corrections into account (solid lines in Fig. 9) is shown to describe correctly the experimental data. The MR of CdSe is negative reflecting the destructive effect of the magnetic field (vector potential) on the interference of scattered waves. A small positive MR visible below 1 K turns out to be due to the influence of spin–orbit scattering that operates in this acentric crystal upon the interference. In the case of $Cd_{0.95}Mn_{0.05}Se$, the MR stemming from the interference is found to be totally masked by the positive MR resulting from the influence of the giant s–d spin splitting upon electron–electron interactions. Since the calculation of $\sigma(H)$ in CdMnSe involves only one adjustable parameter – the effective amplitude of the Coulomb interaction – a good agreement between experimental and theoretical results has to be regarded as significant.

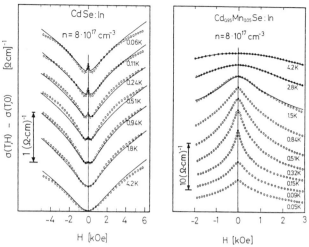

Figure 9. Conductivity changes versus magnetic field in metallic n–CdSe and n–Cd$_{0.95}$Mn$_{0.05}$Se at various temperatures (dots). Solid lines were calculated taking the quantum corrections[15] into account. Calculations for CdMnSe were limited to $T \geq 1.5$ K below which the magnetization, and hence the spin–splitting was unknown (after Sawicki et al.[108]).

Although the present theory[15] is successful in weak magnetic fields, it fails to describe the high–field negative component of the MR, visible in Fig. 7. As argued by Shapira et al.,[103] this negative MR is to be assigned to the spin–splitting–induced redistribution of electrons between the spin subbands; the redistribution increases the kinetic energy of carriers at the Fermi level diminishing their tendency towards Anderson localization.[113] The effect is particularly strong in p-type DMS, as the redistribution of holes between the Γ_8 subbands is connected with a change of the transverse hole mass from a heavy mass to a light one. The redistribution, together with the destructive effect of the magnetic field on the binding energy of bound magnetic polarons, is most probably responsible for the field–induced insulator-to-metal transition observed in magnetic semiconductors[109] as well as in n–Cd$_{0.95}$Mn$_{0.05}$Se[117] and p–Hg$_{0.92}$Mn$_{0.08}$Te.[117,118]

Figure 10 presents zero–temperature transverse and longitudinal conductivity, as well as the inverse Hall coefficient and dielectric constant, as functions of the magnetic field for p–Hg$_{0.92}$Mn$_{0.08}$Te.[118] Similar data for n–Hg$_{0.85}$Mn$_{0.15}$Te[119] are shown in Fig. 11. The data points were obtain from resistivity and capacitance measurements in the temperature range 30 mK$\leq T \leq 0.5$ K. It is seen that p–HgMnTe undergoes an insulator–to–metal transition when the magnetic field gets stronger. The narrow–gap n–HgMnTe undergoes, in turn, a traditional field–induced metal–to–insulator transition. Such a transition takes place because of the diamagnetic effects, which dominate when the electrons are confined at the lowest Landau level, causing the kinetic energy to decrease with increasing magnetic field.

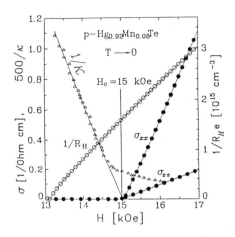

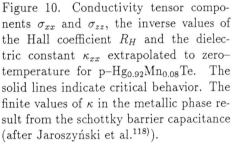

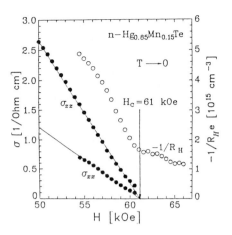

Figure 10. Conductivity tensor components σ_{xx} and σ_{zz}, the inverse values of the Hall coefficient R_H and the dielectric constant κ_{xx} extrapolated to zero–temperature for p–$Hg_{0.92}Mn_{0.08}Te$. The solid lines indicate critical behavior. The finite values of κ in the metallic phase result from the schottky barrier capacitance (after Jaroszyński et al.[118]).

Figure 11. Conductivity tensor components and the inverse Hall coefficient for n–$Hg_{0.85}Mn_{0.15}Te$ near the field–induced localization. Solid lines indicate critical behavior. Above H_c the conductivity seems to be perturb by accumulation or inversion layers (after Wróbel et al.[119]).

Several important conclusions can be drawn from the data displayed in Figs. 10 and 11. Firstly, in both cases the transition is continuous. This underlines the leading role of randomness, as MITs occurring in ordered systems (Mott–Hubbard transition, Wigner crystallization) would be, presumably, discontinuous in the 3d case. Secondly, the Hall coefficient R_H is seen to remain finite in the magnetic fields at which the conductivity vanishes. This contradicts the simple freeze–out picture in the framework of which the carrier concentration should vanish at the same field as the conductivity. The third conclusion comes from the fact that the diagonal components of the conductivity tensor tend to exhibit critical behavior of the form $\sigma_{ii} \sim |H - H_c|^t$, where $t = 1.0 \pm 0.2$. Such a value of t, observed commonly in doped semiconductors,[7] is in apparent disagreement with $t = 1.8 \pm 0.2$ expected from the percolation theory. This indicates that the transition is caused neither by long–range fluctuations of the impurity potentials nor by macroscopic inhomogeneities in the impurity distribution. At the same time $t = 1$ is in accord with theoretical predictions for Anderson's MIT, provided that effects of scattering–modified electron–electron interactions are taken into account.[16] This conclusion is supported by the temperature dependence of the conductivity near the MIT,[117] as well as by the character of the divergence of the localization length ξ at H_c, evaluated from the hopping conductivity[120] and dielectric constant[118] measurements on the insulating side of the MIT. The large values of ξ indicate also that the localization of electrons at the Fermi energy is caused by many scattering centers and must *not* be regarded as a freeze–out of carriers on individual impurities. Finally, in spite of the large field–induced anisotropy, visible in Figs. 10 and 11, the same values of H_c and t are observed for both σ_{xx} and σ_{zz}. This finding confirms the theoretical expectations concerning the quantum localization in anisotropic systems.[121]

The experimental results discussed above substantiate the picture of the MIT as the quantum localization of the Fermi liquid. According to this model, a Fermi–liquid–type description of electronic states remains valid as long as the size of the system is smaller than the localization length ξ. On the other hand, it is well known that deep in the insulating phase the electrons are bound to individual impurities forming, owing to the Hubbard energy U, localized magnetic moments. Obviously, in the latter case the electronic states cannot be described within the Fermi–liquid formalism at any length scale. A question then arises when does a transformation of the Fermi liquid electrons into local moments occurs. There is a growing number of evidences that this transformation begins already on the metallic side of the MIT, leading to the existence of local moments in metallic samples.[122] At the same time, it has been speculated[119] that the divergence of the Hall coefficient, visible in Figs. 10 and 11, marks the point where the Fermi–liquid electrons disappear totally.

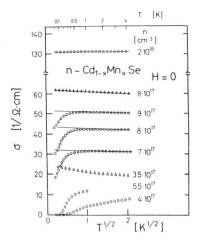

Figure 12. Zero–field conductivity of n–CdSe ($\triangle$) and n–Cd$_{0.95}$Mn$_{0.05}$Se (o). Solid lines were calculated taking quantum corrections into account but ignoring the presence of bound magnetic polarons (after Sawicki et al.[108] and Dietl et al.[123]).

The presence of local moments in metallic samples has been inferred, in particular, from the unusual temperature dependence of the conductivity discovered in Cd$_{0.95}$Mn$_{0.05}$Se:In below 0.5 K,[108] as shown in Fig. 12.[123] In magnetic materials the local moments can polarize, via s–d interaction, the neighboring Mn spins (there are about 200 Mn ions within the Bohr orbit in Cd$_{0.95}$Mn$_{0.05}$Se). The ferromagnetic bubbles (bound magnetic polarons) formed in this way constitute centers of spin–dependent scattering for the itinerant electrons. The efficiency of this scattering increases rather steeply with decreasing temperature, as the degree of the bubble polarization is proportional to the magnetic susceptibility of Mn spins, $\chi(T)$. It has been demonstrated that the drop of conductivity below 0.5 K can be semiquantitatively explained by taking into account the influence of this scattering mode on the quantum corrections to the conductivity.[123] It is also probable that the drop of σ is partly caused by the temperature dependent transformation of the Fermi–liquid electrons into local moments, as the binding energy of bound magnetic polarons becomes greater when the temperature is lowered.

3.4. <u>Conclusions</u>. Results of extensive studies summarized above indicate that the low–field positive magnetoresistance is primarily caused by the influence of the spin–splitting on those quantum corrections to the conductivity which originate from the scattering–modified electron–electron interactions. This interpretation is strongly supported by a quantitative agreement between the measured and calculated conductivities in the weakly localized regime,[28,106,108,124] where the perturbation approach should be valid.[15] The current theory turns out, however, to be less successful in the case of the Hall coefficient[124] as well as at very low temperatures, where effects of the bound magnetic polarons seem to show up.[96,108,123] The high–field negative magnetoresistance results presumably from the spin–splitting–induced repopulation of spin–subbands, and the associated increase in the kinetic energy of majority–spin electrons.[117] Such an increase, in terms of the Anderson model, reduces the relative magnitude of the quantum corrections to the conductivity, leading in this way to the negative magnetoresistance.[113] The relevance of Anderson's localization mechanism is further supported by the critical behavior of the conductivity[117–120] and the dielectric susceptibility[118] near the MIT. At the same time some experimental results, such as the temperature dependent localization[96] and the critical behavior of the Hall coefficient,[118,119] seem to indicate that certain elements of the Hubbard–Mott mechanism may also be of relevance in doped semiconductors.

III. TRANSITION METALS AS RESONANT IMPURITIES

1. Charge correlation in mixed–valence region

The experimental evidence that substitutional Fe impurity acts as a resonant donor in HgSe is presented in Fig. 13, where the low–temperature electron concentration n, as deduced from Hall data, is plotted as a function of the Fe concentration N_{Fe}.[125–128] It is seen that for small values of N_{Fe}, n increases with N_{Fe} reflecting the donor character of Fe. However, for N_{Fe} greater than a critical value $N_{Fe}^* \approx 4 \times 10^{18} \text{cm}^{-3}$, n tends to saturate indicating a pinning of the Fermi level to the resonant d–states. Surprisingly, the electron mobility μ increases sharply at the onset of the mixed valence phase, as shown in Fig. 13. For $N_{Fe} \approx 2 \times 10^{19} \text{cm}^{-3}$, μ attains a value which is about four times larger than that theoretically expected when considering scattering by a system of randomly distributed ionized donors, and neglecting resonant enhancement of the cross section. The high mobility values remain unchanged down to 40 mK[129] indicating a minor influence of the resonant scattering and the Kondo effect even for quasi–particle energies as low as ~ 4 μeV. The above results were substantiated by studies of the Shubnikov–de Haas and de Haas–van Alfen effects.[125,130] The quantum oscillations yielded the cross section of the Fermi sphere (electron concentration) in an accord with Hall data demonstrating the one–band character of the electron transport. Furthermore, the cyclotron effective masses were found to indicate a negligible modification of the conduction–band shape by the resonant states. In turn, the Dingle temperature in HgSe:Fe[125,130] is diminished in comparison to that of HgSe:Ga[67] proving independently that there is a significant reduction in the scattering rate of conduction–band electrons in HgSe:Fe.

These findings are understood at present in terms of a model put forward by J. Mycielski.[131] This model assumes the Fe resonant states to be long living, so that effects

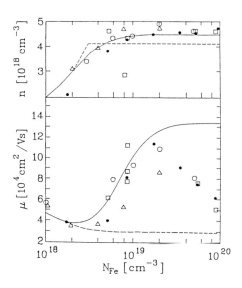

Figure 13. Electron concentration and mobility at 4.2 K for HgSe:Fe versus Fe concentration. Experimental points are taken from (o) Ref. 125; (•) Ref. 126; (□) Ref. 127; (△) Ref. 128. Solid lines were calculated within the short range correlation model; dashed lines ignoring inter–site Coulomb interactions after Wilamowski et al.[132]).

of hybridization can be neglected. If this is the case, the material is to be regarded as an inhomogeneous mixed–valence system, in which the inter–site Coulomb repulsions are the dominant interactions in the system. These interactions will tend to keep ionized donors apart leading, in an extreme case $N_{Fe} \gg N_{Fe}^*$, to a formation of a pinned Wigner crystal. It is the correlation of the positions of donor charges, which is supposed[131] to cause a reduction of the ionized–impurity scattering rate in the mixed–valence regime. Results of recent quantitative calculations,[132] depicted by solid lines in Fig. 13, support this conjecture. In those calculations, the structure factor $S(q)$ (which determines both the efficiency of ionized–donor scattering and the ground–state energy) has been evaluated at $T = 0$ taking into account short–range correlation induced by the screened Coulomb repulsion. In addition, Monte–Carlo simulations of the system have been carried out[132] yielding, among other things, the value of the one–particle density of impurity states. The interactions have been found to lead to a considerable broadening of this density–of–states as well as to the appearance of the so–called Coulomb gap at the Fermi energy ϵ_F.[133,134] It is the lack of impurity states at ϵ_F, which causes the resonant scattering to be suppressed. It is possible, however, that a residual contribution of this scattering is responsible for the mobility drop observed at high doping levels, $N_{Fe} > 2 \times 10^{19} \mathrm{cm}^{-3}$.

Extremely high electron mobilities were also found in zero–gap p–type $Hg_{1-x}Mn_xTe$ at low temperatures with the Fermi level pinned to resonant acceptor states.[34,135] This reduction of scattering efficiency is also though to be due to correlation induces by the inter–impurity Coulomb interactions.[4,136] In this material the ionized donors are likely to be in the vicinity of ionized acceptors owing to their mutual Coulomb attraction. Thus a system of dipoles develops which scatters the band carriers less efficiently then isolated donors do.

To summarize, there is a growing amount of evidences that there exist systems in which the Fermi–liquid and local states coexist. This coexistence, while contradicting the one–electron approach, can occur due to the on–site repulsion and the decoupling effect of the Coulomb gap that appears at ϵ_F in any system of localized charges interacting by means of the Coulomb forces. The same inter–site interactions impose spatial correlations in positions of impurity charges resulting in a dramatic reduction of the scattering rate of conduction–band electrons. This leads to the conclusion, which is of considerable practical significance, that the use of resonant donors or acceptors may result in semi-conducting materials having much greater mobility values than those characteristic of systems incorporating hydrogenic–like impurities.[132]

2. Localization by scattering

Semiconductors in which carriers originate from resonant impurities offer a novel opportunity to study electron localization. This is because in the case when potential of individual impurities has no bound states, the localization by the Mott-Hubbard mechanism should be inoperative. At the same time Anderson's metal–to–insulator transition may occur as a result of the collective action of many scattering centers on the electron motion.

Recently, a study of $Cd_{1-x}Mn_xSe$ doped with Sc, which according to Fig. 1 acts as a resonant donor, has been undertaken.[137] Preliminary results indicate that both $Cd_{1-x}Mn_xSe:Sc$ and $Cd_{1-x}Mn_xSe$ doped with hydrogenic–like impurities, In and Ga, exhibit a similar dependence of the conductivity on the electron concentration, temperature, and magnetic field in the vicinity of the metal–to–insulator transition. This may suggest that localization in doped semiconductors is caused primarily by Anderson's mechanism. There is some work in progress to substantiate this conclusion and, in particular, to find out whether Sc creates resonant states only.

IV. INFLUENCE OF CRYSTAL IMPERFECTIONS

1. Grain boundaries

It is well know that $Hg_{1-x}Mn_xTe$ ingots, grown either by the Bridgman or zone–travelling methods, consist usually of several essentially monocrystalline grains, with a typical disorientation angle of the order of several degrees. Transport studies[138] of samples which contained a single grain boundary (GB) conclusively demonstrated that defects in the GB plane have a donor character. Thus, in p–type material these defects may lead to the appearance of a two–dimensional inversion–layer in the GB vicinity. If such layer is formed, it determines the low–temperature conductivity, as holes in the bulk are usually localized. Studies of the inversion layer conductivity and magneto-conductivity have provided a great deal of information. In particular, the temperature measurements of the Shubnikov–de Haas oscillations revealed an influence of the s–d exchange interaction upon the spin splitting of the two–dimensional electron gas.[139] At the same time, it was experimentally established that the presence of magnetic ions had no significant effect on the precision of the Hall resistivity quantization.[140] Next, it was found that under hydrostatic pressure the concentration of electrons diminishes, whereas their mobility increases.[140–142] As shown in Fig. 14, under these conditions the quantum Hall effect can be observed at a magnetic field as low as 60 kOe and at a temperature as high as 4.2 K.[140] Detailed investigation[141,142] demonstrated that the

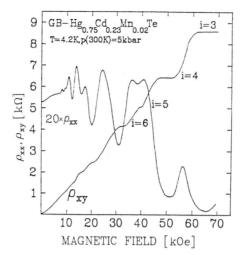

Figure 14. Diagonal and Hall resistivity at 4.2 K and under hydrostatic pressure of 5 kbar (applied at 300 K) versus magnetic field for two–dimensional electron gas adjusted to a grain boundary in p–$Hg_{0.75}Cd_{0.23}Mn_{0.02}Te$ (after Grabecki et al.[140]).

pressure dependence of transport coefficients reflects the pressure–induced change in the charge state of the GB donors accompanied by a strong lattice relaxation. Finally, a comparison was made between experimental and theoretical values of electric–subband occupations yielding information on the shape of the self–consistent potential.[142,143] The result seems to suggest the occurrence of substantial segregation of alloy composition in the GB plane.

The works discussed above are concerned with boundaries between strongly disoriented single–crystalline grains. It is, however, almost a rule in II–VI compounds that the grains themselves contain twinnings or exhibit a mosaic structure with a large number of low angles GBs. The corresponding concentration of GB donors is, presumably, too small to create inversion layers in p–type materials. It was pointed out,[144,145]

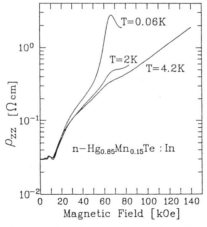

Figure 15. Longitudinal magnetoresistance of n–$Hg_{0.85}Mn_{0.15}Te$ at selected values of temperature (after Wróbel et al.[119]).

however, that these donors might be responsible for the presence of highly conducting accumulation layers that have been observed in n–ZnSe[144] and n–Hg$_{1-x}$Cd$_x$Te.[145,146] Figure 15 presents longitudinal MR of n–Hg$_{0.85}$Mn$_{0.15}$Te, which was interpreted on the same lines. A strong and temperature dependent positive MR is seen below 65 kOe. As argued in the previous Section this MR is caused by the field–induced Anderson localization. Surprisingly, a saturation, or even – at the lowest temperatures – a decrease of the resistivity with the magnetic field is observed between 65 and 80 kOe. Such results suggest the presence of accumulation–layer electrons,[38] or inversion–layer holes,[147] which can undergo the field–induced localization at higher fields than the bulk electrons, and whose conductivity can exhibit a complex field dependence. Since the anomalous magnetoresistance is remarkably isotropic, it was concluded[119] that accumulation layers reside at low–angles GBs rather than on the surface.

In summary, defects at GBs in II–VI compounds appear to be electrically active and, under some circumstances, determine electronic properties of the material. The studies[138–143] of the in–plane conductivity have provided information on both the nature of the GB defects and the properties of the two–dimensional electron gas in semimagnetic semiconductors.

2. Superconductivity by proximity effect

Recent millikelvin resistance and magnetic susceptibility measurements performed on HgSe:Fe have revealed the existence of superconductivity in some samples.[129] As shown in Fig. 16, the transition to a superconducting state occurs in a relatively wide temperature range. Furthermore, the diamagnetic signal and the resistance drop are seen to be quenched by rather small magnetic fields and electric currents.

In order to elucidate the origin of this phenomenon the experiments were carried out for samples prepared in various ways. It was found that the low–temperature drop of the resistance was suppressed in samples originating from stoichiometric melts, in which Fe impurities were compensated by an appropriate amount of Se. On the other hand the low–temperature anomaly appeared in the HgSe sample that contained an excess of Hg. The anomaly did not show–up, however, in those HgSe:Fe samples which were

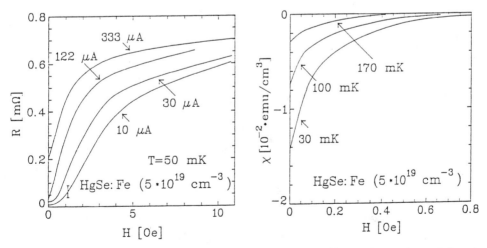

Figure 16. Magnetic field dependence of apparent resistance (for various values of electric current and of diamagnetic susceptibility for a HgSe:Fe sample with Fe concentration 5×10^{19}cm^{-3} (after Lenard et al.[129])).

annealed in saturated Se vapor, the procedure known to release the excess of Hg from the material.

The above series of experiments suggests strongly that the phenomenon is to be related to the presence of superconductive Hg inclusions which, by means of the proximity effect,[148] lead to the apparent macroscopic superconductivity of the bulk. It was estimated that the penetration radius of the Cooper pairs, because of the extremely long mean free path of electrons in HgSe:Fe, can be as large 40 μm at the lowest temperatures and magnetic fields experimentally accessible[129] (30 mK and $\sim$ 50 mOe, respectively). For this radius, 10^8 Hg–related inclusions per ccm may give rise to the development of the superconducting percolative cluster.

The conclusion which can be drawn from the above findings is that even a minute number of superconducting inclusions will result in macroscopic superconductivity of a normal metal provided that the temperature, the external magnetic field, and the electric current will be kept sufficiently small.

V. FINAL REMARKS

Looking once more at the energy diagram in Fig. 1, it is clear that the materials examined so–far by no means exhaust all interesting possibilities that are offered by DMS incorporating transition metals or, possibly, rare earth elements. Future studies of novel systems and their heterostructures will certainly lead to surprising discoveries which will shed new light on various fundamental aspects of electron transport in solids. At the same time the role played by native defects will be further elucidated. Progress in crystal growth and characterization will hopefully result in the reduction of the defect concentration, or even better, in their fruitful utilization for purposes of fundamental research and application.

ACKNOWLEDGEMENTS

I am grateful to W. D. Dobrowolski, G. Grabecki, J. Jaroszyński, J. Kossut, A. Lenard, W. Plesiewicz, M. Sawicki, W. Suski, L. Świerkowski, Z. Wilamowski, T. Wojtowicz, and J. Wróbel for many years of fruitful collaboration and discussions, and to R. R. Gałazka, A. Mycielski, and T. Skośkiewicz for continuous support and encouragement.

REFERENCES

[1] S. Methfessel, and D. C. Mattis, *Magnetic Semiconductors* in: *Handbuch der Physik*, Vol. XVIII (Springer, Berlin 1968); P. Wachter, in: *Handbook on the Physics and Chemistry of Rare Earths*, Vol. 2; Eds. K. A. Gschneidner and LeRoy Eyring (North Holland, Amsterdam 1979) p. 507.

[2] E. L. Nagaev, *Physics of Magnetic Semiconductors* (MIR, Moscow 1983).

[3] see, e.g., R. R. Gałązka and J. Kossut, *Landolt–Borstein, New Series Group III*, Eds. O. Madelung et al. (Springer, Berlin, 1983) Vol. 17b, p. 245; J. K. Furdyna, *J. Appl. Phys.* **64** (1988) R 29.

[4] N. B. Brandt and V. V. Moshchalkov, *Adv. Phys.* **33** (1984) 193; I. I. Lyapilin and I. M. Tsidilkovskii, *Usp. Fiz. Nauk* **146** (1985) 35.

[5] J. Kossut, in: *Diluted Magnetic Semiconductors*, Eds. J. K. Furdyna and J. Kossut, in: *Semiconductors and Semimetals*, Vol. 25, Eds. R. K. Willardson and A. C. Beer (Academic Press, New York, 1988) p. 183.

[6] for review see, e.g., *The Metal Non–Metal Transition in Disordered Systems*, Eds. L. R. Friedman and D. P. Tunstall (SUSSP Publ., Edinburgh 1978).

[7] see, e.g., *Anderson Localization*, Eds. T. Ando and H. Fukuyama (Springer, Berlin 1988).

[8] see, e.g., T. M. Rice, *Phil. Mag.* **B42** (1985) 419.

[9] see, e.g., M. Sundaran, A. C. Gossard, J. H. English, and R. M. Westervelt, *Superlattices and Microstructures* **4** (1988) 683.

[10] G. Nimitz, in: *High Magnetic Fields in Semiconductor Physics*, Ed. G. Landwehr (Springer, Berlin 1987) p. 491, and references therein; R. R. Gerhardts, ibid., p. 482.

[11] see, e.g., C. M. Care and N. H. March, *Adv. Phys.* **24** (1975) 101; I. M. Tsidilkovskii, *Usp. Fiz. Nauk* **152** (1987) 583 [*Sov. Phys. Usp.* **30** (1987) 676].

[12] B. I. Shklovskii and A. L. Efros, *Electronic Properties of Doped Semiconductors*, (Springer, Berlin 1984); R. Zallen, *The Physics of Amorphous Solids* (John Willey & Sons, 1983); B. A. Aronzon and I. M. Tsidilkovskii, *Phys. Stat. Sol. (b)* **157** (1990) 17.

[13] see, e.g., Y. Song, T. W. Noh, S.-I. Lee, and J. R. Gaines *Phys. Rev.* **B33** (1986) 904.

[14] see, e.g., S. Luryi, in Ref. 10, p. 16.

[15] for review see, e.g., B. L. Altshuler, A. G. Aronov, D. E. Khmelnitskii, and A. I. Larkin, in: *Quantum Theory of Solids*, Ed. I. M. Lifshits (Mir Publishers, Moscow 1982) p. 130; P. A. Lee and T. V. Ramakrishnan, *Rev. Mod. Phys.* **57** (1985) 287; B. L. Altshuler and A. G. Aronov, in: *Electron–Electron Interactions in Disordered Systems*, Eds. A. L. Efros and M. Pollak (North-Holland, Amsterdam 1985) p. 1; H. Fukuyama, ibid. p. 155.

[16] A. M. Finkelstein, *Zh. Eksp. Teor. Fiz.* **86** (1984) 367 [*Sov. Phys. JETP* **59** (1984) 212]; C. Castellani, C. Di Castro, P.A. Lee, and M. Ma, *Phys. Rev.* **B30** (1984) 527; C. Di Castro, in Ref. 7, p. 96; G. Kotliar, in Ref. 7, p. 107; T. Dietl, M. Sawicki, J. Jaroszyński, T. Wojtowicz, W. Plesiewicz, and A. Lenard, in: *Proc. 19th Int. Conf. on Physics of Semiconductors, Warsaw 1988*, Ed. W. Zawadzki (Institute of Physics, Warsaw 1988) p. 1189.

[17] E. J. Moore, *Phys. Rev.* **160** (1967) 607 and 618.

[18] see, e.g., P. Vogl and J. M. Baranowski, *Acta Phys. Polon.* **A67** (1985) 133; A. Zunger, in: *Solid State Physics*, Vol. **39**, Eds. F. Seitz and D. Turnbull (Academic Press, New York, 1986) p. 275; V. I. Sokolov, *Fiz. Tverd. Tela* **29** (1987) 1848 [*Soviet Phys.–Solid State* **29** (1987) 1061].

[19] J. M. Langer, C. Delerue, M. Lannoo, and H. Heinrich, *Phys. Rev.* **B38** (1988) 7723; A. Zunger, in Ref. 18.

[20] W. Dobrowolski, K. Dybko, C. Skierbiszewski, T. Suski, E. Litwin-Staszewska , J. Kossut and A. Mycielski, in: *Proc. 19th Intern. Conf. on Physics of Semiconductors*, Ed. W. Zawadzki (Institute of Physics, Warszawa 1988) p. 1247.

[21] T. Dietl and J. Kossut, *Phys. Rev.* **B38** (1988) 10941.

[22] For a review of Mn–based semimagnetics see, e. g., *Diluted Magnetic Semiconductors*, Eds. J. K. Furdyna and J. Kossut, in: *Semiconductors and Semimetals*, Vol. 25, Eds. R. K. Willardson and A. C. Beer (Academic Press, New York 1988).

[23] For review of Fe–based semimagnetics see, e. g., A. Mycielski, *J. Appl. Phys.* **63** (1988) 3279.

[24] A. Twardowski, P. Głód, W. J. M. de Jonge and M. Demianiuk, *Solid State Commun.* **64** (1987) 63.

[25] A. Mycielski, P. Dzwonkowski, B. Kowalski, B. A. Orłowski, M. Dobrowolska, M. Arciszewska, W. Dobrowolski, and J. M. Baranowski, *J. Phys.* **C19** (1986) 3605.

[26] J. A. Gaj, in Ref. 22, p. 276.

[27] Y. Shapira and R. L. Kautz, *Phys. Rev.* **B10** (1974) 4781; see also D. J. Kim and B. B. Schwartz, *Phys. Rev.* **B15** (1977) 377.

[28] M. Sawicki, T. Dietl, and J. Kossut, *Acta Phys. Polon.* A **67** (1985) 399.

[29] G. Bastard, C. Rigaux, Y. Guldner, J. Mycielski, and A. Mycielski, *J. de Phys. (Paris)* **39** (1978) 87.

[30] M. Jaczyński, J. Kossut, and R. R. Gałązka, *Phys. Stat. Sol. (b)* **88** (1978) 73.

[31] J. A. Gaj, J. Ginter, and R. R. Gałązka, *Phys. Stat. Sol. (b)* **89** (1978) 655.

[32] A. Sandauer and P. Byszewski, *Phys. Stat. Sol. (b)* **109** (1982) 167.

[33] A. B. Davydov, B. B. Ponikarov, and I. M. Tsidilkovskii, *Phys. Stat. Sol. (b)* **101** (1981) 127; *Fiz. Tekh. Poluprov.* **15** (1981) 881; B. B. Ponikarov, I. M. Tsidilkovskii, and N. G. Shelushina, *Fiz. Tekh. Poluprov.* **15** (1981) 296.

[34] M. Sawicki, T. Dietl, W. Plesiewicz, P. Sękowski, L. Śniadower, M. Baj, and L. Dmowski, in:*High Magnetic Fields in Semiconductor Physics*, ed G. Landwehr (Springer, Berlin 1983) p. 382.

[35] A. B. Germanenko, L. P. Zverev, V. V. Kruzhaev, G. M. Minkov, and O. E. Rut, *Fiz. Tverd. Tela* **27** (1985) 1857; *Fiz. Tekh. Poluprov.* **20** (1986) 80.

[36] see, e.g., I. M. Tsidilkovskii, W. Giriat, G. I. Kharus, and E. A. Neifeld, *Phys. Stat. Sol. (b)* **64** (1974) 717.

[37] R. T. Delves, *Proc. Phys. Soc.* **87** (1966) 809.

[38] T. Wojtowicz and A. Mycielski, *Physica* **117–118 B** (1983) 475.

[39] W. B. Johnson, J. R. Anderson, and D. R. Stone, *Phys. Rev.* **B29** (1984) 6679; J. R. Anderson, *Physica B* **164** (1990) 67.

[40] A. V. Germanenko, L. P. Zverev, V. V. Kruzhaev, G. M. Minkov, O. E. Rut, N. P. Gavaleshko, and W. M. Frasunyak, *Fiz. Tverd. Tela* **26** (1984) 1754; A. V. Germanenko, V. V. Kruzhaev, G. M. Minkov, and O. E. Rut, *Fiz. Tekh. Poluprov.* **20** (1986) 1662.

[41] J. Jaroszyński and T. Dietl, *Solid State Commun.* **55** (1985) 491.

[42] A. E. Belyaev, Yu. G. Semenov, and N. V. Schevchenko, *Pisma Zh. Eksp. Teor. Fiz.* **48** (1988) 623; see also, ibid. **51** (1990) 164 [*JETP Lett.* **48** (1988) 675; **51** (1990) 186].

[43] M. Chmielowski, T. Dietl, P, Sobkowicz, and F. Koch, in: *Proc. 18th Intl. Conf. Physics of Semiconductors*, Ed. O. Engström (World Scientific, 1987) p. 1787.

[44] see, e. g., F. E. Maranzana, *Phys. Rev.* **160** (1967) 421.

[45] J. N. Chazalviel, *Phys. Rev.* **B11** (1975) 3918.

[46] N. B. Brandt, V. V. Moshchalkov, A. O. Orlov, L. Skrbek, I. M. Tsidilkovskii, and S. M. Chudinov, *Zh. Eksp. Teor. Fiz.* **84** (1983) 1059; N. G. Gluzman, N. K. Lerinman, L. D. Sabirzyanova, I. M. Tsidilkovskii, and V. M. Frazunyak, *Pisma Zh. Eksp. Teor. Fiz.* **43** (1986) 600 [*Sov. Phys. JETP Lett.* **43** (1986) 777].

[47] F. Rys, J. S. Helman, and W. Baltensperger, *Phys. kondens. Materie* **6** (1967) 105.

[48] R. B. Bylsma, J. Kossut, W. M. Becker, U. Dębska, and D. Yodershort, *Phys. Rev.* **B33** (1986) 8207.

[49] J. A. Gaj and A. Golnik, *Acta Phys. Polon.* **A71** (1987) 197.

[50] S. M. Ryabchenko, Yu. G. Semenov, and O. V. Terletskii, *Phys. Stat. Sol. (b)* **144** (1987) 661.

[51] A. K. Bhattacharjee, *Solid State Commun.* **65** (1988) 275.

[52] J. Blinowski and P. Kacman, *Acta Phys. Polon.* **A75** (1989) 215.

[53] C. Haas, *Phys. Rev.* **168** (1968) 531; P. Leroux-Hugo, *J. Mag. Mag. Mat.* **3** (1976) 165.

[54] J. Kossut, *Phys. Stat. Sol. (b)* **72** (1975) 359; **78** (1976) 537.

[55] K. H. Fisher, *Z. Physik* **34** (1979) 45.

[56] A. A. Abrikosov, *Adv. Phys.* **29** (1980) 869.

[57] T. Dietl, M. Sawicki, E. D. Isaacs, M. Dahl, D. Heiman, M. J. Graf, S. I. Gubarev, and D. L. Alov, *Phys. Rev. B*, in press.

[58] J. A. Gaj, A. Golnik, J.–P. Lascaray, D. Coquillat, and M. C. Desjardins–Deruelle, in: *MRS Conf. Proceedings*, Vol. 89, Eds. R. L. Aggarwal et al. (MRS, Pittsburg 1987) p. 59.

[59] M. Dobrowolska, W. Dobrowolski, M. Otto, T. Dietl, and R. R. Gałązka, *J. Phys. Soc. Japan.* **49** (1980) 815, Suppl. A.

[60] H. Pascher, P. Rothlein, G. Bauer, M. von Ortenberg, *Phys. Rev.* **B40** (1989) 10469.

[61] G. Bastard, *Surf. Sci* **142** (1984) 284; *Appl. Phys. Lett.* **43** (1983) 591.

[62] G. Bastard and L. L. Chang, *Phys. Rev.* **B41** (1990) 7899.

[63] for review of spin-glasses, see, e.g., K. Binder and A. P. Young, *Rev. Mod. Phys.* **58** (1986) 801.

[64] see, e. g., P. A. Wiegman, in: *Quantum Theory of Solids*, Ed. I. M. Lifshits (Mir, Moscow 1982) p. 238.

[65] P. Noziers, *Ann. Phys.* **10** (1985) 19.

[66] J. Kossut, *Phys. Stat. Sol. (b)* **86** (1978) 593.

[67] T. Dietl, *J. Physique (Paris)* **39** (1978) C6–1081.

[68] A. Wittlin, M. Grynberg, W. Knap, J. Kossut, and Z. Wilamowski, *J. Phys. Soc. Japan* **49** (1980) 635, Suppl. A.

[69] Y. Ono and J. Kossut, *J. Phys. Soc. Japan* **53** (1984) 1128.

[70] J. A. Gaj, in: *Proc. 17th Intern. Conf. Physics of Semiconductors*, Eds. J. D. Chadi and W. A. Harrison (Springer, New York 1984) p. 1423; J. Warnock, D. Heiman, P. A. Wolff, R. Kershaw, D. Ridgley, K. Dwight, A. Wold, and R. R. Gałązka, *ibid*, p. 1407.

[71] H. Krenn, W. Zawadzki, and G. Bauer, *Phys. Rev. Lett.* **55** (1985) 1510; D. D. Awschalom, J. Warnock, and S. von Molnar, *ibid.* **58** (1987) 812; H. Krenn, K. Kaltenegger, T. Dietl, J. Spałek, and G. Bauer, *Phys. Rev.* **B39** (1989) 10918.

[72] T. Dietl and J. Spałek, *Phys. Rev. Lett.* **48** (1982) 355; *Phys. Rev.* **B28** (1983) 1548.

[73] R. Stępniewski, *Acta Phys. Polon.* **A69** (1986) 1001; *Solid State Commun.* **58** (1986) 19.

[74] V. J. Goldman, H. D. Drew, M. Shayegan, and D. A. Nelson, *Phys. Rev. Lett.* **56** (1986) 968.

[75] see, e.g., P. W. Anderson, in *Solid State Physics*, Eds. F. Seitz and D. Turnbull (Academic Press, New York, 1963) Vol. 14, p. 99.

[76] T. Story, R. R. Gałązka, R. B. Frankel, and P. A. Wolff, *Phys. Rev. Lett.* **56** (1986) 777;

[77] T. Story, L. Świerkowski, G. Karczewski, W. Staguhn, and R. R. Gałązka, in Ref. 20, p. 1567.

[78] A. Sandauer, *Phys. Stat. Sol. (a)* **111** (1989) K 219.

[79] S. M. Ryabchenko and Yu. G. Semenov, *Zh. Eksp. Teor. Fiz.* **84** (1983) 1419 [*Sov. Phys. JETP* **57** (1983) 825]; L. R. Ram–Mohan and P. A. Wolff, *Phys. Rev.* **B38** (1988) 1330; L. Świerkowski and T. Dietl, *Acta Phys. Polon.* **A73** (1988) 431.

[80] C. E. T. Goncalves da Silva, *Phys. Rev.* **B32** (1985) 6962; **33** (1986) 2923; J.-W. Wu, A. V. Nurmikko, and J. J. Quinn, *Solid State Commun.* **57** (1986) 853; *Phys. Rev.* **B34** (1986) 1080; D. R. Yakovlev, W. Ossan, G. Landwehr, R. N. Bicknell–Tassius, A. Waag, and I. N. Uraltsev, *Solid State Commun.*, in press.

[81] M. von Ortenberg, *Solid State Commun.* **52** (1984) 111; *Acta Phys. Polon.* **A69** (1986) 977; A. Golnik and J. Spałek, *J. Mag. Mag. Mat.* **54–57** (1986) 1207.

[82] D. L. Alov, G. I. Gubarev, and V. B. Timofeev, *Zh. Eksp. Teor. Fiz.* **86** (1984) 1124.

[83] see, e. g., A. L. Kuzemsky, D. I. Marvakov, and J. P. Vlahov, *Physica* **B138** (1986) 129.

[84] see, e. g., J. R. Schrieffer, X. G. Wen, and S. C. Zhang, *Phys. Rev. Lett.* **60** (1988) 944.

[85] S. Sadchev, *Phys. Rev.* **B39** (1989) 5297.

[86] See, e. g., M. Pollak and M. Ortuno, in: *Electron–Electron Interactions in Disordered Systems*, Eds. A. L. Efros and M. Pollak (North–Holland, Amsterdam, 1985) p. 409; A. L. Efros and B. I. Shklovskii, *ibid.* p. 483.

[87] V. J. Goldman and H. D. Drew, *Phys. Rev.* **B30** (1984) 6221.

[88] A. Kurobe and H. Kamimura, *J. Phys. Soc. Japan* **51** (1982) 1904.

[89] B. Movaghar and L. Schweitzer, *J. Phys.* **C 11** (1978) 125; see also Y. Osaka, *J. Phys. Soc. Japan* **47** (1979) 729.

[90] M. von Ortenberg, W. Erhardt, A. Twardowski, M. Demianiuk, in *High Magnetic Fields in Semiconductor Physics*, Ed. by G. Landwehr (Springer, Berlin 1987) p. 446.

[91] J. Mycielski and C. Rigaux, *J. de Phys. (Paris)* **44** (1985) 1041.

[92] A. Mycielski and J. Mycielski, *J. Phys. Soc. Japan* **49** (1980) 807, Suppl. A.

[93] T. R. Gawron, *J. Phys. C* **19** (1986) 21; **19** (1986) 29; A. D. Bykhovsky, E. M. Vakhabova, B. L. Gelmont, and A. L. Efros, *Fiz. Tekh. Poluprov.* **18** (1984) 2094; P. Janiszewski, in Ref. 20, p. 1285.

[94] T. Wojtowicz, T. R. Gawron, J. L. Robert. A. Raymond, C. Bousquet, and A. Mycielski, *J. Cryst. Growth* **72** (1985) 385.

[95] J. Mycielski, in Ref. 34, p. 431.

[96] T. Dietl, L. Świerkowski, J. Jaroszyński, M. Sawicki, and T. Wojtowicz, *Physica Scripta* **T14** (1986) 29.

[97] J. Mycielski, A. Witowski, M. Grynberg, and A. Wittlin, *Phys. Rev.* **B40** (1989) 8437; **B41** (1990) 5351.

[98] T. Kasuya and A. Janase, *Rev. Mod. Phys.* **40** (1968) 684; A. Janase and T. Kasuya, *J. Phys. Soc. Japan* **25** (1968) 1025.

[99] see, e. g., P. A. Wolff, in Ref. 22, p. 413.

[100] J. Jaroszyński, T. Dietl, M. Sawicki, and E. Janik, *Physica* **117–118 B** (1983) 473.

[101] T. Dietl, J. Antoszewski, and L. Świerkowski, *Physica* **117–118 B** (1983) 491.

[102] A. C. Ioselevich, *Pisma Zh. Eksp. Teor. Fiz.* **43** (1986) 148.

[103] Y. Shapira, N. F. Oliveira, Jr., D. H. Ridgley, R. Kershaw, K. Dwight, and A. Wold, *Phys. Rev.* **B34** (1986) 4187.

[104] T. Ichiguchi, H. D. Drew, and J. K. Furdyna, *Phys. Rev. Lett.* **50** (1983) 612; J. Stankiewicz, S. von Molnar, and W. Giriat, *Phys. Rev.* **B33** (1986) 3573; J. R. Anderson, W. B. Johnson, D. R. Stone, and J. K. Furdyna, *J. Phys. Chem. Solids* **48** (1987) 481.

[105] D. Heiman, Y. Shapira, and S. Foner, *Solid State Commun.* **45** (1983) 899.

[106] Y. Shapira, N. F. Oliveira Jr., P. Becla, and T. Q. Wu, *Phys. Rev.* **B41** (1990) 5931.

[107] D. M. Finlayson, J. Irvine, and L. S. Peterkin, *Phil. Mag.* **B39** (1979) 253; Y. Zhang, P. Dai, M. Levy, and M. P. Sarachik, preprint.

[108] M. Sawicki, T. Dietl, J. Kossut, J. Igalson, T. Wojtowicz, and W. Plesiewicz, *Phys. Rev. Letters* **56** (1986) 508; M. Sawicki, T. Wojtowicz, T. Dietl, J. Jaroszyński, W. Plesiewicz, and J. Igalson, in: *Proc. 18th Intern. Conf. on Physics of Semiconductors* Stockholm, 1986, Ed. O. Engström (World Scientific, Singapore 1987) p. 1265.

[109] see, e. g., Y. Shapira, S. Foner, and T. B. Reed, *Phys. Rev.* **B5** (1972) 4877; **B8** (1973) 2299; S. von Molnar, J. Flouquet, F. Holtzberg, and G. Remenyi, *Solid State Electron.* **28** (1985) 127; J. Stankiewicz, S. von Molnar, and F. Holtzberg, *J. Mag. Mag. Mat.* **54-57** (1986) 1217.

[110] J. Kubler and D. F. Vigren, *Phys. Rev.* **B8** (1973) 2299.

[111] M. I. Auslander, E. M. Kogan, and S. V. Tretyakov, *Phys. Stat. Sol. (b)* **148** (1988) 289.

[112] P. Leroux-Hugon, *Phys. Rev. Lett* **29** (1972) 939; J. B. Torrance, M. W. Shafer, and T. R. McGuire, *ibid.* **29** (1972) 1168.

[113] H. Fukuyama and K. Yosida, *Physica* **105 B+C** (1981) 132; *J. Phys. Soc. Japan* **46** (1979) 102.

[114] D. Ceperly and B. Alder, *Phys. Rev. Lett.* **45** (1980) 566.

[115] A. P. Grigin and E. L. Nagaev, *Pisma Zh. Eksp. Teor. Fiz.* **16** (1972) 438 [*Sov. Phys. JETP Lett.* **16** (1972) 312].

[116] T. Dietl and L. Świerkowski, *Acta Phys. Polon.* **A71** (1987) 213.

[117] T. Wojtowicz, T. Dietl, M. Sawicki, W. Plesiewicz, and J. Jaroszyński, *Phys. Rev. Letters* **56** (1986) 2419; T. Wojtowicz, M. Sawicki, T. Dietl, W. Plesiewicz, and J. Jaroszyński, in: *High Magnetic Fields in Semiconductor Physics*, Ed. G. Landwehr (Springer, Berlin 1987) p. 442;

[118] J. Jaroszyński, T. Dietl, M. Sawicki, T. Wojtowicz, and W. Plesiewicz, in: *High Magnetic Fields in Semiconductor Physics II*, Ed. G. Landwehr (Springer, Berlin 1989) p. 514.

[119] J. Wróbel, T. Dietl, G. Karczewski, J. Jaroszyński, W. Plesiewicz A. Lenard, M. Dybiec, and M. Sawicki *Semicond. Sci. Technol.* **5** (1990) S 299.

[120] T. Wojtowicz, M. Sawicki, J. Jaroszyński, T. Dietl, and W. Plesiewicz *Physica B* **153** (1989) 357;

[121] R. N. Bhatt, P. Wölfle, and T. V. Ramakrishnan, *Phys. Rev.* **B32** (1985) 569.

[122] R. N. Bhatt, M. A. Paalanen, and S. Sachdev, *J. Physique* **49–C8** (1988) 1179, and references therein; M. Milovanović, S. Sachdev, and R. N. Bhatt, *Phys. Rev. Lett.* **63** (1989) 82; M. Lakner and H. v. Lohneyson, *Phys. Rev. Lett.* **63** (1989) 648; D. Romero, M.–W. Lee, H. D. Drew, M. Shayegan, and B. S. Elman, in Ref. 7, p. 53.

[123] T. Dietl, M. Sawicki, T. Wojtowicz, J. Jaroszyński, W. Plesiewicz, L Świerkowski, and J. Kossut, in Ref. 7, p. 58.

[124] M. Sawicki and T. Dietl, in Ref. 20, p. 1217.

[125] W. Dobrowolski, K. Dybko, A. Mycielski, J. Mycielski, J. Wróbel, S. Piechota, M. Palczewska, H. Szymczak, and Z. Wilamowski, in: *Proc. 18th Intern. Conf. on Physics of Semiconductors*, Stockholm 1986, Ed. O. Engström (World Scientific, Singapore 1987) p. 1743.

[126] F. Pool, J. Kossut, U. Debska, and R. Reifenberger, *Phys. Rev.* **B35** (1987) 3900.

[127] N. G. Gluzman, L. D. Sabirzyanova, I. M. Tsidilkovskii, L. D. Paranchich, and S. Yu. Paranchich, *Fiz. Tekh. Poluprov.* **20** (1986) 94, 1996.

[128] C. Skierbiszewski, T. Suski, E. Litwin–Staszewska, W. Dobrowolski, K. Dybko, and A. Mycielski, *Semicond. Sci. Technol.* **4** (1989) 293.

[129] A. Lenard, T. Dietl, M. Sawicki, K. Dybko, W. Dobrowolski, T. Skośkiewicz W. Plesiewicz, and A. Mycielski, *Acta Phys. Polon.* **A75** (1989) 249; *J. Low. Temp. Phys.* **80** (1989) 15.

[130] M. Miller and R. Reifenberger, *Phys. Rev.* **B38** (1988) 3423 and 4120

[131] J. Mycielski, *Solid State Commun.* **60** (1986) 165.

[132] Z. Wilamowski, K. Świątek, T. Dietl, and J. Kossut, *Solid State Commun.* **78** (1990) 833; J. Kossut, W. Dobrowolski, Z. Wilamowski, T. Dietl, and K. Świątek, *Semicond. Sci. Technol.* **5** (1990) S260.

[133] T. Dietl, *Japan. J. Appl. Phys.* **26**, Suppl. **26-3** (1987) 1907.

[134] Z. Wilamowski, W. Jantsch, and G. Hendorfer, *Semicond. Sci. Technol.* **5** (1990) S266.

[135] C. Morrissy, *Ph. D. Thesis*, Oxford 1973, unpublished.

[136] Y. G. Arapov, A. B. Davydov, and I. M. Tsidilkovskii, *Fiz. Tekh. Poluprov.* **17** (1983) 24.

[137] P. Głód, M. Sawicki, A. Lenard, T. Dietl, and W. Plesiewicz, *Acta Phys. Polon.*, in press.

[138] G. Grabecki, T. Dietl, P. Sobkowicz, J. Kossut, and W. Zawadzki, *Appl. Phys. Letters* **45** (1984) 1214.

[139] G. Grabecki, T. Dietl, P. Sobkowicz, J. Kossut, and W. Zawadzki, *Acta Phys. Polon.* **A67** (1985) 297.

[140] G. Grabecki, T. Suski, T. Dietl, T. Skośkiewicz, and M. Gliński, in: *High Magnetic Fields in Semiconductor Physics*, Ed. G. Landwehr (Springer, Berlin, 1987) p. 127.

[141] T. Suski, P. Wiśniewski, L. Dmowski, G. Grabecki, and T. Dietl, *J. Appl. Phys.* **65** (1989) 1203.

[142] P. Wiśniewski, T. Suski, G. Grabecki, P. Sobkowicz, and T. Dietl, in: *Polycrystalline Semiconductors*, Eds. H. J. Moller et al. (Springer, Berlin 1989) p. 338.

[143] P. Sobkowicz, G. Grabecki, P. Wiśniewski, T. Suski, and T. Dietl, in Ref. [20], p. 611.

[144] T. Ohyama, E. Otsuka, T. Yoshida, M. Isshiki, and K. Igaki, *Surface Sci.* **170** (1986) 491.

[145] V. A. Pogrebnyak, D. D. Khalameida, and V. M. Yakovenko, *Solid State Commun.* **68** (1988) 891.

[146] M. Shayegan, V. J. Goldman, and H. D. Drew, *Phys. Rev.* **B38** (1988) 5585, and references therein.

[147] B. A. Aronzon, A. V. Kopylov, and E. Z. Meilikhov, *Fiz. Tekh. Poluprov.* **21** (1987) 1112; **23** (1989) 294 [*Sov. Phys. Semicond.* **21** (1987) 678; **23** (1989) 294].

[148] G. Deutscher and P. G. de Gennes, in: *Superconductivity*, Ed. R. D. Parks (Dekker, New York, 1969) p. 1005.

EXPERIMENTAL STUDIES OF THE d-d EXCHANGE INTERACTIONS IN DILUTE

MAGNETIC SEMICONDUCTORS

Y. Shapira

Department of Physics
Tufts University
Medford, MA 02155, USA

I. INTRODUCTION

Dilute magnetic semiconductors (DMSs) are compound semiconductors in which a fraction of the cations are magnetic. For many purposes one may regard a DMS as being composed of two subsystems: 1) the magnetic subsystem consisting of the $3d$ magnetic ions, and 2) the electronic subsystem consisting of the s-like electrons and p-like holes near the conduction and valence band edges. The most striking phenomena observed in DMSs are those which arise from the sp-d interaction between the magnetic and electronic subsystems. Because of this interaction, a perturbation of the magnetic subsystem, e.g. by applying a magnetic field H or changing the temperature T, affects the electronic susbsystem. This leads to a host of interesting magneto-optical and magneto-transport phenomena, such as a giant Faraday rotation, bound magnetic polarons, and giant magnetoresistance anomalies.[1-3]

In these lectures I shall focus only on the $3d$ magnetic subsystem. Specifically, I shall review the experimental methods which have been used to study the d-d exchange interactions between the $3d$ magnetic cations. Very little of what I shall say has to do with the sp-d interaction. For purposes of these lectures DMSs are merely examples of dilute magnetic systems with antiferromagnetic interactions.

Two types of methods of determining the d-d exchange interactions will be reviewed. The first uses the Curie-Weiss temperature θ, obtained from susceptibility measurements at high temperatures. The second consists of several techniques of probing the energy-level structure of small clusters of spins, mostly pairs. Because the information obtained from studies of pairs is more accurate, and also because of my own interest, the emphasis will be on "pair spectroscopy." The discussion will be limited to II-VI DMSs containing either manganese or cobalt.

II. CRYSTAL STRUCTURE AND MAGNETIC IONS

The best known examples of DMSs are II-VI compounds, e.g., $Cd_{1-x}Mn_xTe$. (Here, and throughout, x is the fraction of cations which are magnetic.) The crystal structure of these

Semimagnetic Semiconductors and Diluted Magnetic Semiconductors
Edited by M. Averous and M. Balkanski, Plenum Press, New York, 1991

compounds is either zinc-blende or wurtzite. In both of these structures each cation is at the center of a tetrahedron whose corners are occupied by the surrounding four anions. There is a strong evidence that the magnetic ions in DMSs are at the cation sites. Thus, a magnetic ion in either the zinc-blende or wurtzite structures finds itself in a tetrahedral crystal field. In the case of the wurtzite structure there is, in addition, a small uniaxial crystal field.

Looking at the cations alone, the cation sublattice of the zinc-blende structure is fcc, while that of the wurtzite is hcp. For either structure, each cation has 12 nearest-neighbor (NN) cations, and 6 next-nearest-neighbor (NNN) cations. The NNN distance r_{NNN} is larger than the NN distance r_{NN} by a factor of $\sqrt{2}$. The number of NN cations (both magnetic and non-magnetic) is usually designated by z_1, while the number of NNN cation sites is designated by z_2. Thus, $z_1 = 12$, and $z_2 = 6$. Next-nearest-neighbors are also called second neighbors. The numbers and distances of neighbors which are farther away than r_{NNN} (e.g., 3rd, 4th, and 5th neighbors) are different for the zinc-blende and wurtzite structures.[4]

In nearly all the works to date the magnetic ions in the II-VI DMSs were either Mn^{++} or Co^{++} or Fe^{++}. Among these, DMSs with Mn^{++} were studied much more extensively. For a free Mn^{++} ion the $3d^5$ electronic configuration leads, via Hund's rules, to an $^6S_{5/2}$ ground level with a zero orbital angular momentum, a spin S = 5/2, and a g factor of 2.00. If the Mn^{++} ion is placed in a crystal then the crystal field should not split the ^6S ground level because this level is orbitally non-degenerate. No spin-orbit splitting of the ground level is expected either, because L = 0. Experimentally, EPR studies[5,6] show some splittings of the ground level, but these splittings are very small, much less than 0.1 K. The g factors of Mn^{++} in II-VI DMSs differ from 2.00 by less than 1%. Thus, for most purposes the Mn^{++} ion may regarded as an ideal spin with S = 5/2 and g = 2.

The free Co^{++} ion has a $3d^7$ electronic configuration, which leads to a $^4F_{9/2}$ ground level. The value J = 9/2 is determined by the spin-orbit interaction. In a crystal, the magnetic ion is subjected to a crystal field which is stronger than the spin-orbit interaction. Accordingly, the energy levels in the crystal are obtained by starting from the ^4F level and applying the crystal-field perturbation first, followed by the spin-orbit interaction. The theory[7,8] indicates that in the zinc-blende structure the ground level is 4-fold degenerate (at H=0), and is well separated from the first excited level, by about 3500 cm^{-1} (equivalent to 5×10^3 K). The ground level can then be described by an effective spin S = 3/2. The g factor of this ground level is typically ~15% higher than the spin-only value g = 2.0. In the wurtzite structure a small axial crystal field splits the 4-fold ground level into two doublets. The splitting for Co^{++} in CdS and CdSe is of order 1 K.[9] The lowest four states then behave like those of an effective spin S = 3/2 which is subjected to a small uniaxial anisotropy of the form DS_z^2. The g factor in this case is slightly anisotropic.

For Fe^{++}, the $3d^6$ electronic configuration leads to a $^5D_4$ ground level for the free ion. The splittings of the ^5D level in a crystal with the zinc-blende or the wurtzite structures was discussed theoretically in Refs. 8, 10, 11. The crucial result is that the ground level in the crystal is a singlet. Such a ground level is magnetically inactive because it does not split when a magnetic field is applied. At low temperatures, where only the ground level is occupied, the susceptibility is due to a field-induced admixture of the ground state with excited states. This is the well known Van Vleck paramagnetism, which is temperature independent at low T. Because the first excited state of Fe^{++} in these materials is only ~20 K above the ground state, such a temperature independence is achieved only in the liquid-helium range.

The singlet ground state of Fe^{++} makes the physics of the magnetism of this ion quite different from that of Mn^{++} or Co^{++}. Here we shall be concerned only with the latter two ions. A review of the magnetism of Fe^{++} in DMSs was recently given by Twardowski.[12]

III. MAGNETIC INTERACTIONS OF THE 3d IONS

Our starting point are the individual 3d ions. As discussed above, a Mn^{++} ion may be regarded as an ideal spin S = 5/2, and a Co^{++} ion may be regarded as an effective spin S = 3/2. The latter is subjected to a weak single-ion anisotropy, DS_z^2, if the ion is in the wurtzite structure.

From general experience with 3d ions one expects that the dominant magnetic coupling *between* these ions is the isotropic exchange interaction

$$\mathcal{H}_{exch} = -2 \sum_{<ij>} J_{ij} \mathbf{S}_i \cdot \mathbf{S}_j , \tag{1}$$

where J_{ij} is the exchange constant between spins $\mathbf{S}_i$ and $\mathbf{S}_j$, and the sum is on all pairs $<ij>$ of magnetic ions. Usually the exchange constants J_{ij} decrease rapidly with distance. Therefore one expects that the exchange constant $J_1 \equiv J_{NN}$, between NN magnetic ions, will be the largest.

Theoretical calculations of the exchange constants for Mn^{++} ions in II-VI DMSs were carried out by Larson et al.[13] They showed that superexchange is the dominant exchange mechanism, and that J_1 is the largest exchange constant. This J_1 is antiferromagnetic (negative, in our notation), with a typical value of −10 K. The second-neighbor exchange constant $J_2 \equiv J_{NNN}$ is also antiferromagnetic, and is an order of magnitude smaller than J_1.

Two other types of exchange interactions, besides the isotropic exchange, may be present.[14] The usual (symmetric) anisotropic exchange is that part of the exchange interaction which depends on the directions of the spins relative to the lattice, and which is symmetric with respect to the interchange of the two interacting spins. One typical term, sometimes called the pseudodipolar exchange, has the form

$$D'(3S_{iz}S_{jz} - \mathbf{S}_i \cdot \mathbf{S}_j) .$$

In the present materials one expects this anisotropic exchange to be very small compared to the isotropic exchange.[15] We shall ignore the symmetric anisotropic exchange entirely.

A second type of anisotropic exchange is the Dzyaloshinski-Moriya (DM) interaction, which is antisymmetric with respect to the interchange of the two interacting spins. For a pair of spins, *i* and *j*, it has the form

$$\mathcal{H}_{DM} = -2\mathbf{D}_{ij} \cdot (\mathbf{S}_i \times \mathbf{S}_j) . \tag{2}$$

The largest $\mathbf{D}_{ij}$ is expected to be for nearest neighbors. Its magnitude D_1 for NN Mn^{++} ions was calculated by Larson and Ehrenreich.[16] The somewhat surprising result is that D_1 depends mainly on the spin-orbit coupling *in the anion*. As a result, D_1 increases with the atomic number of the anion, i.e. largest for tellurides and smallest in sulfides. Even for the tellurides the DM term is small, i.e., $|D_1/J_1| \cong 0.05$. For this reason we shall ignore the DM term in much of our discussion. Its effects will be considered separately, as a perturbation.

In the presence of a magnetic field **H** there is a Zeeman interaction

$$\mathcal{H}_{Zeeman} = \sum_i g\mu_B \mathbf{S}_i \cdot \mathbf{H} , \tag{3}$$

where μ_B is the Bohr magneton. Here, the tiny anisotropy of the g tensor was ignored, and this tensor was replaced by a scalar g factor. In the case of Co^{++}, an additional small H-dependent term should be added to the R.H.S. of Eq. (3). This term arises from the admixture of the ground state with higher orbital states.[8,9] The effects of this term are usually unimportant except at high temperatures. This term will be included only when needed.

IV. EXCHANGE CONSTANTS FROM THE CURIE-WEISS TEMPERATURE

The magnetic susceptibility of DMSs was discussed by Spalek et al.[17] At high temperatures the suscpetibility per unit volume χ obeys the Curie-Weiss law

$$\chi = \frac{C}{T - \theta}.$$ (4)

The Curie constant C is given by

$$C = xNg^2\mu_B^2 S(S+1)/3k_B,$$ (5)

where N is the total number of cations (magnetic and nonmagnetic) per unit volume and k_B is the Boltzmann constant. The Curie-Weiss temperature θ is given by

$$\theta = [2xS(S+1)/3k_B] \sum_i z_i J_i,$$ (6)

where J_i is the exchange constant between a central magnetic ion and a magnetic ion which is on the ith coordination sphere (sphere of ith distant neighbors), and z_i is the number of cation sites on that sphere. Assuming that J_1 is much larger than all other exchange constants, Eq. (6) gives

$$\theta \cong 2z_1 x S(S+1)J_1/3k_B,$$ (7)

where $z_1 = 12$ for the present materials.

Equations (6) and (7) are similar to the standard expressions for θ in an ordinary undiluted paramagnet ($x = 1$), except that in a DMS z_i is replaced by $z_i x$. The crucial assumption behind this replacement is that the magnetic ions are randomly distributed over the cation sites. For such a random distribution the average number of magnetic ions on the ith coordination sphere is $z_i x$.

Experimental data for θ are used to obtain information about the J_i's, via Eqs. (6) or (7). The most common practice has been to use Eq. (7) to estimate J_1. Another use of θ is to check that the magnetic ions are, in fact, randomly distributed. If this is the case then, from Eq. (6), θ should be proportional to x. Here, it is assumed that the J's are independent of x, which is expected to be true if the range of x is not too wide.

In extracting θ from the experimental data, several precautions must be exercised. The measured susceptibility must be corrected, to account for the (negative) diamagnetic susceptibility of the lattice, χ_d. In the case of Co^{++} one must also subtract a small temperature independent term which arises from the admixture of higher orbital states.[9] In all cases, it is

Table I. Values of J_1 deduced from the Curie-Weiss temperature θ.

Material	J_1/k_B (K)	Ref.
$Cd_{1-x}Mn_xTe$	-6.9	17
$Cd_{1-x}Mn_xSe$	-10.6	17
$Zn_{1-x}Mn_xTe$	-11.9	17
	-12	18
$Zn_{1-x}Mn_xSe$	-13.7	19
$Cd_{1-x}Co_xSe$	-37	9
$Zn_{1-x}Co_xSe$	-54	20
$Zn_{1-x}Co_xS$	-47	20

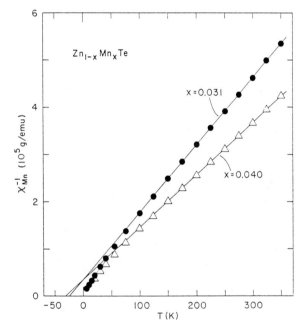

Fig. 1. Temperature dependence of the inverse of the low-field susceptibility χ_{Mn} due to the Mn spins in $Zn_{1-x}Mn_xTe$. Straight line are best fits of the data at $T \geq 175$ K to a Curie-Weiss law (Ref. 18).

necessary to ascertain that the temperature is sufficiently high for Eq. (4) to hold. A good way to do this is to estimate the error in θ by performing calculations using high-temperature series expansions. This procedure is discussed in Ref. 18.

Figure 1 shows an example of susceptibility data[18] used to extract θ. Table I lists some values of J_1 estimated from θ using Eq. (7).

V. NEAREST-NEIGHBOR CLUSTER MODEL (J_1 MODEL)

The most accurate methods of determining J_1 are based on studies of pairs, each consisting of two magnetic ions which are NN of each other. The NN cluster model (J_1 model) is a simple model which brings out the physics behind these methods. The main assumption of the model is that all exchange interactions except those between NN's can be ignored. This is a reasonable assumption in view of the fact that J_1 is the dominant exchange constant. Of course the NN cluster model will fail in some situations. For example, the magnetization calculated from this model may be seriously in error when J_2 is larger than both k_BT and the Zeeman energy per spin. Departures from the NN cluster model will be considered later. For now we consider a Hamiltonian which includes only NN isotropic exchange interactions [Eq. (1)] and the Zeeman energy [Eq. (3)]. Distant-neighbor exchange interactions and anisotropies of all kinds are ignored.

The NN exchange interactions can be regarded as "bonds" connecting NN spins. The magnetic ions (or spins) in a DMS are then viewed as belonging to clusters of different sizes. The smallest cluster is a single, consisting of a magnetic ion which has no NN magnetic neighbors. The next type of cluster is the NN pair (J_1 pair), consisting of two NN magnetic ions which have no other magnetic NN's. Following the pairs there are two types of triplets: closed and open. A closed J_1 triplet (CT) consists of three magnetic ions any two of which are coupled by J_1. In an open J_1 triplet (OT) there are J_1 bonds only between the first and second, and between the second and third spins (but not between the first and third spins). These clusters are sketched in Fig. 2. Besides these small-size clusters there are larger clusters, such as various types of quartets and quintets.

The NN cluster model has been employed in magnetism for many years. The model is useful only when the great majority of spins are in small-size clusters, usually not larger than triplets. This is the case only when the fraction x of magnetic ions is small, say $x < 0.05$ or, at most, 0.1.

For each small cluster it is possible to obtain an exact solution for the magnetic behavior as a function of T and H. This can be done because (in the J_1 model) the individual clusters are independent of each other. Having obtained these solutions, one can predict several effects. First, there are resonance or Raman-like excitations between energy levels of clusters of various types. Second, by summing the contributions of all the clusters one can calculate macroscopic quantities such as the susceptiblity, the magnetization, or the specific heat.

The solution for a single isolated spin is trivial. The energy levels resulting from the Zeeman splittings can be probed by EPR, but this only yields the g factor (in this simple model). The magnetization of the singles follows the Brillouin function, and their susceptibility obeys the Curie law.

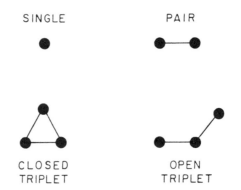

Fig. 2. The various types of small clusters (up to triplets) in the NN cluster model. The three members of an open triplet may or may not be on a straight line.

The Hamiltonian for a J_1 pair is

$$\mathcal{H} = -2J_1 \mathbf{S}_1 \cdot \mathbf{S}_2 + g\mu_B H(S_{1z} + S_{2z}) . \qquad (8)$$

This Hamiltonian is diagonal in the magnitude S_T of total spin of the pair, and in the projection m of the total spin S_T along $\mathbf{H}$. The energy levels can be obtained by noting that $\mathbf{S}_T = \mathbf{S}_1 + \mathbf{S}_2$ obeys the relation

$$\mathbf{S}_T \cdot \mathbf{S}_T = \mathbf{S}_1 \cdot \mathbf{S}_1 + \mathbf{S}_2 \cdot \mathbf{S}_2 + 2\mathbf{S}_1 \cdot \mathbf{S}_2 . \qquad (9)$$

With $S_1 = S_2 = S$ this leads to the energies

$$E = <S_T m \mid \mathcal{H} \mid S_T m> = -J_1[S_T(S_T + 1) - 2S(S + 1)] + g\mu_B m H . \qquad (10)$$

The energy-level diagram for a pair at $H = 0$ is shown in Fig. 3. (This figure is for Mn^{++}. The diagram for Co^{++} is identical except that there are no levels higher than $S_T = 3$.) For the ground level the total spin is zero, corresponding to an antiferromagnetic alignment of $\mathbf{S}_1$ and $\mathbf{S}_2$. For each successive level the total spin S_T increases by one unit, and the energy increases by an integer multiple of $2|J_1|$. Experimental probes of the energy levels in Fig. 3 will be considered later.

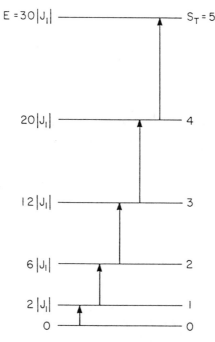

Fig. 3. Energy-level diagram for a J_1 pair of Mn^{++} spins at $H = 0$. Arrows show the allowed "Stokes" transitions for inelastic scattering of neutrons. The inverse "anti-Stokes" transitions between the same levels are also allowed.

The energy levels of open and closed J_1 triplets are known,[21] but will not be discussed here in detail. For either type of triplet the total spin S_T and its projection m along $\mathbf{H}$ are good quantum numbers. For a CT of either Mn^{++} or Co^{++} the ground level at $H = 0$ has $S_T = 1/2$. The first excited level is $3|J_1|$ higher. For an OT the ground level at $H = 0$ has $S_T = S$, i.e. $5/2$ for Mn^{++}, and $3/2$ for Co^{++}. The first excited level is several $|J_1|$ higher.

The number of clusters of a given type depends on the probability P_i that a magnetic ion is in this type of cluster. These probabilities were first calculated by Behringer for various crystal structures.[22] A random distribution of the magnetic ions was assumed. Let P_1, P_2, P_3, P_4 be the probabilities that a magnetic ion will find itself in a cluster which is a single, a J_1 pair, an open J_1 triplet (OT), or a closed J_1 triplet (CT), respectively. Then for the zinc-blende and wurtzite structures, $P_1 > P_2 > P_3 > P_4$ when $x < 0.1$.

The probability P_2 for being in a pair reaches a maximum of 24% at $x = 0.05$. The maximum probabilities of being in an OT or in a CT are 11% and 2.5%, respectively. Usually the calculations based on the NN cluster model consider only clusters with three spins or less. The probability $P_{>3}$ that a magnetic ion is in a cluster larger than a triplet increases with x. At $x = 0.03$, 0.05, and 0.10 the values of $P_{>3}$ are 3%, 11%, and 41%, respectively.

VI. DIRECT PROBES OF THE ENERGY LEVEL DIAGRAM FOR J_1 PAIRS

Studies of the energy levels of pairs in dilute magnetic materials predate the comparatively recent interest in DMSs. Two of the major experimental techniques have been: EPR and inelastic scattering of neutrons. EPR studies of pairs were reviewed by Owen and Harris.[14] When the anisotropic interactions are very small compared to the isotropic exchange, the EPR method is not optimal for obtaining J_1. One reason is that for a vanishing anisotropy, transitions between states with different S_T are forbidden. In that case J_1 cannot be obtained directly from the resonance frequencies. An estimate for J_1 can be obtained, however, based on the temperature dependence of line intensities, which reflect the Boltzmann factors of levels with different S_T. Thus far no accurate values for J_1's in DMSs were obtained by EPR, possibly because the anisotropy in DMSs with Mn^{++} and Co^{++} is small. EPR has been used recently to estimate the (small) anisotropic exchange in $Cd_{1-x}Mn_xTe$.[23]

Inelastic scattering of neutrons from pairs was reviewed by Furrer and Güdel.[24] This scattering is governed by the selection rules $\Delta S_T = 0, \pm 1$ and $\Delta m = 0, \pm 1$. Thus, the only allowed transitions at $H = 0$ are those between adjacent levels in Fig. 3. The energies of such transitions yield J_1 directly. At temperatures $k_B T \ll 2|J_1|$, only the ground level with $S_T = 0$ is occupied so that inelastic scattering can occur only between this level and the level with $S_T = 1$. The neutron energy in this case decreases by $2|J_1|$. In optical terminology the energy loss corresponds to a "Stokes" process. At higher temperatures, Stokes processes between higher levels, with energy losses of $4|J_1|$, $6|J_1|$, etc. become possible. In addition, "anti-Stokes" processes, e.g. from the $S_T = 1$ level to the ground level, are also possible at high temperatures.

The first inelastic scattering of neutrons from pairs in a DMS was carried out by Corliss et al. on $Zn_{1-x}Mn_xTe$.[25] An example of their data, showing the Stokes transition from the ground level to the first excited level, is given in Fig. 4. Extensive neutron studies of pairs in DMSs were carried out later by Giebultowicz and coworkers.[26–28] At present, neutron diffraction and magnetization steps (discussed later) are the leading techniques for measuring J_1. One limitation of the neutron method is the large neutron capture cross section for the naturally abundant Cd isotope. For this reason all neutron studies of pairs in DMSs performed thus far used DMSs whose cation in the parent compound is zinc. Values of J_1 obtained by neutron studies of pairs are listed in Table II.

Transitions between adjacent levels in Fig. 3 were also observed by Raman scattering of light.[29] Thus far the Raman experiments on pairs were successful only in $Cd_{1-x}Mn_xS$ and

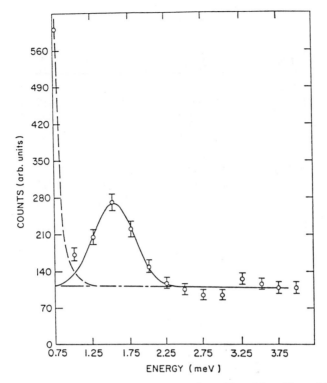

Fig. 4. Inelastic scattering of neutrons from $Zn_{0.97}Mn_{0.03}Te$ at 4.2 K. The peak is the "Stokes" transition from the ground state to the first excited state of a NN pair (Ref. 25).

$Cd_{1-x}Mn_xSe$, both of which have the wurtzite structure. The details of the Raman process are still unclear. The results for J_1 are included in Table II.

VII. TECHNICAL SATURATION AND MAGNETIZATION STEPS IN THE NN CLUSTER MODEL

The magnetization of a DMS at a temperature $k_BT \ll |J_1|$ was discussed in Refs. 18 and 30. These early discussions were based on the the NN cluster model (J_1 model). They explain the salient features which are observed in many experiments.

A. Technical saturation

For any cluster the magnitude S_T of the total spin is a good quantum number. The value of S_T in the zero-field ground level of the cluster will be called S_T(ground). At the other extreme, at very high magnetic fields where all the spins in the cluster are parallel, the cluster is in a state with $S_T = S_T$(ferro). The designation "ferro" ephasizes the ferromagnetic (parallel) alignment of the spins. For all clusters except singles,

$$S_T(\text{ground}) < S_T(\text{ferro}) .$$

Table II. Values of J_1 obtained from neutron and Raman studies of J_1 pairs.

Material	J_1/k_B (K)	Method	Ref.
$Zn_{1-x}Mn_xTe$	-9.35	Neutron	28
	-8.79	Neutron	25
$Zn_{1-x}Mn_xSe$	-12.3	Neutron	28
$Zn_{1-x}Mn_xS$	-16.1	Neutron	28
$Zn_{1-x}Co_xTe$	-38.0	Neutron	28
$Zn_{1-x}Co_xSe$	-49.5	Neutron	28
$Zn_{1-x}Co_xS$	-47.5	Neutron	28
$Cd_{1-x}Mn_xSe$	-8.1	Raman	29
$Cd_{1-x}Mn_xS$	-10.6	Raman	29

This is a consequence of the antiferromagnetic interaction.

For a single, $S_T(\text{ground}) = S_T(\text{ferro}) = S$, i.e., 5/2 for a single of Mn^{++} and 3/2 for Co^{++}. For a pair, $S_T(\text{ground}) = 0$, and $S_T(\text{ferro}) = 2S$. For an OT, $S_T(\text{ground}) = S$, which is only a third of the value of $S_T(\text{ferro})$. A CT of Mn^{++} or Co^{++} has $S_T(\text{ground}) = 1/2$. For a CT of Mn^{++} this corresponds to 1/15 of the value of $S_T(\text{ferro})$, whereas for a CT of Co^{++} $S_T(\text{ground})$ is 1/9 of $S_T(\text{ferro})$. For larger clusters, such as the various types of quartets, $S_T(\text{ground})$ is expected to be only a fraction of $S_T(\text{ferro})$. This is a consequence of the antiferromagnetic interaction.

At a temperature $k_B T << 2|J_1|$, and when $H = 0$, all clusters are in their zero-field ground levels. The magnetization process at relatively low magnetic fields, $g\mu_B H \lesssim k_B T$, corresponds to the alignment of the ground-level moment of each cluster. For clusters of a given type this alignment follows the Brillouin function for an effective spin of magnitude $S_T(\text{ground})$. When the ground-level moments of all the clusters become aligned, the magnetization M exhibits an apparent, or "technical," saturation. The technical saturation value M_S is lower than the true saturation value M_0. (M_0 is the value of M at $H = \infty$.) The difference between M_S and M_0 is that M_S corresponds to the alignment of $S_T(\text{ground})$, whereas M_0 corresponds to the alignment of $S_T(\text{ferro})$.

The ratio between M_S and M_0 depends on: 1) the ratios $S_T(\text{ground})/S_T(\text{ferro})$ for the various clusters, and 2) the probabilities P_i that a magnetic ion is in a cluster of a given type. If clusters containing more than 3 spins are neglected then, for a DMS with Mn^{++} ions

$$M_S/M_0 = P_1 + (P_3/3) + (P_4/15) , \tag{11}$$

where the various P_i's are defined in Sec. V. Note that the probability P_2 for being in a pair does not enter into Eq. (11). The reason is that pairs, for which $S_T(\text{ground}) = 0$, do not contribute to M_S. The largest contribution to M_S comes from the singles, which are represented in Eq. (11) by P_1. When the magnetic ions in a DMS are Co^{++} rather than Mn^{++}, Eq. (11) still holds except that $P_4/15$ should be replaced by $P_4/9$. This change (first noted in Ref. 31) reflects the different ratios $S_T(\text{ground})/S_T(\text{ferro})$ for CT's of Mn^{++} and Co^{++}.

The contribution of clusters with more than 3 spins to M_S has not been calculated thus far. This contribution is very small for $x < 0.04$. For larger x, up to ~0.10, clusters with more than 3 spins can be included by adding an empirical correction term to the R.H.S. of Eq. (11). The empirical correction is $(1/5)P_{>3}$, where $P_{>3}$ is the probability that a magnetic ion is in a cluster with more than 3 spins. The emprical correction is equivalent to an assumption that the average value of $S_T(\text{ground})/S_T(\text{ferro})$ for clusters with more than 3 spins is 1/5.

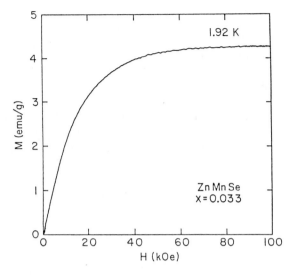

Fig. 5. Magnetization of $Zn_{0.967}Mn_{0.033}Se$ at 1.92 K. Note the apparent saturation above ~70 kOe [after Heiman *et al.*, *Solid State Commun.* **51**, 603 (1984)].

An example of technical saturation is shown in Fig. 5. The value $M_S = 4.3$ emu/g in this case is about 30% lower than M_0. Figure 6 compares the measured ratio M_S/M_0 for several DMSs with theory. The solid curve represents Eq. (11), whereas the dashed line includes the empirical correction due to large clusters. Both curves are calculated using the probabilities

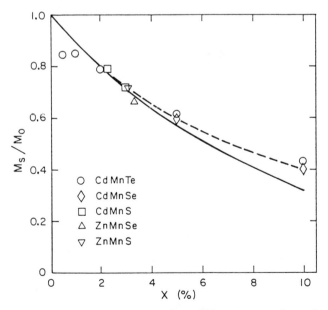

Fig. 6. Ratio M_S/M_0 as a function of Mn concentration x for various DMSs. The solid curve represents Eq. (11), with the probabilities P_i for a random Mn distribution. The dashed curve includes the empirical correction due to clusters larger than triplets (Ref. 30).

131

P_i which follow from a random distribution of the magnetic ions over the cation sites. There are no adjustable parameters. The good agreement between experiment and theory (particulary for $x < 0.05$, where the empirical correction is not needed) is a strong evidence in favor of a random distribution.

B. Magnetization steps due to J_1 pairs

In a magnetic field **H**, the energy levels of a J_1 pair exhibit a Zeeman splitting. This is shown in Fig. 7b. The only level which is not split is the zero-field ground level, with $S_T = 0$. One consequence of the Zeeman splitting is that the ground state of a pair changes with increasing H. These changes are caused by level crosssings which occur at magnetic fields H_1, H_2, H_3, etc. The first change, at H_1, is from the zero-field ground state with $S_T = m = 0$ to the state with $S_T = 1$, $m = -1$. Because of the jump in $|m|$, the low-temperature magnetization will exhibit a step at H_1. Subsequent level crossings at H_2, H_3 etc. also increase $|m|$ by one unit at a time, and they too will result in magnetization steps (MSTs).

The MSTs due to J_1 pairs are sketched in Fig. 7c. This figure is for Mn^{++} pairs, in which case there are five MSTs. For Co^{++} pairs there are only three MSTs. Once the last MST is completed, the magnetization of the pair is saturated, i.e., $|m| = S_T(ferro) = 2S$, where S is the

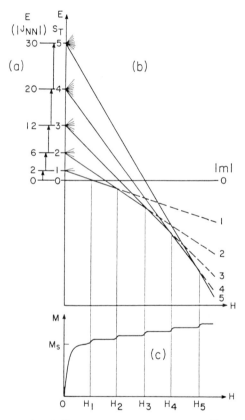

Fig. 7. (a) Energy level diagram for a pair of NN Mn^{++} ions at H = 0. (b) Zeeman splitting of these levels in a magnetic field H. Note the level crossings at H_n, which change the ground state. (c) Schematic of the magnetization curve at low T, showing the apparent saturation ($M = M_S$), followed by magnetization steps due to the NN pairs. The MSTs due to triplets are not shown.

spin of an individual magnetic ion. As Fig. 7c indicates, the MSTs are superimposed on M_S. At the low temperatures which are being considered, $k_B T \ll 2|J_1|$, the technical saturation responsible for M_S is achieved before the onset of the first MST.

In the NN cluster model the magnetic fields at the centers of the MSTs are given by

$$g\mu_B H_n = 2|J_1|n \tag{12}$$

where n = 1,2,3,...,2S. This equation will be modified slightly later, to take into account exchange interactions with distant neighbors. For a typical value $J_1/k_B \cong -10$ K in DMSs with Mn^{++}, the first MST is near 150 kOe (15 T). Such fields are accessible in a reasonable number of laboratories. For DMSs with Co^{++} the NN exchange constant, and hence also H_1, are much higher (Tables I and II), which is why the MSTs due to J_1 pairs are yet to be observed in these materials.

The rise of the magnetization associated with each MST is independent of n and is given by

$$\delta M = P_2 M_0/5 \quad \text{for } Mn^{++} \text{ pairs}, \qquad \delta M = P_2 M_0/3 \quad \text{for } Co^{++} \text{ pairs}. \tag{13}$$

Physically, the saturation magnetization for pairs, which is equal to $P_2 M_0$, is achieved by five equal steps for Mn^{++} pairs, and by three equal steps for Co^{++} pairs. This result is not affected by distant-neighbor interactions, provided that they are weak compared to the NN interaction. For the zinc-blende and wurtzite structures the probability P_2 has a peak near $x = 0.05$. At this peak, δM is about 5% of M_0 (or 8% of M_S) if the magnetic ions are Mn^{++}, and about 8% of M_0 if the magnetic ions are Co^{++}.

In the NN cluster model the shape of a MST is controlled by the Boltzmann factors which govern the populations of the two levels which are crossing. Assuming that the separation beteen two adjacent MSTs is much larger than the width of a single MST, the shape of the nth MST is given by

$$M_n = \frac{\delta M}{1 + \exp[g\mu_B(H_n - H)/k_B T]} \tag{14}$$

Sometimes one measures the derivative dM/dH rather than M. Each MST gives rise to a peak in this derivative. If the line shape is given by Eq. (14) then the peak in dM/dH is symmetric, and its width at half height is given by

$$\delta H = 3.53 k_B T/g\mu_B . \tag{15}$$

It should be stressed that Eqs. (14) and (15) hold only in the NN cluster model and when thermal equilibrium prevails. Otherwise, these equations must be modified, as discussed later. The width δH given by (15) is purely thermal in origin, and will be called the thermal width. At a typical temperature of 1.4 K the thermal width is $\delta H = 37$ kOe (3.7 T), assuming $g = 2$.

The MSTs due to J_1 pairs were discovered in 1984.[30,32] Since then the MSTs have been studied extensively in laboratories in the U.S., France, and Japan. The MSTs can be observed not only in measurements of M or dM/dH, but also in measurements of optical properties which depend on the magnetization.[32,33] A list of works on the MSTs can be found in a recent review.[34] Some value of J_1 which were obtained from data on MSTs are given in Sec. IX.

Open and closed J_1 triplets also give rise to MSTs.[30] However, these MSTs are harder to observe. We shall not discuss them here.

VIII. EFFECTS OF DISTANT-NEIGHBOR INTERACTIONS ON THE MSTS ARISING FROM J_1 PAIRS

The exchange interactions between distant neighbors are weak compared to the NN exchange interaction. Nevertheless, they affect the MSTs arising from J_1 pairs in two ways: 1) the fields H_n where the steps occur are shifted, and 2) the steps are broadened. These effects were treated by Larson et al. using an effective-field method.[4] Before summarizing their results we shall discuss a simple example using a different approach. This example brings out the main qualitative effects which are obtained from the more general method of Larson et al.

A. The J_1-J_2 model for low x

Consider a DMS with $x = 0.01$. To find out how the NNN interaction affects the MSTs associated with J_1 pairs we introduce the J_1-J_2 model. In this model both J_1 and J_2 are non-zero, but all other J's vanish. We also assume that both J_1 and J_2 are antiferromagnetic, and that $|J_2| \ll |J_1|$. Because x is low, it is useful to consider a cluster model. There are now three categories of clusters: 1) pure J_1 clusters, with J_1 bonds only, 2) pure J_2 clusters, and 3) mixed clusters, with both J_1 and J_2 bonds. Each of these three categories contains clusters of different types. The various types of clusters which occur in the J_1-J_2 model (up to triplets), as well as their probabilities, were discussed by Kreitman and Barnett.[35]

In the NN cluster model (J_1 only), 10.0% of the spins in the present example are in J_1 pairs. When J_2 is included, 92.3% of these original pairs still remain as pure J_1 pairs, but 6.6% of the original pairs are now in mixed open triplets (each with a J_1 and a J_2 bonds). The remaining 1.1% of the original J_1 pairs are in larger mixed clusters. We ignore these larger clusters and focus on the two main groups: pure J_1 pairs, and mixed open triplets.

The fields at which the MSTs due to the pure J_1 pairs occur are still given by Eq. (12). On the other hand, the fields at the MSTs due to the open mixed triplets are shifted, i.e.,

$$g \, \mu_B \, H_n = 2 \, | \, J_1 \, | \, n + \varepsilon \qquad (16)$$

The shift ε can be found by considering the energy-level diagram of a mixed open triplet. This diagram was obtained by Okada[36] for the case $|J_1| \gg |J_2|$. Using his results one can show that ε is independent of n and is given by

$$\varepsilon = (5/2) \, | \, J_2 \, | \quad \text{for Mn}^{++}, \qquad \varepsilon = (3/2) \, | \, J_2 \, | \quad \text{for Co}^{++}, \qquad (17)$$

At ultra low temperatures, $k_B T \ll |J_2|$, each of the original MSTs (in the NN cluster model) will split into two resolved steps: a magnetization step arising from the pure J_1 pairs, and a smaller step arising from the open mixed triplets. The smaller step occurs at a magnetic field which is higher by $\delta = \varepsilon/g\mu_B$.

Usually the temperature T is not sufficiently low to resolve the two steps which are separated by δ. In that case each of the observed magnetization steps will be a superposition of the two unresolved steps. The weighted center of the observed magnetization step will then be at a slightly higher field than that given by Eq. (12) for pure J_1 pairs. In addition, the unresolved splitting will broaden the MST, and will cause the peak in dM/dH to be asymmetric. The shift in the average position of the MST, the increased broadening, and the asymmetry are independent of n.

B. Effective field treatment

Larson et al.[4] used an effective field method to treat the influence of distant-neighbor interactions (J_2, J_3, etc.) on the MSTs arising from J_1 pairs. Each J_1 pair sees an effective field h_e due to the distant neighbors. This h_e is usually different for different J_1 pairs, because the positions of the distant neighbors relative to a pair are different. For example, some J_1 pairs have no NNNs, some have one NNN, and others have two or more NNNs. Thus, in any given crystal there is a distribution of h_e's.

Consider a particular J_1 pair, consisting of spins 1 and 2, and let H be along the z axis. The z component of h_e, acting on this pair, is given by

$$g\mu_B h_{ez} = -\sum_r J_{1r} < S_{rz} > - \sum J_{2s} < S_{sz} >, \qquad (18)$$

where r and s are the distant neighbors of spins 1 and 2, respectively. In DMSs it is expected that J_{1r} and J_{2s} will be negative (antiferromagnetic). Also, at the high fields where the MSTs occur, $<S_{rz}>$ and $<S_{sz}>$ are negative (the magnetic moments are parallel to H, but the spins are antiparallel). Thus, h_{ez} is negative, i.e. antiparallel to the applied field. The net effective field seen by the pair therefore has the magnitude

$$H_{eff} = H - |h_{ez}|. \qquad (19)$$

The fields at the MSTs are obtained by replacing H in Eq. (12) by H_{eff}. This gives

$$g\,\mu_B\,H_n = 2\,|J_1|\,n + g\,\mu_B\,|h_{ez}|, \qquad (20)$$

for this particular pair. Thus, H_n is shifted upward by $|h_{ez}|$.

For low x most distant spins are fully aligned at the high fields where the MSTs occur. Thus, Larson et al. take $<S_{rz}>$ and $<S_{sz}>$ to be independent of H, and equal to $-S$. In this approximation the shift $|h_{ez}|$ is independent of n. We note in passing that for the previously-considered example of an open mixed J_1-J_2 triplet, Eq. (18) gives $g\mu_B|h_{ez}| = S|J_2|$. This is the same as ε in Eq. (17). The shift $|h_{ez}|$ in H_n is then the same as the shift $\delta = \varepsilon/g\mu_B$ obtained previously by treating this cluster explicitly.

Because the h_e's are different for different J_1 pairs, the shift in the positions of the MSTs is different for different pairs. An *observed* MST will be a superposition of many ministeps, each occuring at its own shifted position. This results in an additional broadening of the MST, which is not included in Eqs. (14) and (15). In general, the distribution of the shifts $|h_{ez}|$ is not symmetric, so that the observed peak in dM/dH will be asymmetric. In the approximation that h_{ez} is indepedent of n, the broadening and asymmetry remain the same for all the MSTs.

Because an observed MST is not symmetric, it is necessary to choose some special point on it as the value of H_n. In practice this point is often either the maximum of dM/dH or the midpoint in the rise of M. Given this choice, H_n is given by

$$g \, \mu_B \, H_n = 2 \, | \, J_1 \, | \, n + \Delta \qquad (21)$$

where Δ depends on the distribution of the shifts $|h_{ez}|$, and is approximately independent of n. The magnitude of Δ is smaller for smaller x (fewer distant neighbors). In typical experiments with $x \leq 0.05$, the ratio $\Delta/2|J_1|$ is less than ~0.1.

The value of J_1 can be obtained by measuring the difference in the positions of two steps, e.g.,

$$g \mu_B (H_2 - H_1) = 2 \, | \, J_1 \, | \, . \qquad (22)$$

Here, Δ was treated as independent of n, which usually leads to an error of less than 1% in J_1. This tiny error may be reduced by assuming that Δ is proportional to the magnetization M. Such a relation is expected because $<S_{rz}>$ and $<S_{sz}>$, which govern the shift $|h_{ez}|$ via Eq. (18), should be roughly proportional to M. There is also empirical evidence that Δ is proportional to M.[37] Until now the usual practice has been to treat Δ, in a given sample, as a constant and use Eq. (22).

In some experiments only the first MST can be reached, and the experimenter is faced with the problem of extracting J_1 from H_1 alone. The simplest way is to ignore Δ and use Eq. (12). Typically, this will result in an overestimate of J_1 by up to ~10%. A much better estimate can be obtained by using a procedure suggested by Barilero et. al.[38]

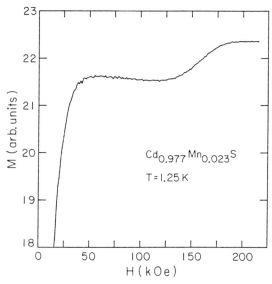

Fig. 8. Upper portion of a magnetization curve for $Cd_{0.977}Mn_{0.023}S$, showing the first MST due to J_1 pairs (Ref. 39).

IX. VALUES OF J_1 FROM MSTS

The MSTs due to J_1 pairs were observed in virtually all II-VI DMSs containing Mn^{++}, but as yet not in DMSs containing Co^{++}. Most experiments were carried out on samples with several percent of manganese, at $T \cong 1.3$ K. Data were taken both in dc magnetic fields (up to ~30 T), and in pulsed fields (up to ~60 T).

Figure 8 shows the first of the five expected MSTs in $Cd_{1-x}Mn_xS$. These data were taken in dc fields.[39] From this single step a value $J_1/k_B = -10.4$ K was obtained using the method of Barilero et al.[38]

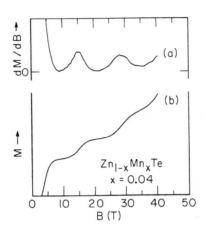

Fig. 9.   (a) Differential susceptibility dM/dB for $Zn_{0.96}Mn_{0.04}Te$, measured in pulsed fields. The bath temperature is 1.4 K. (b) The magnetization curve, as obtained by integrating the data in part (a). (After Ref. 40.)

Figure 9 shows the first two MSTs, and the beginning of the third, in $Zn_{1-x}Mn_xTe$. These data were taken by Foner et al. using pulsed fields.[40] Part (a) shows the data for dM/dB, while part (b) shows the magnetization obtained by integrating the data in part (a). From the positions of the two MSTs, a value $J_1/k_B = -9.0\pm0.2$ K was deduced. Similar pulsed-field data on the same material were obtained earlier by another group.[41] All five MSTs due to J_1 pairs were observed in $Cd_{1-x}Mn_xTe$ using pulsed fields.[40]

Much of the work on the MSTs arising from J_1 pairs was summarized in Refs. 34 and 42. Table III gives selected values of of J_1 obtained by this method.

X. DZYALOSHINSKI-MORIYA INTERACTION

The DM interaction for a J_1 pair has the form

$$\mathcal{H}_{DM} = -2\mathbf{D}_{12} \cdot \mathbf{S}_1 \times \mathbf{S}_2 . \tag{23}$$

Table III. Values of J_1 obtained from MSTs. Values in parentheses were obtained from a single MST using the method suggested in Ref. 38.

Material	J_1/k_B (K)	Ref.
$Cd_{1-x}Mn_xTe$	-6.1±0.2	39,43
	-6.2±0.2	40
$Cd_{1-x}Mn_xSe$	-7.7±0.3	43
	-7.6±0.2	40
$Cd_{1-x}Mn_xS$	(-10.4)	39
	-9.65±0.2, -11.0±0.2	44
$Zn_{1-x}Mn_xTe$	-8.8±0.1	41
	-9.0±0.2	40
	(-9.25)	38
$Zn_{1-x}Mn_xSe$	-12.2±0.3	40
	(-12.2)	39
$Zn_{1-x}Mn_xS$	(-16.9)	44
$Hg_{1-x}Mn_xTe$	-5.1±0.5	42
$Hg_{1-x}Mn_xSe$	-6.0±0.5	42

For two NNs in the zinc-blende structure the direction of the axial vector $\mathbf{D}_{12}$ is perpendicular to the plane containing the two NNs and the intervening anion.[16] The magnitude of $\mathbf{D}_{12}$ is sometimes called D_1.

Consider the NN cluster model, but now with the DM term. The Hamiltonian of a NN pair is the sum of R.H.S.s of Eqs. (8) and (23). Because $|D_1/J_1| \ll 1$, it is convenient to start from the eigenstates $|S_Tm\rangle$ of the Hamiltonian (8). The DM term then mixes states which differ by $\Delta S_T = \pm 1$, $\Delta m = 0, \pm 1$. Despite the fact that $S_1 \times S_2$ is a pseudovector, states with the same S_T are not mixed, as can be seen from the following arguments.

The eigenstates $|S_Tm\rangle$ are linear combinations of the eigenstates $|m_1m_2\rangle$ in which the z component of spin 1 is m_1, and that of spin 2 is m_2. That is

$$| S_Tm > = \Sigma \, | m_1m_2 >< m_1m_2 | S_Tm > . \tag{24}$$

In general the Clebsch-Gordan coefficients have the property

$$< j_1j_2m_1m_2 | j_1j_2jm > = (-1)^{j-j_1-j_2} < j_2j_1m_2m_1 | j_2j_1jm > . \tag{25}$$

In the present case, with $j_1 = j_2 = S$, and $j = S_T$,

$$< m_2m_1 | S_Tm > = (-1)^{S_T - 2S} < m_1m_2 | S_Tm > . \tag{26}$$

This means that for a fixed S_T the Clebsch-Gordan coefficients are either all even or all odd with respect to an interchange of the values of m_1 and m_2. Thus, for a given S_T the states $|S_Tm\rangle$ are all either symmetric or antisymmetric with respect to an interchange of m_1 and m_2. The symmetric states are linear combinations of terms of the form $(|m_\alpha m_\beta\rangle + |m_\beta m_\alpha\rangle)$. The antisymmetric states are linear combinations of terms of the form $(|m_\alpha m_\beta\rangle - |m_\beta m_\alpha\rangle)$. The values of S_T which lead to symmetric states are $S_T = 2S, 2S-2, ...,$ while the antisymmetric states are for $S_T = 2S-1, 2S-3, ...$ These properties are illustrated by Table V of Ref. 14, which gives the explict expansions of $|S_T m\rangle$ in terms of $|m_1 m_2\rangle$ for $S = 5/2$.

The DM interaction in Eq. (23) is antisymmetric with respect to an interchange of the values of spins 1 and 2. When $\mathcal{H}_{DM}$ operates on a symmetric (antisymmetric) state $|S_T m\rangle$ it produces antisymmetric (symmetric) combinations. That is, the symmetry is reversed. The combinations produced by $\mathcal{H}_{DM}|S_T m\rangle$ are orthognal to any state $|S_T m'\rangle$ with the original S_T, because of the opposite symmetry. Thus, $\langle S_T m' | \mathcal{H}_{DM} | S_T m\rangle = 0$.

Figure 10 illustrates the effect of $\mathcal{H}_{DM}$ on the first MST.[44] The dashed lines show the energy levels in the absence of the DM term. The first MST, at H_1 is then caused by the level crossing of the states $|S_T m\rangle = |0,0\rangle$ and $|1,-1\rangle$. When the DM term is added, it mixes some states. In the vicinity of H_1 it is sufficient to consider the mixing of $|0,0\rangle$ and $|1,-1\rangle$ only. The energy levels are then obtained by diagonalizing the 2×2 matrix

$$\begin{pmatrix} 0 & \gamma \\ \gamma^* & 1-h \end{pmatrix},$$

(27)

where all energies are normalized to $2|J_1|$, $h = g\mu_B H/2|J_1|$, and γ is the (normalized) matrix element connecting $|0,0\rangle$ with $|1,-1\rangle$. The diagonal terms in (27) are not affected by the DM interaction. The matrix element γ depends on the orientation of $\mathbf{H}$ relative to $\mathbf{D}_{12}$, and is therfore different for different NN pairs in the crystal. When $\mathbf{H}$ is not nearly parallel to $\mathbf{D}_{12}$, γ is of order $D_1/2J_1$.

Diagonalization of the matrix (27) leads to the anticrossing shown by the solid curves in Fig. 10. The lower solid branch starts as a nearly pure $|0,0\rangle$ state and ends up as a nearly pure $|1,-1\rangle$ state. For the upper solid branch the situation is just the oppostite. For either branch most of the change in the spin component along $\mathbf{H}$ occurs in the anticrossing region. It can be shown that at H_1 the average spin component along $\mathbf{H}$ for either branch is $-1/2$. Thus, the

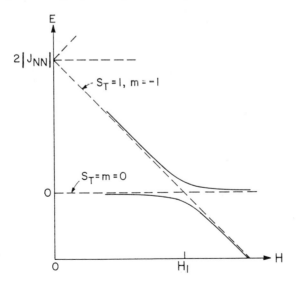

Fig. 10. Energy levels for a NN pair near the first MST. The dashed lines are from the NN cluster model. The solid curves show the anticrossing resulting from the addition of the DM interaction. The anticrossing is exaggerated for clarity (Ref. 44).

center of the MST is not shifted by the DM interaction. Therefore, the DM term does not spoil the determination of J_1 from the positions of the MSTs. However, the anticrossing produced by the DM term leads to an extra broadening of the MSTs.

Using this model (NN and DM interactions only) it can be shown[45] that at $T = 0$ the width at half height of the first MST is $\delta h \cong 3\gamma$. This width is due to the anticrossing. In an actual experiment at a finite temperture the width should be a combination of the thermal width, the broadening due to distant neighbors, and the broadening due to the anticrossing. Here, we have implicitly assumed the existence of thermal equilibrium, leading to a definite thermal width. This assumption holds in dc fields but often not in pulsed fields. Non-equilibrium situations in pulsed fields will be discussed later.

Although Fig. 10 is for the first MST at H_1, a similar behavior is expected for the other MSTs at higher H_n.

XI. NARROWING AND FINE STRUCTURE IN PULSED FIELDS

In typical pulsed-field experiments on MSTs the pulse duration is several ms. Thermal equilibrium does not prevail in many of these experiments, at least not during some portions of the pulse. For example, there is strong evidence that in some experiments the sample warms in the initial portion of the pulse (due to the rapid magnetization), approaches thermal equilibrium in the high-field portion of the pulse (where the magnetization is changing slowly), and cools in the last portion of the pulse (rapid demagnetization).[40] The MSTs at the lower fields are then better resolved in the down-portion of the pulse, because the sample is colder.

A different, and much more striking, non-equilibrium phenomenon was observed in $Cd_{1-x}Mn_xS$ and $Zn_{1-x}Mn_xS$ crystals with relatively low x (2–3%).[44] The MSTs in these pulsed experiments were much narrower, by a factor of 3-to-4, than the thermal width at the starting temperature (bath temperature). Moreover, the width was the same during the up- and down-portions of the pulse.

A plausible model for the narrowing, which assumes a long spin-lattice relaxation time for the J_1 pairs, was proposed.[44] In this model the pairs are effectively decoupled from the lattice for a period longer than that required to sweep through a single MST (about 0.5 ms). During the pulse the pairs then remain on the lower solid branch of Fig. 10. The thermal width then does not play a role, and the observed width is determined only by the width of the anticrossing region and by distant-neighbor interactions. It was suggested that the low value of D_1 in the sulfides was crucial for the narrowing. First because it produced a very narrow anticrosssing region, and second because it was responsible for the long spin-lattice relaxation time.

Another model for the narrowing of the MSTs (not entirely different from the first model) was developed very recently.[46] The basic ideas in this model are: 1) the existence of a phonon bottleneck for the relaxation of the spin system, and 2) a significant cooling of the spins as the pulsed field approaches H_n from either side.

Usually, when spin-lattice relaxation takes place, the lattice temperature can be regarded as constant. At low temperatures, however, the lattice specific heat may be too small for the lattice to act as a thermal reservoir. The observed effective spin-lattice relaxation time T_1^{eff} may then be controlled by the relatively slow rate at which phonons can exchange energy between the spin system and the bath in which the sample is immersed. This is the "phonon bottleneck".[47] One indication of a phonon bottleneck is a dependence of the measured T_1^{eff} on sample size. Such a size dependence was observed recently in the samples which exhibited the narrow MSTs.[46] Although these observations of a phonon bottleneck were made only at $H << H_1$, it is quite possible that such a bottleneck exists also at fields where the MSTs occur. In the model for the narrow MSTs it is assumed that the relaxation time between the spins and the bath is longer than the time in which $g\mu_B H$ changes by $k_B T_B$, where T_B is the temperature of the bath. In that case the temperature of the spins can differ appreciably from that of the bath.

The cooling of the spins as the pulsed field approaches H_n is analogous to the familiar cooling accompanying the adiabatic demagnetization of a dilute paramagnet. For an ordinary paramagnet the cooling is a consequence of the fact that *for singles* the ground level is degenerate at $H = 0$, apart from small crystal-field splittings. The situation *for pairs* is similar except that the ground level is degenerate at H_n, apart from the small splitting due to the DM term. It can be shown that when the sample is in poor thermal contact with the bath, cooling will occur (due to the pairs) as the pulsed field approaches H_n. As a result of the cooling, the MSTs are narrower. Computer simulations, using reasonable parameters for the samples in which narrow MSTs were observed, support this conclusion. It should be noted that in the special case where 1) there is no relaxation between the spin system and the bath, and 2) the lattice specific heat can be ignored, the new model gives the same results as the old model.

The narrowing of the MSTs (regardless of its cause) allows a fine structure to be resolved. One possible origin of a fine structure is the existence of two inequivalent types of nearest-neighbor sites in the wurtzite structure. One type are the six "in plane" NN cation sites. The other type are the six "out of plane" NN cation sites (three above and three below). The two types of NNs can have slightly different J_1's, which would result in a splitting of the MSTs. The expected signature of this type of splitting is: 1) the splitting is proportional to n, and 2) the two peaks in dM/dH resulting from the splitting of a single MST should have equal intensities. These characteristics were observed in the splitting of the first two MSTs in $Cd_{1-x}Mn_xS$, shown in Fig. 11. The splitting of the first MST in this sample was also observed in dc experiments at 0.56 K, but it was not apparent at the higher temperature of Fig. 8. Earlier Raman experiments also showed the same splitting.[29] The two values of J_1 deduced from Fig. 11 are given in Table III. The 14% difference between the two J_1's is close to a recent theoretical estimate for a similar material with the wurtzite structure.[48]

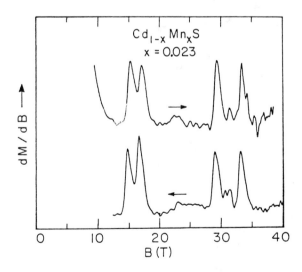

Fig. 11.   dM/dB versus B for $Cd_{0.977}Mn_{0.023}S$, measured in pulsed fields. The two traces are for increasing and decreasing B. The bath is at 1.4 K. Note the splitting of the first two MSTs (Ref. 44). The samples in this figure and in Fig. 8 are from the same boule.

Another type of fine structure may arise from different configurations of distant neighbors surrounding J_1 pairs. For example, in Sec. VIII-A we considered a case in which the largest group of J_1 pairs had no NNNs, but a smaller group had one NNN. The MSTs for these two groups occur at fields which differ by δ, which would lead to a fine structure. Because the splitting δ is independent of n, such a fine structure can be distinguished from that due to different J_1's. For this purpose, however, it is necessary that the fine structure is observed in more than a single MST. It has been suggested that a fine structure which was observed in $Zn_{1-x}Mn_xS$ might have been due to different distant-neighbor environments.[44] However, because only a single MST was observed in this case, this suggestion could not be verified.

XII. MAGNETIZATION STEPS DUE TO J_2 PAIRS AND J_3 PAIRS

In Sec. VIII-A the J_1-J_2 model was used to study the effect of the NNN interaction on the MSTs arising from J_1 pairs. The particular clusters which were of interest there were the pure J_1 pairs, and the mixed J_1-J_2 open triplets. We now turn our attention to the pure J_2 clusters, such as J_2 pairs and J_2 open triplets. These clusters too give rise to MSTs, but at fields which are much lower than H_n for MSTs from J_1 pairs. The most prominent MSTs due to J_2 clusters are those arising from J_2 pairs. One of the requirements for observing such MSTs (at thermal equilibrium) is that $k_B T \ll |J_2|$. For Mn^{++} in II-VI DMSs, $|J_2| \lesssim 1$ K so that the required temperatures are quite low. The much larger d-d exchange constants for Co^{++} make the temperature requirement much less stringent for these magnetic ions. The MSTs due to J_2 pairs were recently observed in $Zn_{1-x}Co_xSe$, using ^3He at 0.6 K.[49]

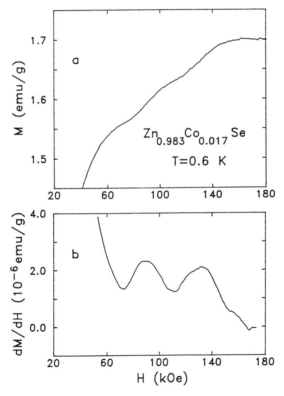

Fig. 12. (a) Upper portion of the magnetization curve for $Zn_{0.983}Co_{0.017}Se$ at 0.6 K. The two MSTs are attributed to J_2 pairs. (b) The derivative dM/dH obtained from the data in part (a). (After Ref. 49.)

The theory of the MSTs arising from J_2 pairs is completely analogous to that for MSTs from J_1 pairs. Assuming that the exchange interactions decrease rapidly with distance ($J_1 \gg J_2 \gg J_3, ...$), the MSTs due to J_2 pairs are obtained from the J_1-J_2 model in the same manner that the MSTs due to J_1 pairs are derived from the J_1 model. Interactions between third neighbors, and more distant neighbors, are then added in a manner similar to that in Sec. VIII. The result is that the positions of the MSTs are given by Eq. (21), except that: 1) J_1 is replaced by J_2 and, 2) the shift Δ is now due to J_3, J_4, etc. The size of the MSTs due to J_2 pairs is governed by the probability of finding a magnetic ion in a J_2 pair. For the zinc-blende structure the maximum size of the MSTs from J_2 pairs is only a fifth of the maximum size of the MSTs due to J_1 pairs.

Figure 12 shows some of the data for the MSTs from J_2 pairs in $Zn_{1-x}Co_xSe$. The two steps correspond to the second and third MSTs in the series. (There are only three MSTs in a series when the magnetic ion is Co^{++}.) From the difference (H_3-H_2) an exchange constant J_2/k_B = -3.04±0.1 K was obtained in this case. This should be compared with J_1/k_B = -49.5±1 K.[28]

MSTs can also arise from J_3 pairs. The theory for these MSTs uses the J_1-J_2-J_3 model. Some of the recent data in Ref. 49 are consistent with MSTs from J_3 pairs.

MSTs from J_1 pairs are easily identified as such, because an approximate value for J_1 is usually already known from the Curie-Weiss θ. In addition, the predicted size of the MSTs from J_1 pairs is usually much larger than those for other types of MSTs, so that a comparison with the observed size leads to a conclusive identification. In contrast, the identification of the MSTs associated with distant-neighbor pairs, such as J_2 pairs or J_3 pairs, is much more difficult. The most effective identification tool thus far has been the value of the magnetization in fields slightly above a given series of MSTs. If the series is due to J_2 pairs then, for low x, this magnetization should be close to M_S, as calculated in Sec. VII. If the MSTs are due to J_3 pairs, or still more distant pairs, then the magnetization should be lower. The reason is that the J_2 pairs are still not magnetized, and the J_2 open triplets contribute only a third of their saturation magnetization.

It remains to be seen how far the MSTs method can be pushed to obtain various d-d exchange constants. The results thus far are encouraging.

ACKNOWLEDGMENTS

I am grateful to my collaborators: S. Foner, D. Heiman, P.A. Wolff, N.F. Oliveira, Jr., V. Bindilatti, E.J. McNiff, Jr., C.R. McIntyre, T.Q. Vu, A. Wold, K. Dwight, R. Kershaw, C-M. Niu, and P. Becla. This work was supported by NSF Grant No. DMR-89-00419. The Francis Bitter National Magnet Laboratory, where our group performs the experiments on the MSTs, is supported by NSF.

REFERENCES

1.   *Diluted Magnetic Semiconductors*, Vol. 25 of *Semiconductors and Semimetals*, edited by J.K. Furdyna and J. Kossut (Academic, New York, 1988).

2.   *Diluted Magnetic (Semimagnetic) Semiconductors*, edited by R.L. Aggarwal, J.K. Furdyna and S. von Molnar (Materials Research Society, Pittsburgh, 1988).

3.   J.K. Furdyna, *J. Appl. Phys.* **64**, R29 (1988).

4.   B.E. Larson, K.C. Hass, and R.L. Aggarwal, *Phys. Rev.* **B 33**, 1789 (1986).

5.   S.P. Keller, I.L. Gelles, and W.V. Smith, *Phys. Rev.* **110**, 850 (1958); P.B. Dorain, *Phys. Rev.* **112**, 1058 (1958); J. Lambe and C. Kikuchi, *Phys. Rev.* **119**, 1256 (1960); H.H. Woodbury and G.W. Ludwig, *Bull. Am. Phys. Soc.* **6**, 118 (1961).

6.   Additional refences to EPR studies may be found in a review by S. Oseroff and P.H. Keesom (Ref. 1).

7.   H.A. Weakliem, *J. Chem. Phys.* **36**, 2117 (1962).

8.   M. Villeret, S. Rodriguez, and E. Kartheuser, *Physica* **B 162**, 89 (1990), and *Phys. Rev.* **B 41**, 10028 (1990).

9.   A. Lewicki, A.I. Schindler, I. Miotkowski, and J.K. Furdyna, *Phys. Rev.* **B 41**, 4653 (1990).

10.   0. J. P. Mahoney, C.C. Lin, W. Brumage, and F. Dorman, *J. Chem. Phys.* **53**, 4286 (1970).

11.   W. Low and M. Weger, *Phys. Rev.* **118**, 1119 (1960); G.A. Slack, S. Roberts, and J.T. Vallin, *Phys. Rev.* **187**, 511 (1969).

12.   A. Twardowski, *J. Appl. Phys.* **67**, 5108 (1990). See also A. Twardowski, and C. Benoit à la Guillaume, in this volume.

13.   B.E. Larson, K.C. Hass, H. Ehrenreich, and A.E. Carlsson, *Phys. Rev.* **B 37**, 4137 (1988); **38**, 7842 E (1988). For a more physical discussion of these results see B.E. Larson and H. Ehrenreich, *J. Appl. Phys.* **67**, 5084 (1990).

14.   J. Owen and E.A. Harris in *Electron Paramagnetic Resonance*, edited by S. Geschwind (Plenum, New York, 1972).

15.   T.M. Giebultowicz, P. Klosowski, J.J. Rhyne, T.J. Udovic, J.K. Furdyna, and W. Giriat, *Phys. Rev.* **B 41**, 504 (1990).

16.   B.E. Larson and H. Ehrenreich, *Phys. Rev.* **B 39**, 1747 (1989).

17.   J. Spalek, A. Lewicki, Z. Tarnawski, J.K. Furdyna, R.R. Galazka and Z. Obuszko, *Phys. Rev.* **B 33**, 3407 (1986).

18.   Y. Shapira, S. Foner, P. Becla, D.N. Domingues, M.J. Naughton, and J.S. Brooks, *Phys. Rev.* **B 33**, 356 (1986).

19.   A. Lewicki, J. Spalek, J.K. Furdyna, and R.R. Galazka, *Phys. Rev.* **B 37**, 1860 (1988).

20.   A. Lewicki, A.I. Schindler, J.K. Furdyna, and W. Giriat, *Phys. Rev.* **B 40**, 2379 (1989).

21.   S. Nagata, R.R. Galazka, D.P. Mullin, H. Akbarzadeh, G.D. Khattak, J.K. Furdyna, and P.H. Keesom, *Phys. Rev.* **B 22**, 3331 (1980).

22.   R.E. Behringer, *J. Chem. Phys.* **29**, 537 (1958).

23.   L.M. Claessen, A. Wittlin, and P. Wyder, *Phys. Rev.* **B 41**, 451 (1990).

24.   A. Furrer and H.U. Güdel, *J. Magn. Magn. Mater.* **14**, 256 (1979). See also E.C. Svensson, M. Harvey, W.J.L. Buyers, and T.M. Holden, *J. Appl. Phys.* **49**, 2150 (1978).

25.   L.M. Corliss, J.M. Hastings, S.M. Shapiro, Y. Shapira, and P. Becla, *Phys. Rev.* **B 33**, 608 (1986).

26.   T.M. Giebultowicz, J.J. Rhyne, and J.K. Furdyna, *J. Appl. Phys.* **61**, 3537 (1987).

27.   T.M. Giebultowicz, P. Klosowski, J.J. Rhyne, T.J. Udovic, J.K. Furdyna, and W. Giriat, *Phys. Rev.* **B 41**, 504 (1990).

28.   T.M. Giebultowicz, J.J. Rhyne, J.K. Furdyna, and P. Klosowski, *J. Appl. Phys.* **67**, 5096 (1990).

29.   D.U. Bartholomew, E.-K. Suh, S. Rodriguez, A.K. Ramdas, and R.L. Aggarwal, *Solid State Commun.* **62**, 235 (1987).

30. Y. Shapira, S. Foner, D.H. Ridgley, K. Dwight, and A. Wold, *Phys. Rev.* **B 30**, 4021 (1984).

31. D.U. Bartholomew, E.-K. Suh, A.K. Ramdas, S. Rodriguez, U. Debska, and J.K. Furdyna, *Phys. Rev.* **B 39**, 5865 (1989).

32. R.L. Aggarwal, S.N. Jasperson, Y. Shapira, S. Foner, T. Sakakibara, T. Goto, N. Miura, K. Dwight, and A. Wold, in *Proceedings of the 17th International Conference on the Physics of Semiconductors, San Francisco, 1984*, edited by J.D. Chadi and W.A. Harrison (Springer, New York, 1985), p. 1419.

33. R.L. Aggarwal, S.N. Jasperson, P. Becla, and R.R. Galazka, *Phys. Rev.* **B 32**, 5132 (1985).

34. Y. Shapira, *J. Appl. Phys.* **67**, 5090 (1990).

35. M.M. Kreitman and D.L. Barnett, *J. Chem. Phys.* **43**, 364 (1965).

36. O. Okada, *J. Phys. Soc. Jpn.* **48**, 391 (1980).

37. X. Wang, D. Heiman, S. Foner, and P. Becla, *Phys. Rev.* **B 41**, 1135 (1990).

38. G. Barilero, C. Rigaux, Nguyen Hy Hau, J.C. Picoche, and W. Giriat, *Solid State Commun.* **62**, 345 (1987).

39. Y. Shapira and N.F. Oliveira, Jr., *Phys. Rev.* **B 35**, 6888 (1987).

40. S. Foner, Y. Shapira, D. Heiman, P. Becla, R. Kershaw, K. Dwight, and A. Wold, *Phys. Rev.* **B 39**, 11793 (1989).

41. J.P. Lascaray, A. Bruno, M. Nawrocki, J.M. Broto, J.C. Ousset, S. Askenazi, and R. Triboulet, *Phys. Rev.* **B 35**, 6860 (1987).

42. R.R. Galazka, W. Dobrowolski, J.P. Lascaray, M. Nawrocki, A. Bruno, J.M. Broto, and J.C. Ousset, *J. Magn. Magn. Mater.* **72**, 174 (1988).

43. E.D. Isaacs, D. Heiman, P. Becla, Y. Shapira, R. Kershaw, K. Dwight, and A. Wold, *Phys. Rev.* **B 38**, 8412 (1988).

44. Y. Shapira, S. Foner, D. Heiman, P.A. Wolff, and C.R. McIntyre, *Solid State Commun.* **71**, 355 (1989).

45. C.R. McIntyre (unpublished).

46. V. Bindilatti, T.Q. Vu, and Y. Shapira, to be published.

47. For a brief discussion of the phonon bottleneck see S. Geschwind in *Electron Paramagnetic Resonance*, edited by S. Geschwind (Plenum, New York, 1972).

48. B.E. Larson, *J. Appl. Phys.* **67**, 5240 (1990).

49. Y. Shapira, T.Q. Vu, B.K. Lau, S. Foner, E.J. McNiff, Jr., D. Heiman, C.L.H. Thieme, C-M. Niu, R. Kershaw, K. Dwight, A. Wold, and V. Bindilatti, *Solid State Commun.* **75**, 201 (1990).

BOUND MAGNETIC POLARONS IN DILUTED MAGNETIC SEMICONDUCTORS

P. A. Wolff

NEC Research Institute
4 Independence Way
Princeton, NJ 08540

ABSTRACT

These notes develop the semi-classical theory of bound magnetic polarons (BMP) and use it to calculate their thermodynamic properties, such as energy, susceptibility, etc. The theory demonstrates that the polaron evolves continuously from a high temperature, fluctuation-dominated regime to a low temperature, collective state with large net moment (20–$50\mu_B$). In the simplest cases (donor - BMP in materials such as $Cd_{1-x}Mn_xSe$ with small x), the theory is in good agreement with optical experiments that determine microscopic properties of the polaron. To date, there is no theory of the acceptor - BMP that fully incorporates valence band degeneracy. However, hydrogenic models are in fair agreement with experiment, and demonstrate saturation of the acceptor - BMP below 20K.

The semiclassical theory correctly predicts thermodynamic averages, but is inadequate for describing polaron kinetics or the detailed energy level structure. For that purpose, a fully quantum mechanical theory is required. Simplified, soluble, quantum mechanical models of the polaron have been developed by several authors. We discuss them using Feynman's variational principle for the free energy. Optimized models will be compared with experiment and the semi-classical theory.

Finally, the behavior of polarons in materials with large Mn^{2+} concentration will be discussed. Experimentally, in $Cd_{1-x}Mn_xTe$ the polaron binding energy peaks at x=0.2 and decreases thereafter. This effect is ascribed to competition between the ferromagnetic, polaron-forming interaction (mediated by the carrier) and the direct, antiferromagnetic Mn^{2+}–Mn^{2+} interaction. A phenomenological polaron theory, that incorporates Mn^{2+}–Mn^{2+} interactions via the measured susceptibility, accounts for the observed saturation. To avoid saturation, which reduces polaron energies by a factor of five compared to those that would otherwise be attainable, we propose *ordered* DMS compounds. Growth of such crystals provides an important challenge to materials scientists.

INTRODUCTION

In semiconductors containing magnetic ions, there can be a sizable exchange interaction between carrier spins and those of the magnetic ions. Exchange causes novel spin-dependent phenomena in these materials including giant spin-splittings of the bands, large Faraday rotations, and magnetic polarons. Free magnetic polarons - magnetization clouds associated with unbound carriers - have not yet been conclusively demonstrated.

However, there is abundant evidence that carriers localized at impurities can induce sizable magnetizations in their vicinity. These complexes, termed bound magnetic polarons (BMP), often have moments exceeding $25\mu_B$.

BMP were first studied in europium chalcogenides, mainly by transport and magnetization experiments; there is an extensive literature on the subject. However, none of these experiments unambiguously determines microscopic parameters, such as binding energy, moment, or size of the BMP. Direct observation of BMP in Eu-chalcogenides is impeded by their complicated band structure, relatively poor electronic and optical properties, and by competing magnetic interactions. The situation is more favorable in DMS[1] with large, direct bandgaps and simple band structures; examples include $Cd_{1-x}Mn_xS$, $Cd_{1-x}Mn_xSe$, $Cd_{1-x}Mn_xTe$, $Zn_{1-x}Mn_xS$, $Zn_{1-x}Fe_xSe$, $Zn_{1-x}Mn_xSe$, etc. These crystals, in contrast to the EuX compounds, have highly structured luminescence spectra whose features can be associated with well-characterized centers. Luminescence has been used to study polaron interactions in the acceptor, the acceptor-bound exciton (A^0X), and the donor-bound exciton (D^0X). DMS also favor light scattering and Faraday rotation experiments, since both processes are enhanced by large factors as the laser frequency approaches the direct bandgap. This resonance facilitates spin-flip Raman scattering studies which provide a direct measure of the internal, exchange field in BMP.

2. THE BMP MODEL

Consider the donor-BMP[2] in a DMS of moderate (x<0.1) magnetic ion concentration. The single electron in such a complex moves in a nondegenerate conduction band and has an extended orbit well-described by the effective mass approximation. Moreover, for small x, the magnetic ion susceptibility is nearly Curie-like for T>4K. These facts simplify the analysis. Thus, the donor case will be used to develop the theory; it will later be extended, in a less rigorous way, to acceptor-BMP.

The Hamiltonian of the donor-BMP is:

$$H = \frac{p^2}{2m^*} - \frac{e^2}{\varepsilon_0 r} - \alpha\Sigma[(s \cdot S_j)\delta(r-R_j)], \qquad (2.1)$$

where m* is the effective mass, ε_0 the dielectric constant, and α the exchange constant of the conduction band. A typical value is $\alpha N_0 = 0.25ev$ in $Cd_{1-x}Mn_xSe$. In applying Eq. (2.1) to wurtzite crystals, such as CdMnS or CdMnSe, we will ignore the small (5%) anisotropy of m* and ε_0. A more serious approximation is the neglect of direct $Mn^{2+}-Mn^{2+}$ interactions. Experimentally, $Cd_{1-x}Mn_xX$ crystals with small x-values have nearly Curie-like magnetic susceptibilities, but with magnitudes corresponding to an effective Mn^{2+} concentration $(\bar{x})$ smaller than x. This behavior is due to magnetic ion pairs that are coupled antiferromagnetically and make no contribution to the low field susceptibility[3]. The model quantitatively explains the observed variation of $\bar{x}$ with x if Mn^{2+} ions are assumed randomly distributed on the cation sublattice. From the BMP viewpoint, Mn^{2+} doublets are "locked out" of the problem by antiferromagnetic pairing. This argument is only valid for x<0.1; but in that range it enables one, via the $\bar{x}$ trick, to treat the BMP problem as if there were no direct $Mn^{2+}-Mn^{2+}$ interactions. The resulting model Hamiltonian contains the essential physics of the BMP in its purest form.

The exchange interaction in the Schrodinger equation mixes orbital and spin degrees of freedom; for example, exchange can scatter an electron from one orbital to another while flipping its spin. In practice, such matrix elements are small. The Schrodinger equation then has a solution of the form:

$$\Psi(r,s;S_j) \equiv \phi(r)\chi(s;S_j). \qquad (2.2)$$

With the aid of Eq. (2.2) one can derive separate, coupled Schrodinger equations for ϕ and χ by integrating Eq. (2.1) against χ^* or ϕ^*. They are

$$\left\{ \frac{p^2}{2m^*} - \frac{e^2}{\varepsilon_0 r} - \alpha\Sigma_j \left[<s \cdot S_j> \delta(r-R_j) \right] \right\} \phi = E\phi,$$ (2.3)

and

$$-\alpha\Sigma_j \left[(s \cdot S_j) |\phi(R_j)|^2 \right] \chi = (E - E_0)\chi,$$ (2.4)

where

$$<(s \cdot S_j)> = <\chi | (s \cdot S_j) | \chi>$$ (2.5)

and

$$E_0 = \int \phi^*(r) \left[\frac{p^2}{2m^*} - \frac{e^2}{\varepsilon_0 r} \right] \phi(r) d^3 r.$$ (2.6)

Generally speaking, Eqs. (2.3) and (2.4) must be solved self-consistently to determine ϕ and χ. However, in the donor-BMP case, the orbital wave functions are hardly modified. Thus, we will use the standard hydrogenic donor wave functions to evaluate the spin Hamiltonian:

$$H_{spin} = -\alpha\Sigma_j \left[(s \cdot S_j) |\phi(R_j)|^2 \right].$$ (2.7)

3. THE BMP PARTITION FUNCTION

The spin problem defined by Eq. (2.7) has $2(6)^N$ degrees of freedom, where $N \approx 100$ is the number of Mn^{2+} spins within a donor orbit of radius 40A. This number is huge; even at low temperatures an enormous number of spin states is accessible to the donor-BMP system. Thus, the individual eigenfunctions and eigenvalues of H_{spin} are of less interest than their thermodynamic average. The primary task of the theory is to calculate the spin partition function,

$$Z = Tr \left[e^{-\beta H_{spin}} \right].$$ (3.1)

Subsequent analysis is simplified by rewriting the spin Hamiltonian in the form

$$H_{spin} = -\alpha\Sigma_j [(s \cdot S_j) |\phi(R_j)|^2] = -\Sigma_j [K_j (s \cdot S_j)] = -(s \cdot \Gamma)$$ (3.2)

where

$$K_j = \alpha |\phi(R_j)|^2,$$ (3.3)

and

$$\Gamma = \Sigma_j (K_j S_j).$$ (3.4)

The two factors determining K_j, α and $|\phi(R_j)|^2$, are both larger in acceptor-BMP than in donor-BMP. When combined they imply that polaron interactions in acceptors are an order of magnitude stronger than in donors. This difference precludes use of an unperturbed acceptor wave function in studying the acceptor-BMP; instead, a variational calculation is required to determine the optimum acceptor radius in the presence of strong exchange interactions.

The quantity Γ is proportional to the effective field experienced by the electron in the BMP. It is a vector operator whose components do not commute with one another; that fact prevents an exact evaluation of the trace in Eq. (3.1). On the other hand, if Γ were a classical variable the trace over electron spin variables (s) could easily be evaluated via the standard identity;

$$e^{\beta(s \cdot \Gamma)} = \left[\cosh\left[\frac{\beta\Gamma}{2}\right] + \frac{2(s \cdot \Gamma)}{\Gamma} \sinh\left[\frac{\beta\Gamma}{2}\right] \right]. \tag{3.5}$$

In the Mn^{2+} case, the individual spins are fairly large (S = 5/2) and combine in the BMP to give a net moment of many Bohr magnetons. Thus, it should be a good approximation to treat Γ as a classical variable by assuming the validity of Eq. (3.5).

Before proceeding to the evaluation of Eq. (3.5) we make a change which considerably simplifies the following analysis. The BMP system, though large, is finite. Thus, its partition function is an analytic function of β in the finite complex plane (this fact also implies that the BMP cannot exhibit a phase transition). In the following, therefore, we evaluate Z at an imaginary temperature ($\beta \rightarrow i\beta$), and later determine the physical $Z(\beta)$ by analytic continuation. With the aid of Eq. (3.5) one finds:

$$Z(i\beta) = Tr\left[e^{i\beta(s \cdot \Gamma)}\right]$$

$$= 2\ Tr_{\{S_j\}} \int \cos\left[\frac{\beta\gamma}{2}\right] \delta(\gamma - \Gamma) d^3\gamma$$

$$= 2\ Tr_{\{S_j\}} \left\{ \iint \frac{d^3\lambda d^3\gamma}{(2\pi)^3} \cos\left[\frac{\beta\gamma}{2}\right] e^{i\lambda \cdot [\gamma - \Sigma_j (K_j S_j)]} \right\}$$

$$= 2(6)^N \iint \frac{d^3\lambda d^3\gamma}{(2\pi)^3} \cos\left[\frac{\beta\gamma}{2}\right] e^{i\lambda \cdot \gamma} \prod_j \left[\mathcal{F}_{5/2}(-i\lambda K_j) \right], \tag{3.6}$$

where

$$\mathcal{F}_{5/2}(x) \equiv 1/6 \left[e^{5/2x} + e^{3/2x} + e^{1/2x} + e^{-1/2x} + e^{-3/2x} + e^{-5/2x} \right]. \tag{3.7}$$

Integrating over angles gives the result

$$Z(i\beta) = \frac{2(6)^N}{\pi} \int_0^\infty \lambda d\lambda \int_0^\infty \gamma d\gamma \left\{ \cos\left[\frac{\beta\gamma}{2}\right] \left[e^{i\lambda\gamma} - e^{-i\lambda\gamma} \right] \prod_j \left[\mathcal{F}_{5/2}(-i\lambda K_j) \right] \right\}$$

$$= -2(6)^N \frac{d}{d\beta} \left\{ \int_0^\infty \lambda\, d\lambda \int_{-\infty}^\infty d\gamma \sin\left[\frac{\beta\gamma}{2}\right] \left[e^{i\lambda\gamma} - e^{-i\lambda\gamma} \right] \prod_j \left[\mathcal{F}_{5/2}(-i\lambda K_j) \right] \right\}$$

$$= 2(6)^N \frac{d}{d\beta} \left\{ (i\beta) \prod_j \left[\mathcal{F}_{5/2}\left(-\frac{i\beta K_j}{2} \right) \right] \right\}. \tag{3.8}$$

Analytically continuing gives finally:

$$Z(\beta) = 2(6)^N \frac{d}{d\beta} \left\{ \beta \prod_j \left[\mathcal{F}_{5/2}\left(\frac{\beta K_j}{2} \right) \right] \right\}. \tag{3.9}$$

This is the central result of the semiclassical theory. Eq. (3.9) can also be derived directly (without the analytic continuation trick) from Eqs. (3.1) and (3.5). It is interesting to note that the product, $2(6)^N \prod_j \left[\mathcal{F}_{5/2}\left(\frac{\beta K_j}{2} \right) \right]$, appearing in Eq. (3.9), is the partition function for an Ising model of the BMP with Hamiltonian:

$$H_{Ising} = -\sum_j \left[K_j (s_z S_{jz}) \right]. \tag{3.10}$$

Eq. (3.9) is a complicated function of temperature because the alignment of each Mn^{2+} ion varies nonlinearly with the local exchange field (K_j), and saturates when $(\beta K_j/2) \gg 1$. In acceptor-BMP saturation begins below T=10K; much lower temperatures are required in donor- BMP. In the unsaturated regime the expression can be considerably simplified. For small γ, the distribution of internal fields appearing in Eq. (3.6) is Gaussian;

$$P(\gamma) \equiv \underset{\{s_j\}}{Tr} \left[\delta(\gamma - \Gamma) \right] = (6)^N \int \frac{d^3\lambda}{(2\pi)^3} \left\{ e^{i\lambda\cdot\gamma} \prod_j \left[\mathcal{F}_{5/2}(-i\lambda K_j) \right] \right\}$$

$$\cong 6^N \int \frac{d^6\lambda}{(2\pi)^3} \left[e^{i\lambda\cdot\gamma - 35/24\lambda^2\Sigma(K_j^2)} \right] = \frac{6^N e^{-\gamma^2/2W_0^2}}{(2\pi W_0^2)^{3/2}}, \tag{3.11}$$

where

$$W_0^2 = \frac{35}{12} \sum_j (K_j^2). \tag{3.12}$$

Substitution of Eq. (3.11) into Eq. (3.9) yields the following formula for the partition function in the Gaussian limit:

$$Z = 8\pi(6)^N \int \frac{\gamma^2 d\gamma}{(2\pi W_0^2)^{3/2}} \left[\cosh\left[\frac{\beta\gamma}{2}\right] e^{-\gamma^2/2W_0^2} \right]$$

$$=2(6)^N \left[1+\frac{\beta^2 W_0^2}{4}\right] e^{\beta^2 W_0^2/8}.$$ (3.13)

It is interesting to compare this expression with the corresponding result for the Ising model (3.10). In the Ising case:

$$Z_{Ising} = Tr \left[e^{\beta \Sigma(K_j s_z S_{jz})}\right] = Tr \left[e^{\beta s_z \Gamma_z}\right]$$

$$= 2 \, Tr_{\{S_j\}} \left\{\int_{-\infty}^{\infty} d\gamma \left[\cosh\left[\frac{\beta\gamma}{2}\right] \delta(\gamma-\Gamma_z)\right]\right\}$$

$$\cong 2(6)^N \int \frac{d\gamma}{\sqrt{2\pi W_0^2}} \left[\cosh\left[\frac{\beta\gamma}{2}\right] e^{-\gamma^2/2W_0^2}\right] = 2(6)^N e^{\frac{\beta^2 W_0^2}{8}}.$$ (3.14)

The integrals determining Z and Z_{Ising} are similar in form. Note, however, that the former involves an extra factor of γ^2 as compared to the latter. This factor accounts for the many possible orientations of the BMP moment, whereas in the Ising model the direction of the moment is uniquely determined by the external z-axis. Practically speaking, this factor of γ^2 profoundly alters the spin-flip light scattering spectrum. We will argue later that this spectrum is given by the integrand of Eq. (3.13). The $\gamma^2 \cosh\left[\frac{\beta\gamma}{2}\right] e^{-\gamma^2/2W_0^2}$ form implied by Eq. (3.13) is quite different from that one would infer from a true Ising model [Eq. (3.14)]. Experiment confirms the spectrum of Eq. (3.13).

Finally, note that $Z(\beta)$ and $Z_{Ising}(\beta)$ do, in fact, satisfy Eq. (3.9).

4. THERMODYNAMIC PROPERTIES OF BMP

Much of BMP behavior is determined by statistical mechanics, and can be calculated from the partition function [Eqs. (3.9) or (3.13)]. In particular, thermodynamic functions clearly demonstrate the gradual transition from the fluctuation- dominated regime to the collective regime mentioned above.

Consider the internal energy, U, of the BMP. It is an especially important thermodynamic variable, directly measured by luminescence studies of acceptor-BMP and spin-flip Raman studies of donor-BMP. U can be calculated from the usual statistical mechanical relation:

$$U = \frac{-\partial(lnZ)}{\partial\beta} = -<s\cdot\Gamma>$$ (4.1)

Differentiation of Eq. (3.9) produces a complicated, not-too- meaningful formula. The result simplifies, however, in the unsaturated limit. One there finds from Eq. (3.13):

$$U = -\left[\frac{\beta W_0^2}{4}\right]\left[\frac{12+(\beta W_0)^2}{4+(\beta W_0)^2}\right].$$ (4.2)

This expression can be seen to be identical to that of Dietl and Spalek [their Eq. (4.11)] by making the replacement $\beta W_0^2 \rightarrow 4\varepsilon_p$ and assuming a Curie susceptibility.

Eq. (4.2) has different temperature variations in the high ($\beta W_0 \ll 1$) and low ($\beta W_0 \gg 1$) temperature regimes. To understand the meaning of these two regimes, we note that there are two ways in which the BMP can lower its internal energy. At high temperature, though $<\Gamma>=0$, there are sizable fluctuations of Γ about its mean value because the BMP contains a *finite* number of Mn^{2+} spins. As temperature is reduced the carrier spin, which relaxes rapidly compared to those of the Mn^{2+} ions, aligns with instantaneous fluctuations of Γ and reduces U. As will be shown below, the magnitude of Γ remains essentially unchanged throughout this spin alignment process. It is solely determined by the statistics of Mn^{2+} spin orientation; hence the term "fluctuation regime."

When alignment of the carrier spin is complete, for $\beta W_0 \cong 2$, the system can further reduce its energy by forcing the Mn^{2+} spins to adopt statistically unfavorable, but energetically favorable, configurations with larger values of $\sqrt{<\Gamma^2>}$. $\sqrt{<\Gamma^2>}$ gradually increases throughout this BMP formation regime where $\beta W_0 > 2$. Ultimately, at sufficiently low temperatures the Mn^{2+} spins may saturate, $<s \cdot S_j> \rightarrow 5/4$. This interpretation of the temperature variation of U can be tested by rewriting the internal energy as the product of two factors, one describing the carrier spin alignment process, the other the gradual lengthening of Γ :

$$U = - \left[\frac{2<s \cdot \Gamma>}{\sqrt{<\Gamma^2>}} \right] \cdot \left[\frac{\sqrt{<\Gamma^2>}}{2} \right]. \tag{4.3}$$

$<\Gamma^2>$ is calculated from Eq. (3.13) with the aid of the classical approximation for $(s \cdot \Gamma)^2$. One finds:

$$<\Gamma^2/4> \equiv <(s \cdot \Gamma)^2> = \frac{1}{Z} \frac{\partial^2 Z}{\partial \beta^2}.$$

$$= \left[3W_0^2/4 \right] \left[\frac{1+(\beta W_0)^2/2+(\beta W_0)^4/48}{1+(\beta W_0)^2/4} \right], \tag{4.4}$$

This equation implies $\sqrt{<\Gamma^2>}=\sqrt{3}W_0$ when $(\beta W_0) \ll 1$. As βW_0 increases, $\sqrt{<\Gamma^2>}$ remains essentially constant until $\beta W_0 \cong 2-3$; thereafter it increases as βW_0. Conversely, the spin orientation factor, $2<s \cdot \Gamma>/\sqrt{<\Gamma^2>}$, grows linearly throughout the fluctuation regime ($\beta W_0 < 2$) and has value near unity in the low temperature range. Thus, though the transition is a gradual one, it has two distinct regimes - a higher temperature one dominated by magnetic ion spin fluctuations, and a lower temperature one within which the net moment of the complex ($\sqrt{<\Gamma^2>}$) gradually increases.

Not surprisingly, in the high temperature regime ($\beta W_0 \ll 1$) there are large percentagewise fluctuations of the BMP energy about its mean value. In the Gaussian limit

$$<(\Delta H)^2> = <H^2>-<H>^2 = \frac{\partial^2 (lnZ)}{\partial \beta^2}$$

$$= \frac{W_0^2}{4} \left[1 + \frac{8}{(4+\beta^2 W_0^2)} - \left[\frac{4\beta W_0}{4+\beta^2 W_0^2} \right]^2 \right], \qquad (4.5)$$

and $<(\Delta H)^2>/<H^2> \rightarrow 1$ as $\beta W_0 \rightarrow 0$. On the other hand, in the BMP regime the energy is quite sharply defined since

$$<(\Delta H)^2>/<H^2> \rightarrow \left[\frac{4}{\beta^2 W_0^2} \right] \ll 1 \text{ as } \beta W_0 \rightarrow \infty.$$

In donor-BMP, the exchange interaction is barely strong enough to force the system into the collective regime at T = 4K. It is not surprising, therefore, that early SFRS experiments on CdMnSe showed only faint indications of BMP formation in the temperature range T> 2K. For a time thereafter it was believed that antiferromagnetic, next nearest neighbor $Mn^{2+}-Mn^{2+}$ interactions would preclude stronger polaron effects below T=2K. Isaacs[5] realized, however, that some statistically favored Mn^{2+} ions in a randomly alloyed sample would have no nearest or next nearest neighbor Mn^{2+} ions, and would continue to increase the tendency for polaron formation below 2K. To test this idea he performed low temperature SFRS experiments - to 130mK, a world's record - on two CdMnSe samples. Fig. 1 shows the temperature variation of the zero magnetic field Zeeman splitting (E_0) that he observed in a $Cd_{0.9}Mn_{0.1}Se$ crystal. The data clearly show two distinct regimes of polaron behavior; a fluctuation regime above 3K where E_0 is independent of T, and a collective regime below 3K within which the polaron energy increases rapidly with decreasing temperature. The transition between them is continuous, though relatively abrupt. In the high temperature range E_0 = 0.75 meV; theoretically, the peak of the Raman spectrum is expected to be at $E_0 = \sqrt{2} W_0$; one calculates E_0 = 0.70 meV from Eq. (3.12).

Fig. 2 compares the measured SFRS spectrum to the theoretical spectrum:[2]

$$S(E) \approx E^2 e^{\frac{-E^2}{2W_0^2}} \cdot e^{\frac{E}{2k_B T}}. \qquad (4.6)$$

This spectrum is essentially the integrand of the partition function [Eq. (3.13)]; thus, it directly reflects the distribution of internal fields experienced by the electron.

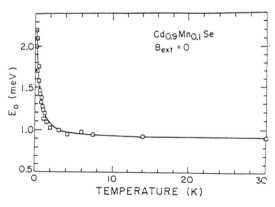

Fig. 1   BMP energy, E_0, vs. temperature for $Cd_{0.9}Mn_{0.1}Se$ in zero magnetic field. Solid line is theoretical fit.

154

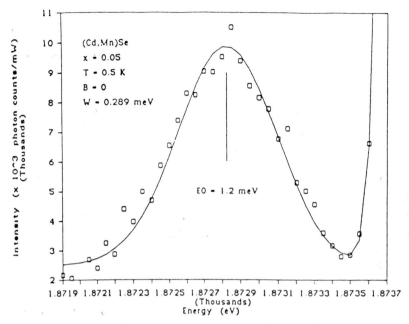

Fig. 2   Spin-flip Raman scattering spectrum for (Cd,Mn)Se with x = 0.05 and T = 0.5 K.
The data are fit (solid line) to Eq. (4.6) with W_0=0.29 meV.

Isaac estimates that at T=130mK in the x=0.1 sample the donor-BMP has magnetic
moment $\mu = 111\mu_B$. Though large, this value is still well below the saturation moment of
$750\mu_B$. Saturation will be exceedingly difficult to achieve in donor-BMP. However, it can
easily be attained in strongly-coupled acceptor- BMP.

Acceptor-BMP are more interesting than donor-BMP because they exhibit the full
range of possible BMP behavior, from the fluctuation-dominated regime at high
temperatures to the fully saturated collective regime at low temperatures. In a typical wide
gap semimagnetic crystal, the coupling constant for acceptor-BMP is $W_0 \cong 6 meV$, as
compared to 0.7meV in the corresponding donor-BMP. Thus, in the acceptor case, the
transition from the fluctuation-dominated regime to the collective regime occurs at $T \cong 30K$,
where the Mn^{2+} susceptibility is Curie-like and the theoretical ideas of Section 3 should
apply.

Unfortunately, it is not easy to study the properties of the simple acceptor-BMP.
SFRS of holes has not yet been seen in DMS, and most luminescence features are produced
by three-body complexes (A^0X, D^0X) whose wave functions are exceedingly difficult to
calculate. The donor-acceptor pair (DAP) line is an exception; it involves the recombination
of electrons and holes bound to fairly well separated donors and acceptors. Moreover, since
the coupling constant of acceptor-BMP is much larger than that of donor-BMP, any polaron
effects observed in DAP luminescence can be attributed to the acceptor.

The time resolved DAP luminescence technique, developed by Nhung and Planel,[6]
has made possible direct optical measurements of the acceptor-BMP binding energy in
CdMnTe. Their results for a series of alloys are shown in Fig. 3. These data have two
important features: (1) acceptor-BMP binding energies are substantially larger, at high
temperatures, than that of the acceptor in CdTe and (2) there is a gradual increase in binding

energy with decreasing temperature. Nhung and Planel ascribe the high temperature energy difference to Mn^{2+} spin fluctuations and suggest that BMP formation is responsible for the increase in binding energy at lower temperatures. In their original publication, they tested this picture with a calculation that made several assumptions concerning the acceptor-BMP, namely:

1. Mn^{2+} magnetization is a continuous function of position, M(r).

2. The hole wave function is hydrogenic.

3. Mn^{2+} magnetization is linear in effective field to saturation, and constant thereafter.

4. The magnetic entropy can be calculated in the classical, Gaussian approximation.

Two groups have developed formalisms which do not require assumptions (1), (3), and (4). The analysis parallels that of Section 3. For the valence band one postulates, in the hydrogenic approximation, a spin Hamiltonian of the form:

$$H = -J\Sigma_j \left[(s \cdot S_j) |\phi(R_j)|^2 \right] = -\Sigma_j \left[K_j (s \cdot S_j) \right]. \tag{4.7}$$

where J is the valence band exchange constant ($JN_o \cong 0.9ev$), the hole spin s = 3/2, and $\phi(r) = e^{-r/a}/\sqrt{\pi a^3}$ is the acceptor wave function. The radius a is determined variationally. The spin partition function, in the classical approximation, is then

$$Z = \underset{\{s,S_j\}}{Tr} \int \delta \left[\gamma - \Sigma_j (K_j S_j) \right] e^{\beta(s \cdot \gamma)} d^3\gamma$$

$$= \underset{\{S_j\}}{Tr} \int \delta \left[\gamma - \Sigma_j (K_j S_j) \right] \left[e^{3/2\beta\gamma} + e^{1/2\beta\gamma} + e^{-1/2\beta\gamma} + e^{-3/2\beta\gamma} \right] d^3\gamma$$

$$= 6^N \int \frac{d^3\lambda d^3\gamma}{(2\pi)^3} e^{i\lambda\gamma} \underset{j}{\Pi} \left[\mathcal{F}_{5/2}(-i\lambda K_j) \right] \left[e^{3/2\beta\gamma} + e^{1/2\beta\gamma} + e^{-1/2\beta\gamma} + e^{-3/2\beta\gamma} \right] \tag{4.8}$$

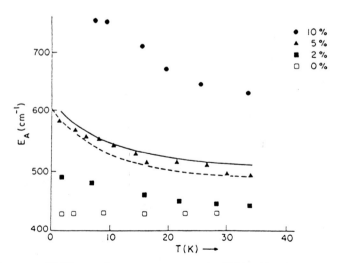

Fig. 3   Acceptor-BMP energies vs. temperature in (Cd,Mn)Te alloys. Dashed line; theory of Nhung and Planel: solid line; theory of Warnock and Wolff (both for 5% Mn).

This expression is the analogue for a spin 3/2 carrier, of Eq. (18). The integrals can again be evaluated as outlined in Part IV, Section 1. The result is:

$$Z = 2(6)^N \frac{d}{d\beta} \left\{ \beta\Pi_j \left[\mathcal{F}_{5/2} \left[\frac{3\beta K_j}{2} \right] \right] + \beta\Pi_j \left[\mathcal{F}_{5/2} \left[\frac{\beta K_j}{2} \right] \right] \right\}, \qquad (4.9)$$

a formula quite similar to Eq. (3.9). Wolff and Warnock[7] calculated BMP energies from this expression by making the continuum approximation, though in principle the discrete sum could have been evaluated. Their results are indicated in Fig. 3 along with earlier calculations of Nhung and Planel.[6] The only adjustable parameter (a=13A) in this fit was determined from the measured acceptor binding energy of CdTe. Other quantities (JN_0, x) are known from independent measurements. Both of the calculations illustrated in Fig. 3 imply that Mn^{2+} spins within the acceptor orbit are fully aligned with the hole spin at low temperatures. The net moment of the BMP is then about $20\mu_B$. The radius of the acceptor wave function could, in principle, vary with temperature via the changing exchange interaction. However, this effect is found to be small.

Calculations of Nhung et al.[8] are compared with their experiments in Fig. 4. These results were obtained by numerically integrating an expression for the internal energy derived from Eq. (4.9):

$$U = -\frac{\partial(lnZ)}{\partial\beta}$$

$$= -\frac{\int d^3\gamma\gamma P(\gamma) \left[3/2e^{3/2\beta\gamma} + 1/2e^{1/2\beta\gamma} - 1/2e^{-1/2\beta\gamma} - 3/2e^{-3/2\beta\gamma} \right]}{\int d^3\gamma P(\gamma) \left[e^{3/2\beta\gamma} + e^{1/2\beta\gamma} + e^{-1/2\beta\gamma} + e^{-3/2\beta\gamma} \right]}. \qquad (4.10)$$

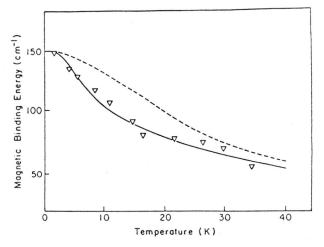

Fig. 4   Comparison of theory and experiment for acceptor-BMP energies. The solid curve is the exact, numerical calculation; the triangles are experimental points for $Cd_{0.95}Mn_{0.05}Te$. The dashed curve corresponds to a truncated Gaussian approximation.

A finite number of Mn^{2+} ions was randomly distributed throughout the BMP to give the correct x-value, and the sum over j was cut off at $R_c = 1.5a = 15A$. Physically, R_c is the distance beyond which the hole exchange interaction is no longer strong enough to break up antiferromagnetic, $Mn^{2+}-Mn^{2+}$ spin alignments.

Though the calculations outlined above differ in detail, they contain similar physics. It is remarkable, and gratifying, that first principles calculations can predict the E vs. T curve for a center as complicated as the acceptor-BMP. Both calculations use hydrogenic wave functions for the acceptor. Errors in the binding energy caused by this approximation are minimized by choosing the acceptor radius to give the correct energy in CdTe. However, other properties of the acceptor-BMP may be poorly described by a wave function that does not incorporate the degeneracy of the valence band.

5. EXACTLY SOLUBLE POLARON MODELS

Several authors have discussed an exactly soluble model of the BMP. Though unrealistic, it provides a valuable test of the semiclassical approximation. The model is derived by replacing the hydrogenic wave function in Eq. (2.7) by a normalized step function:

$$<\phi(r)>^2 \rightarrow (3/4\pi R_0^3)\theta(R_0 - r), \tag{5.1}$$

where

$$\theta(x) = 1 \quad (x > 0)$$

$$= 0 \quad (x < 0), \tag{5.2}$$

and R_0 is a variational parameter that will later be chosen to optimize the free energy. With this assumption the spin Hamiltonian [Eq. (2.7)] takes the form:

$$H_{spin} = -\alpha(3/4\pi R_0^3)\Sigma_i \left[\theta(R_0 - |R_i|)(s \cdot S_i) \right]$$

$$= -\alpha(3/4\pi R_0^3) \sum_{i=1}^{N} \left[(s \cdot S_i) \right]$$

$$= -\alpha(3/4\pi R_0^3)(s \cdot S), \tag{5.3}$$

where N is the number of Mn^{2+} ions within radius R_0 of the donor. N is determined by the conditions;

$$N = \Sigma_i \theta(R_0 - |R_i|) \equiv \bar{x}N_0 \int \theta(R_0 - r)d^3r = \bar{x}N_0 \left[\frac{4\pi R_0^3}{3} \right]. \tag{5.4}$$

It is convenient to define a constant $K = \alpha \left[3/4\pi R_0^3 \right]$. Note that $KN = \bar{x}(\alpha N_0)$. The spin Hamiltonian then takes the form

$$H_{spin} = -K(s \cdot S), \tag{5.5}$$

and can be immediately diagonalized since

$$H_{spin} = -\frac{K}{2}\left[\left[\vec{S}+\vec{s}\right]^2 - S^2 - s^2\right]$$

$$= -\frac{K}{2}\left[J^2 - S^2 - s^2\right]. \tag{5.6}$$

Here $J = (\vec{S}+\vec{s})$ is the total angular momentum of the spin system. J^2 and J_z are constants of the motion, as are S^2 and s^2. Thus, H_{spin} has eigenvalues

$$E = -\frac{K}{2}[J(J+1) - S(S+1) - 3/4]. \tag{5.7}$$

For a given value of S, J can take on values $(S \pm 1/2)$ corresponding to parallel or antiparallel alignment of the electron spin relative to the total Mn^{2+} spin. The corresponding energies are

$$E_+(S) = -\frac{KS}{2}$$

$$E_-(S) = \frac{K(S+1)}{2}. \tag{5.8}$$

Within this simple model, SFRS corresponds to transitions $E_+(S) \rightarrow E_-(S)$ with S conserved.

A typical donor orbit of radius 40A in a DMS crystal with x = .03 contains about 100 Mn^{2+} spins. That spin system has an enormous number of possible spin configurations, $0(6^{100})$. Thus, we focus on the partition function of the model. It is given by

$$Z_m = 2\Sigma_S \left\{ D(S)\left[(S+1)e^{\beta KS/2} + Se^{-\beta K(S+1)/2}\right]\right\}, \tag{5.9}$$

where $D(S)$ is the number of Mn^{2+} spin configurations, of fixed S_z, with total spin S. Yanase and Kasuya[9] describe a technique for calculating $D(S)$. When $1 << S << \frac{5N}{2}$, it can be shown that:

$$D(S) \cong \frac{6^N}{\sqrt{\pi}}\left[\frac{6}{35N}\right]^{3/2}(2S+1)e^{-6S^2/35N}. \tag{5.10}$$

The summation over S is thus similar in form to the integration over γ in Eq. (3.13). Replacing this sum by an integration gives the result:

$$Z_m \cong 2(6)^N\left[1 + \frac{35NK^2\beta^2}{48}\right]e^{\frac{35NK^2\beta^2}{96}}. \tag{5.11}$$

This formula is similar in form to Eq. (3.13), and becomes identical to it if one chooses $35NK^2/12 = W_0^2$.

The preceding discussion implies that, with proper choice of the coupling constant, the partition function and thermodynamic properties of the model quantum mechanical problem [Eq. (5.5)] can be made identical to the semiclassical limit of the true BMP problem. this coincidence only holds when $P(\gamma)$ [Eq. (3.11)] is Gaussian; nevertheless, it is still a striking result because the energy level spectrum of the model problem is quite different, in detail, from that of the BMP. The energy levels of the model have degeneracy $D(S)$ (which is enormous), whereas these degeneracies are removed, by the spatial variation of K_j, in the true BMP problem. Yet, thermodynamically, the two systems are equivalent.

To further explore the relationship between the BMP and the step function model, it is helpful to invoke a variational principle developed by Feynman.[10] He considers a thermodynamic system described by Hamiltonian H, that one wants to approximate by a model Hamiltonian, H_m. Feynman's theorem states that

$$F \leq F_m + <H - H_m>_m \equiv \mathcal{F}, \tag{5.12}$$

where F is the true free energy, F_m the free energy of the model, and

$$<H - H_m> = \frac{Tr\left[(H - H_m)e^{-\beta H_m}\right]}{Tr\left[e^{-\beta H_m}\right]}. \tag{5.13}$$

For our model the right hand side of Eq. (5.12) can be analytically evaluated. The result is quite simple in the low and high temperature limits; one finds

$$\mathcal{F}(\beta, N) \cong 1 (or\ 3) \left[\frac{35\beta(\bar{x}\alpha N_0)^2}{96N} - \frac{35\beta(\bar{x}\alpha^2 N_0)}{48N} \sum_{j=1}^{N} |\phi(R_j)|^2\right]. \tag{5.14}$$

Here K has been eliminated via the relation $KN = \bar{x}(\alpha N_0)$. $\phi(R)$ is the true orbital wave function of the BMP. Minimizing $F(\beta, N)$ with respect to N gives the interesting equation,

$$4\pi \int_0^{R_0} |\phi(r)|^2 r^2 dr = \frac{4\pi R_0^3}{3} |\phi(R_0)|^2 + 1/2, \tag{5.15}$$

which was previously derived by Ryabchenko[11] via perturbation theory. We see here that it is an exact result obtained by minimizing the Feynman expression. With a hydrogenic wave function, Ryabchenko finds $R_0 = 1.82$ a from Eq. (5.15).

On the other hand, with a hydrogenic wave function, the condition

$$\frac{35NK^2}{12} = \frac{35\bar{x}(\alpha N_0)^2}{12}\left[\frac{3}{4\pi N_0 R_0^3}\right] = W_0^2 \cong \frac{35\bar{x}\alpha^2 N_0}{12} \int |\phi(r)|^4 d^3r \tag{5.16}$$

implies $R_0 = (6)^{1/3} a = 1.82a$. We conclude that the optimal step function model Hamiltonian (in the Feynman sense) has the same partition function (and thermodynamics) as the semiclassical approximation to the BMP. Experimentally, W_0 is somewhat higher than the theoretical values. Heiman measures $W_0 = 0.66 \pm 0.07 meV$, compared to the theoretical $W_0 = 0.56 meV$; Isaacs finds 0.75 meV vs. 0.70 meV.

The model Hamiltonian of Eq. (5.5) is a gross over-simplification of the correct, spatially-varying BMP interaction [Eq. (3.2)]. Thus, it is pertinent to ask if there are other, exactly soluble quantum mechanical models that better approximate H_{spin}. Work of Kasuya and Yanase[12] suggests that the two-step spin Hamiltonian,

$$H_2 = -K_1(s \cdot S_1) - K_2(s \cdot S_2), \tag{5.17}$$

can also be exactly diagonalized[13]. Here the spatial variation of K_j is better approximated by a two-step $|\phi(r)|^2$. Squaring Eq. (5.17) gives, after some manipulation, the result:

$$H_2^2 = \left\{ \left[\frac{K_1 + K_2}{2} \right] H_2 + \right.$$

$$\left. \left[\frac{K_1 K_2 J^2}{4} + \frac{K_1(K_1 - K_2)S_1^2}{4} + \frac{K_2(K_2 - K_1)S_2^2}{4} - \frac{3K_1 K_2}{16} \right] \right\}, \tag{5.18}$$

where $J^2 = (\vec{S}_1 + \vec{S}_2 + \vec{s})^2$, S_1^2, S_2^2, s^2 and J_z are constants of the motion. Thus, the two-step model has eigenvalues:

$$E_2 = \left[\frac{K_1 + K_2}{2} \right] \pm$$

$$\sqrt{\frac{K_1 K_2 J(J+1)}{4} + \frac{K_1(K_1 - K_2)S_1(S_1 + 1)}{4} + \frac{K_2(K_2 - K_1)S_2(S_2 + 1)}{4} - \frac{3K_1 K_2}{16}} \cdot$$

$$\tag{5.19}$$

In the limit $S_1 \gg 1$, $S_2 \gg 1$, E_2 is approximated by

$$E_2 \cong \pm \left| \frac{K_1 \vec{S}_1}{2} + \frac{K_2 \vec{S}_2}{2} \right|. \tag{5.20}$$

Note that this two-step model removes some of the artificial degeneracy of the one-step model. Numerical calculations to optimize the two-step model are in progress; these results will be reported elsewhere.[13]

6. BMP IN CONCENTRATED DMS

The preceding discussion has focussed on DMS with small magnetic ion concentration ($x < 0.1$) since these materials exhibit BMP behavior in its purest form. Polaron formation has clearly been observed in them, but requires temperatures in the 100mk-10K range. For practical and scientific reasons it would be desirable to increase BMP formation temperatures to the liquid nitrogen range or higher. There are several possible approaches to this goal:

i.   Seek ferromagnetic DMS with $T_c > 77K$.

ii. Increase carrier-magnetic ion exchange constants.

iii. Increase magnetic ion concentration.

Unfortunately, ferromagnetism seems to be forbidden in tetrahedral semiconductors. Magnetic ion exchange constants have been measured in Mn, Fe, Co, and Ni substituted II-VI's and are all antiferromagnetic. Ferrimagnetism is not precluded in mixed alloys, such as Zn(Fe,Mn)Se, but will require difficult-to- fabricate , ordered DMS structures as discussed below. To our knowledge, no group has yet grown them.

Carrier-magnetic ion exchange interactions are large in p- type DMS; usually $JN_0 \cong 0.8$–$1.0e$ v. CdMnS is an exception with $JN_0 \cong 3ev$. Unfortunately, it cannot be made p-type; the exchange constant of holes was inferred from the Zeeman splitting of exciton transitions. The JN_0 of CdMnS is large because this crystal has a small anion-cation spacing;

sp-d hybridization is expected to increase rapidly with decreasing anion-cation distance (see notes of K. Hass). This argument suggests that exchange constants larger than 1ev cannot be achieved in the narrower gap II- VI's, such as CdMnTe, that can be made p-type.

Finally, there is the possibility of increasing BMP energies by increasing the magnetic ion concentration. Naively, one might expect a large increase of E_{BMP} with x, since full saturation of the $(s \cdot S_j)$ correlation in Eq. (2.1) implies, $E_{BMP} \cong 5/4\bar{x}(JN_0)$. For acceptor - BMP this expression gives, for example, $E_{BMP} = 250meV$ when x=0.25. Experimentally, however, the acceptor- BMP energy at low temperatures in $Cd_{0.75}Mn_{0.25}Te$ is only 40meV. The large discrepancy is attributed to competing, antiferromagnetic interactions between Mn^{2+} spins. Detailed studies of the BMP energy in $Cd_{1-x}Mn_xTe$ have been performed by Bugajski et al.[14], using the free electron-A^0 luminescence line as a monitor of BMP energy vs. x and T. At low temperatures they find that the BMP energy initially increases with x according to the relation $E_{BMP} = 5/4x (JN_0)$, but soon departs from it and eventually decreases with x beyond x=0.22. Their low temperature data are illustrated in Fig. 5.

To confirm the role of $Mn^{2+}-Mn^{2+}$ interactions in suppressing polaron energies, we have developed[15] a theory of BMP that includes them in a phenomenological way. An ab initio solution of the interacting Mn^{2+} spin problem is a formidable task, especially when x>0.2 and the system is in the spin glass range. To avoid it, we have used experimental measurements of magnetization vs. field (to 50T) to determine the response of the interacting Mn^{2+} spin system to the exchange field generated by the carrier.

Consider initially the case of noninteracting Mn^{2+} spins. In the semiclassical approximation, the partition function of the BMP has the form $Z = \dfrac{\partial}{\partial \beta}(\beta Z_{Ising})$, where

$$Z_{Ising} = Tr\left[e^{-\beta H_{Ising}}\right], \tag{6.1}$$

and

$$H_{Ising} = J\Sigma_j (s_z S_{jz}). \tag{6.2}$$

The trace of Eq. (6.1) over s gives

$$Z_{Ising} = Tr\left[e^{\beta \Sigma_j (K_j s_z S_{jz})}\right]$$

$$= \underset{\{S_j\}}{Tr}\left[e^{\beta/2\Sigma_j (K_j S_{jz})} + e^{-\beta/2\Sigma_j (K_j S_{jz})}\right]$$

$$\equiv 2Tr\left[e^{-\beta H_M}\right], \tag{6.3}$$

with

$$H_M = -\frac{J}{2}\Sigma_j\left[S_{jz} |\phi(R_j)|^2\right]. \tag{6.4}$$

We can view H_M as describing the alignment of Mn^{2+} spins in an internal, spatially varying magnetic field:

$$\mu g B_{eff}(R_j) = \frac{K_j}{2} = \frac{J |\phi(R_j)|^2}{2}. \tag{6.5}$$

162

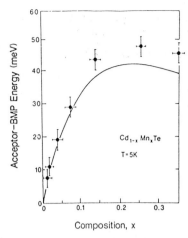

Fig. 5   Composition dependence of acceptor-BMP binding energy in low-temperature
limit in $Cd_{1-x}Mn_xTe$ alloys. Solid line represents the values calculated according
to phenomenological polaron theory.

In the noninteracting case this alignment is well understood; it is described by the function
$F_{5/2}(\beta K_j/2)$. This result, combined with Eq. (2.16), leads immediately to Eq. (2.16) for Z.
To extend these ideas to the interacting Mn^{2+} problem, we must first show (via the
semiclassical approximation) that it can be reduced to an equivalent Ising problem. Then we
must determine the response of the Mn^{2+} spin system. That is a difficult problem when the
spins interact, but can be avoided by the using experimental magnetization data to determine
M vs. B_{eff}. Note, however, that this local approximation assumes that the spatial variation of
the internal field is relatively slow, a condition that is barely satisfied in the most interesting,
acceptor-BMP case.

Before turning to the interacting Mn^{2+} problem, we briefly compare the Ising free
energy with the classical free energy of spins in a spatially varying magnetic field. The Ising
free energy is

$$F_I = kT \ln (Z_I)$$

$$= -kT \left\{ \ln(2) + N \ln(6) + \sum_j \left[\ln F_{5/2}(\beta K_j/2) \right] \right\}$$

$$\equiv -kT \left\{ \ln(2) + N \ln(6) + xN_0 \int d^3r \ln F_{5/2} \left[\beta \mu g B_{eff}(r) \right] \right\}. \tag{6.6}$$

On the other hand, the classical free energy is

$$F_c = -\int d^3r \int_0^{B_{eff}(r)} M(B')dB', \tag{6.7}$$

where

$$M(B) = \mu g x N_0 \frac{F'_{5/2}(\beta \mu g B)}{F_{5/2}(\beta \mu g B)} = kT \mu g x N_0 \frac{d}{dB} \left[\ln F_{5/2}(\beta \mu g B) \right]. \tag{6.8}$$

163

The two free energies have the same magnetic field variation. This relationship will be important in our phenomenological treatment of the spin glass regime below.

Next consider the case of *interacting Mn^{2+}* spins. The BMP Hamiltonian takes the form

$$H \cong -J_{NN} \sum_{\{NN\}} \left[S_i \cdot S_j \right] - \Sigma K_j \left[s \cdot S_j \right]$$

$$= \left[H_0 + H_x \right].$$
(6.9)

In the semiclassical approximation

$$Z = Tr \left[e^{-\beta(H_0 + H_x)} \right] \cong Tr \left[e^{-\beta H_0} e^{-\beta H_x} \right]$$

$$= Tr \int d^3 \gamma e^{-\beta H_0} \delta(\gamma - \Gamma) e^{\beta s \cdot \gamma},$$
(6.10)

where Γ is defined by Eq. (3.4). The trace over s is evaluated with Eq. (3.5) and the delta-function replaced by its integral representation

$$\delta(\gamma - \Gamma) = \int \frac{d^3 \lambda}{(2\pi)^3} e^{i\lambda \cdot (\gamma - \Gamma)}.$$
(6.11)

After integrating over the γ-angular variables one finds

$$Z = \frac{1}{2\pi^2} \frac{d}{d\beta} \left\{ Tr_{\{S_j\}} \int \frac{d^3 \lambda}{\lambda} \int_{-\infty}^{\infty} d\gamma e^{-\beta H_0} sinh(\beta \gamma / 2) sin(\lambda \gamma) e^{-i\lambda \cdot \Gamma} \right\}.$$
(6.12)

As before, we evaluate the γ integral for imaginary β and then analytically continue back to the real axis. The result is

$$Z(\beta) = \frac{1}{4\pi} \frac{d}{d\beta} \left\{ Tr_{\{S_j\}} \int d\Omega_\lambda \beta e^{-\beta(H_0 + \frac{\lambda \cdot \Gamma}{2|\lambda|})} \right\}$$

$$= \frac{1}{4\pi} \frac{d}{d\beta} \left\{ \int d\Omega_\lambda \beta Z_M \left[\beta; B_{eff}(R_j) \right] \right\},$$
(6.13)

where $Z_M [\beta; B_{eff}(R_j)]$ is the partition function of the *interacting Mn^{2+}* spin system in the spatially varying magnetic field

$$\mu g B_{eff}(R_j) = \frac{\vec{\lambda} K_j}{2|\lambda|} = \frac{\vec{\lambda} J |\phi(R_j)|^2}{2\lambda}.$$
(6.14)

Moreover, since this system is isotropic, Z_M is independent of the directions of λ. It then follows, as before, that

$$Z(\beta)=\frac{d}{d\beta}\left[\beta Z_M(\beta,B_{eff})\right].\tag{6.15}$$

To evaluate Z_M we use the classical formula for the free energy:

$$Z_M=e^{-F_c/kT}$$

$$\mathcal{F}_c=-\int d^3r \int_0^{B_{eff}(r)} M(B')dB',\tag{6.16}$$

and the experimentally measured M vs. B relationship for the Mn^{2+} spin system. Heiman *et al.*[16] have found that M vs. B for $Cd_{1-x}Mn_xTe$ crystals with $0.1\leq x\leq0.5$ can be accurately described at low temperatures by the formula:

$$M(B)=\left[\mu g x_{eff}N_0\frac{\mathcal{F}'_{5/2}(\beta\mu gB)}{\mathcal{F}_{5/2}(\beta\mu gB)}+cB\right].\tag{6.17}$$

The data from which this relation is derived are illustrated in Fig. 6.

Here x_{eff} is the effective number of "free" Mn^{2+} spins, whose magnetization saturates below 10T at 4K, and the cB term in Eq. (6.17) is the magnetization of large, spin glass clusters. Note that in the more heavily alloyed samples, the saturation field of the latter exceeds 100T. Thus, the magnetization of these spins is severely depressed. Moreover, as x increases the fraction of easily magnetizable spins *decreases*. $x_{eff}=.038$ when x=0.11, but has fallen to $x_{eff}=.014$ when x=0.49. Both effects act to reduce the BMP energy.

Eq. (6.16) can now be used to evaluate the free energy; the result is

$$\mathcal{F}=-kT\int d^3r\left\{x_{eff}N_0\ln\left[\mathcal{F}_{5/2}\left[\beta\mu gB_{eff}(r)\right]\right]+\frac{cB_{eff}^2(r)}{2}\right\}.\tag{6.18}$$

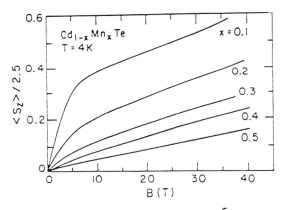

Fig. 6   Spin alignment $<S_z>$ of Mn^{2+} ion, normalized to $\frac{5}{2}$, vs. applied magnetic field B, for $Cd_{1-x}Mn_xTe$, with x=0.11, 0.20, 0.30, 0.38, 0.49, at liquid-He temperature.

165

Finally, the effective exchange potential appearing in the Schrodinger equation of the BMP is:

$$V_x = -kT \frac{\delta (lnZ)}{\delta \phi^*} = \frac{\delta F}{\delta \phi^*}$$

$$= -\frac{5JN_0 x_{eff}}{4} B_{5/2} \left[\frac{J|\phi(r)|^2}{2kT} \right] + C \left(\frac{J}{2\mu g} \right)^2 |\phi(r)|^2. \qquad (6.19)$$

Note that V_x is a function of $|\phi(r)|^2$, so the resulting Schrodinger equation:

$$\left\{ \frac{p^2}{2m^*} - \frac{e^2}{E_0 r} - V_x[|\phi(r)|^2] \right\} \phi(r) = E \phi(r) \qquad (6.20)$$

is highly nonlinear. Eq. (6.20) is a generalization of a nonlinear Schrodinger equation originally derived by Spalek.[17]

All quantities appearing in Eqs. (6.18) - (6.20) are known experimentally. The Schrodinger equation was solved numerically for a hydrogenic model of the acceptor-BMP in CdMnTe, and used to determine total acceptor binding energies and magnetic energies vs. x. A comparison of the latter with measurements of Bugajski et al. is illustrated in Fig. 5. The agreement is better than the many approximations of the theory would lead one to expect. In particular, it supports our previous contention that AF Mn-Mn interactions are responsible for the reduction of BMP energies as x increases.

The preceding arguments preclude large BMP energies in random DMS alloys, but need not do so in *ordered* DMS. To achieve large polaron energies in DMS one requires that the Mn^{2+} spins respond paramagnetically to the carrier exchange field when x is sizable (say x=0.25). They do not do so - as evidenced by the experiments of Heiman et al. - when Mn^{2+} ions occupy nearest neighbor sites in the fcc cation sublattice of a random DMS alloy. On the other hand, with appropriate ordering of the magnetic ions a II-VI crystal can accommodate 25% of them with no magnetic ions being nearest neighbors of one another. Since next nearest neighbor interactions are weak ($J_{NNN} < 0.2J_{NN}$), they can be overcome by

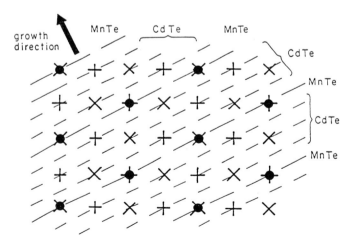

Fig. 7   Cation locations for (210) ALE growth of Cd_3MnTe_4 in the famatinite structure. Dots are Mn atoms; +'s are the top layer cations, x's are next layer.

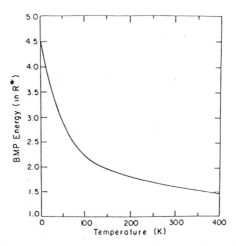

Fig. 8   Calculated acceptor-BMP energy, in units of the Rydberg (R*=53 meV), as a function of temperature in $(Cd_3Mn)Te_4$.

the carrier-generated internal field to produce a strongly bound polaron. Ordering of the type required to achieve this goal occurs naturally in stannite crystals that are tetrahedrally bonded, diamond-type semiconductors with Mn^{2+} spacing $\sqrt{2}$ times that of nearest neighbor Mn^{2+} pairs in $Cd_{1-x}Mn_xTe$. Cu_2MnGeS_4 is a typical stannite-type crystal (there are many such materials) that has been extensively studied by Shapira[18] and his collaborators. They find that the saturation field of $(Cu_2MnGe)S_4$ is 17T, compared to 130T in the random alloy, $Cd_{0.75}Mn_{0.25}Te$, which has the *same* magnetic ion concentration. The large difference is a direct reflection of the fact that $(Cu_2MnGe)S_4$ contains no nearest neighbor Mn pairs.

These results suggest that stannite-type materials should be optimal for BMP formation. In practice, the natural stannites are too difficult to grow and control to ever be useful. However, there is an alternate possibility - of growing *ordered* alloys with 25% Mn concentration. Fig. 7 shows a possible scheme for achieving stannite Cd_3MnTe_4 by ALE or MBE growth in the (120) direction. The resulting structure, if one alternates three layers of CdTe with one layer of MnTe, is a stannite - like form of $(Cd_3Mn)Te_4$ known as famatinite. Detailed solutions[15] of the nonlinear Schrodinger equation for this structure confirm our initial hypothesis that acceptor-BMP energies would increase about 5X when nearest neighbor AF Mn-Mn interactions are eliminated. These calculations are summarized in Fig. 8.

To our knowledge, famatinite Cd_3MnTe has not yet been grown. The task of doing so is a difficult one, but could lead to a material with unique magnetic properties. They could be tested on quite thin samples (<1μ) via magneto-optic experiments.

BIBLIOGRAPHY

[1]   The properties of DMS are reviewed in the book *Diluted Magnetic Semiconductors, Semiconductors and Semimetals, Vol. 25*, Jacek K. Furdyna and Jasek Kossut, editors (Academic Press, San Diego, 1988).

[2]   The theory of BMP in DMS was developed by T. Dietl and J. Spalek, Phys. Rev. Letts. *48*, 355 (1982); Phys. Rev. *B28*, 1548 (1983) and by D. Heiman, P. Wolff, and J. Warnock, Phys. Rev. *B27*, 4848 (1983). The two groups use different formalisms

with identical conclusions. In these notes we use the HWW formalism which is more microscopic in nature, and naturally handles the problem of spin saturation. A comparison of the two formalisms is given in Ref. 1.

[3] Y. Shapira, S. Foner, D. Ridgley, K. Dwight and A. Wold, Phys. Rev. *B30,* 4021 (1984).

[4] M. Nawrocki, R. Planel, G. Fishman, and R. Galazka, *Proc. XV Intl. Conf. Phys. of Semiconductors, Kyoto,* J. Phys. Soc. Japan Suppl. *A49,* 823 (1980). Also see the paper of HWW in Ref. 2.

[5] E. D. Isaacs, D. Heiman, M. J. Graf, B. B. Goldberg, R. Kershaw, D. Ridgley, K. Dwight, A. Wold, J. Furdyna and J. S. Brooks, Phys. Rev. *B37,* 7108 (1988).

[6] T. Nhung and R. Planel, *Proc. XVI Intl. Conf. Phys. of Semiconductors, Montpellier;* Physics 117B-11813, 488 (1983).

[7] P. Wolff and J. Warnock, J. Appl. Phys. *55,* 2300 (1984).

[8] T. Nhung, R. Planel, C. Benoit a la Guillaume and A. Bhattacharjee, Phys. Rev. *B31,* 2388 (1985).

[9] A. Yanese and T. Kasuya, J. Phys. Soc. Japan *25,* 1025 (1968).

[10] R. P. Feynman, *Statistical Mechanics; A Set of Lectures,* (W. A. Benjamin, Inc., Reading, MA, 1972).

[11] S. M. Ryabchenko and Uy. G. Semenov, Sov. Physics JETP *57,* 825 (1983).

[12] T. Kasuya and A. Yanase, Rev. Mod. Phys. *40,* 684 (1968).

[13] C. McIntyre and P. A. Wolff (to be published).

[14] M. Bugajski, P. Becla, P. A. Wolff, D. Heiman and L. R. Ram-Mohan, Phys. Rev. *B38,* 10512 (1988).

[15] L. R. Ram-Mohan and P. A. Wolff, Phys. Rev. *B38,* 1330 (1988).

[16] D. Heiman, E. D. Isaacs, P. Becla and S. Foner, Phys. Rev. *B35,* 3307 (1987).

[17] J. Spalek, Phys. Rev. *B30,* 5345 (1984).

[18] Y. Shapira, E. J. McNiff, Jr., N. F. Oliveira, E. D. Honig, K. Dwight and A. Wold, Phys. Rev. *B37,* 411 (1988).

MAGNETOOPTIC PROPERTIES OF WIDE GAP

$II_{1-x}Mn_xVI$ SEMIMAGNETIC SEMICONDUCTORS

J. P. LASCARAY

Groupe d'Etude des Semiconducteurs, URA CNRS 357
Université de Montpellier II, Place Eugène Bataillon
34095 Montpellier cedex 5, France

I - INTRODUCTION

Since the first observation of a giant exchange splitting of the free exciton in $Cd_{1-x}Mn_xTe$ by Komarov et al in 1977 [1], the magnetooptical properties of semimagnetic semiconductors have been extensively studied [2]. The origin of these remarkable properties (giant Faraday rotation and Zeeman effect, Spin Flip Raman Scattering...) is in the large splitting of conduction and valence bands in magnetic field which arises through the exchange interaction of band carriers with the 3d electrons localized on Mn^{++} ions. Magnetoreflectivity and magnetotransmission measurements were commonly used to determine the Zeeman splitting of $Cd_{1-x}Mn_xTe$, [1,3 to 11], $Zn_{1-x}Mn_xTe$ [12 to 16], $Cd_{1-x}Mn_xSe$ [18 to 22] and $Cd_{1-x}Mn_xS$ [23 to 27].

An important part of these lecture notes will be devoted to the study of the exchange induced bands splitting, including the deviation from the classical mean field model for concentrated alloys and the effects away from the center of the Brillouin zone. The "magnetooptical" effects observed in absence of magnetic field will be also presented.

II - BAND SPLITTING INDUCED BY MAGNETIC FIELD AT THE CENTER OF THE BRILLOUIN ZONE

II-1- Measurements

The magnetoreflectivity is the most commonly used. A classical experimental set up is presented in Figure 1. The sample, generally either cut with a wire saw from an ingot, polished and etched in a solution of bromine in methanol or cleaved, is placed in a magnetooptical cryostat (superconductive coil, low temperatures and optical access) The measurements are performed in Faraday (Voigt) configuration and the reflected beam is analyzed in circular σ^+ and σ^- (linear Π) polarization. In presence of magnetic field, the valence and conduction bands are split as reported in figure 2 in the case of zinc blende SMSC. According to the selections rules $\Delta mj = +1$ and -1 respectively for the two polarizations σ^+ and σ^- and $\Delta mj = 0$ for the Π linear polarization, six optical transitions are allowed. The figure 3 shows an example of magnetoreflectivity spectra.

Semimagnetic Semiconductors and Diluted Magnetic Semiconductors
Edited by M. Averous and M. Balkanski, Plenum Press, New York, 1991

169

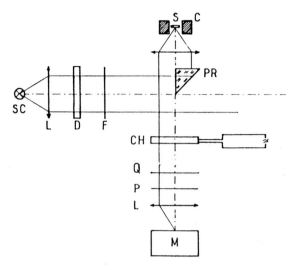

Figure 1. *Experimental set up for magnetoreflectivity measurements :
SC , source ; L , lenses ; PR , Reflecting Prims ; S , Sample ; C,
Superconducting Coil ; CH , Chopper; Q, Quaterwave plate ; P, Linear
Polarizer ; M , Monochromator.*

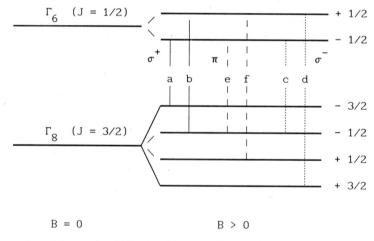

Figure 2. *Schematic illustration of the allowed transitions between
the spin-split valence band maxima and conduction band minima.*

A typical magnetotransmission spectra is presented in figure 4. This
technique allows to obtain more detailed information, for example the
observation of the excited states of the exciton, than the
magnetoreflectivity. Nevertheless, it has not been often used, mainly in
reason of the requested thickness of the samples (below 1 μm) due to the
important absorption coefficient in the excitonic transition region.

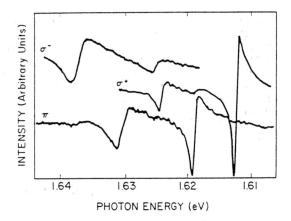

Figure 3. Magnetoreflectivity spectra of Cd$_{0.98}$Mn$_{0.02}$Te at T = 1.4K and B = 1.1 T, from [4].

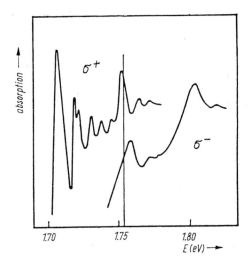

Figure 4. Magnetoabsorption spectra of Cd$_{0.9}$Mn$_{0.1}$Te at T = 1.8K for B = 4T, from [7]

II-2- **Mean field model**

The magnetic part of the Hamiltonian for a band electron in a magnetic field consists in two terms : H = Hm + Hex.
The first,

$$\text{Hm} = - \frac{e}{mc} \, \vec{p}.\vec{A} + g \, \mu_B \, \vec{\sigma}.\vec{B},$$

describes the direct influence of the magnetic field on the carriers states. It is responsible for the Landau quantization and the band splitting of the Landau levels. Its effect can be neglected in wide gap SMSC, compared to effect of Hex. This later term corresponds to Mn 3d ion – carrier exchange interaction. It can be expressed in the form of an Heizenberg Hamiltonian [28,29] :

$$\text{Hex} = - \sum J \, (r - Rn) \, \vec{S}n.\vec{\sigma} \qquad (1)$$

where Sn is the spin of the Mn^{++} localized at the Rn site, α is the spin of the band electron (hole) and J $(r_{++} -$ Rn) is the exchange integral. The summation is extended over all the Mn^{++}. Two approximations can be used to simplify the resolution of (1) : the molecular field and the crystal field approximations. In the first one, all the spin operators are replaced by their mean value < Sz >. This is justified by the spatial extension of the carrier wave function. The second one attributed to every cation site of the crystal a fraction x < Sz >, where x is the Mn concentration, of the mean spin value. In presence of a magnetic field applied in the z direction, (1) becomes :

$$\text{Hex} = - \text{x } \sigma_z < S_z > \sum_i \text{J (r} - \text{Ri)}$$

where the summation is extended over all the cation sites.

a) SMSC with a zinc blende structure
‾‾‾‾‾‾‾‾‾‾‾‾‾‾‾‾‾‾‾‾‾‾‾‾‾‾‾‾‾‾‾‾‾

The following approximations can be used to determine the effect of magnetic field on the valence and conduction bands near the Γ point of the Brillouin zone.
- parabolic bands approximation
- Hm negligible compared to Hex
- large spin orbit splitting of valence band compared to the ions-carrier exchange energy.

Using the basic vector proposed by Bir and Pikus [30] for Γ_6 conduction band and Γ_7 and Γ_8 valence bands :

$\Psi\Gamma_6$: $|3/2, \; 1/2 > = \; S\uparrow$

$|1/2, \; -1/2 > = S\downarrow$

$\Psi\Gamma_8$: $|3/2, \; 3/2 > = (1/\sqrt{2}) \; (X + iY)\uparrow$

$|3/2, \; 1/2 > = (i/\sqrt{6}) \; ((X + iY)\downarrow - 2 \; Z\uparrow)$

$|3/2, \; -1/2 > = (1/\sqrt{6}) \; ((X - iY)\uparrow + 2 \; Z\uparrow)$

$|3/2, \; -3/2 > = (i/\sqrt{2}) \; (X - iY)\downarrow$

$\Psi\Gamma_7$: $|1/2, \; 1/2 > = (1/\sqrt{3}) \; ((X + iY)\downarrow + Z\uparrow)$

$|1/2, \; -1/2 > = (i/\sqrt{3}) \; ((X - iY)\uparrow - Z\downarrow)$

The matrix of Hex are diagonals :

$$< \Psi\Gamma_6 \; |\text{Hex} \; |\Psi\Gamma_6 > \; = \; \begin{vmatrix} 3A & 0 \\ 0 & -3A \end{vmatrix}$$

$$< \Psi\Gamma_8 \; |\text{Hex} \; |\Psi\Gamma_8 > \; = \; \begin{vmatrix} 3B & 0 & 0 & 0 \\ 0 & B & 0 & 0 \\ 0 & 0 & B & 0 \\ 0 & 0 & 0 & -3B \end{vmatrix}$$

and

172

$$< \Psi\Gamma_7 |Hex| \Psi\Gamma_7 > = \begin{vmatrix} -B & 0 \\ 0 & B \end{vmatrix}$$

where

$$A = - (1/6) \, No \, \alpha \, x < Sz >$$

and

$$B = - (1/6) \, No \, \beta \, x < Sz >.$$

No denotes the number of unit cells per unit volume and $\alpha = < S|J|S >$ and $\beta = < X|J|X >$ are the exchange integrals for the conduction and valence band respectively.

The energy of the observed interband transitions in circular polarization are :

σ^+: $Ea = Eo + 3B - 3A$

  $Eb = Eo + B + 3A$

σ^+: $Ec = Eo - B - 3A$

  $Ed = Eo - 3B + 3A$

In this model, the splitting between the two strong components Ea and Ed is given by :

$$\Delta E = - No \, (\alpha - \beta) \, x < Sz > \qquad (2)$$

b) <u>SMSC with hexagonal structure</u>

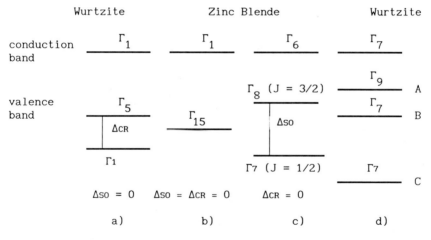

Figure 5.

The figure 5 gives a schematic representation of the band structure of zinc blende and wurtzite semiconductors at the Γ point of the Brillouin zone. c and d correspond to the real case. Because the presence of an anisotropic crystal field, the situation of wurtzite SMSC (eg. Cd1-xMnxSe and Cd1-xMnxS) is more complicated than zinc blende SMSC. The exchange splittings are thus comparable with the interband energy between the A, B and C valence subbands and strongly mix their states. Furthermore, the existence of the anisotropic crystal field is responsible of a strong angle dependence of the spin splitting. The theoretical description of the observed excitonic structure is possible using the complete Hamiltonian obtained by adding an exchange term Hex to the original wurtzite Hamiltonian Hwur [23,21,19,20]. Using the notation of Ref. [20], and in the mean field approximation, the splitting of the conduction band in two subbands is given by :

$$2A = No\ \alpha\ x < Sz >.$$

In the simplest geometry where the external magnetic field is parallel to the C axis of the crystal, the eigenvalues and eigenfunctions can be solved analytically, given six energy values for the A, B and C split subbands :

$$A : E(3/2 \pm 3/2) = \Delta_1 + \Delta_2 \pm Bx$$

$$B : E(3/2 \pm 1/2) = ((\Delta_1 - \Delta_2)/2) \pm ((Bz - Bx)/2) + E^{\pm}$$

$$C : E(1/2 \pm 1/2) = ((\Delta_1 - \Delta_2)/2) \pm ((Bz - Bx)/2) - E^{\pm}$$

with

$$E^{\pm} = (([\Delta_1 - \Delta_2 \pm (Bx + Bz)^2]/2) + 2\ \Delta_3^2)^{1/2}$$

where Δ_1, Δ_2 and Δ_3 are the wurtzite band structure parameters describing the crystal field and spin orbit interactions and

$$Bx,z = No\ \beta_{x,y}\ x < Sz >$$

here

$$\beta_x = < X|J|X > \quad \text{and} \quad \beta_z = < Z|J|Z >.$$

Using the basis which diagonalizes Hwur,

$$A : |3/2, \pm 3/2 > = (1/\sqrt{2})|(X + iY)\uparrow, \downarrow>$$

$$B : |3/2, \pm 1/2 > = \pm ((1 - b\pm^2)^{1/2}/\sqrt{2})|(X \pm iY)\downarrow, \uparrow> + b\pm |Z\downarrow, \downarrow>$$

$$C : |1/2, \pm 1/2 > = (b\pm /\sqrt{2})|(X \pm iy)\downarrow, \uparrow> \mp (1 - b\pm^2)^{1/2}|Z\uparrow, \downarrow>$$

where the wave function mixing the coefficients b$\pm$ are given by :

$$b\pm = ((1 - [(\Delta_1 - \Delta_2) \pm (\beta_x + \beta_z)]) / 2E^{\pm})/ 2$$

the matrix of Hex are diagonal and the allowed transitions energy from excitons A, B, C and conduction b and are :

$$E_A^{\pm} = Eo \mp A - \Delta_1 - \Delta_2 \pm Bx$$

$$E_B^{\pm} = Eo \pm A - ((\Delta_1 - \Delta_2)/2) \pm (Bz - Bx)/2 - E^{\pm}$$

$$E_C^{\pm} = Eo \pm A - ((\Delta_1 - \Delta_2)/2) \pm (Bz - Bx)/2 + E^{\pm}$$

(3)

for Faraday configuration and :

$$E_B^{\pi} = Eo \mp A - ((\Delta_1 - \Delta_2)/2) \pm (Bz - Bx)/2 - E^{\pm}$$

$$E_C^{\pi} = Eo \mp A - ((\Delta_1 - \Delta_2)/2) \pm (Bz - Bx)/2 + E^{\pm}$$

for Voigt configuration.
Eo - Δ_1 - Δ_2 is the energy of the transition for the A exciton in absence of external magnetic field. A schematic representation of allowed transitions is reported in the figure 6, in the case of $Cd_{1-x}Mn_xSe$. The figure 7 represents the fit of exciton lines obtained by magnetoabsorption measurements, by the formula (3) [20].

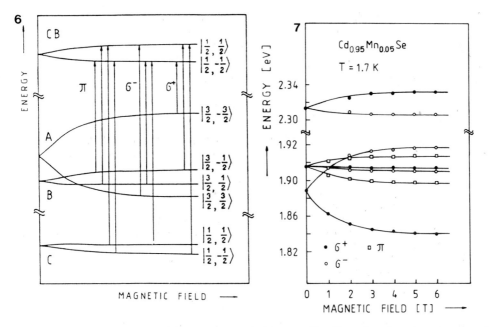

Figure 6. *Schematic representation of the band structure of Cd₁-ₓMnₓSe* [20].

Figure 7. *Energy of the exciton lines versus magnetic field in Cd₀.₉₅Mn₀.₀₅Se, from* [20].

II-3- **Mn-band carriers exchange parameters α and β**

a) Détermination of α and β from band splitting measurements

For low concentrated samples, it is easy to derive α and β from mean field model. In the case of SMSC with zinc blende structure, magnetoreflectivity (or magnetotransmission) measurements performed in Faraday configuration with circular polarized light gives the energy of the four allowed transitions E_a, E_b, E_c and E_d. Magnetization measurements performed on the same samples and in the same conditions of temperature and magnetic field gives :

$$x < S_z > = M_n \, m \, (x) \, / g \, \mu_B \, N_a$$

where M_n is the measured magnetization per unit mass, N_a the Avogadro number, $m(x)$ the mole mass of the compound and $g \sim 2$. It is then possible to determine α and β from :

$$N_o \, (\alpha - \beta) = (E_d - E_a) \, x < S_z >$$

$$\beta/\alpha = (E_c - E_b) \, / \, (E_d - E_a)$$

For crystals with wurtzite structure, when all the transitions corresponding to the splitting of the excitons A, B and C can be measured, the fitting of the experimental data following equ. (3) gives :
 - the band parameters Δ_1, Δ_2, Δ_3, E_{AB}, E_{BC} and b^2
 - the value of the exchange constants α and β.
The values of α an β for Mn based wide gap SMSC are found in the following ranges: $0.18 < N_o \, \alpha < 0.26$ eV and $-0,9 < N_o \, \beta < -1.4$ eV exepted for

$Cd_{1-x}Mn_xS$ which is distinguishable from other because β appears to varies with Mn concentration (N β = - 5.6 eV for x =0.001 [23] ; Noβ = - 1.8 eV for x = 0.013 [27]). The α and β values for Fe based SMSC and for Hg alloys are presented in [38] and [39] of the present volume. The first determination of both α and β for a Co based SMSC $Cd_{1-x}Co_xSe$ gives No α = 279 ± 29 meV and No β = - 1873 ± 42 meV [41].

b) <u>Origin of the Mn-band carriers exchange</u>

The difference in both sign and magnitude of α and β were first explained by Bhattacharjee et al [32]. They have noted that the Mn 3d-sp bands exchange results in two competing mechanism. The first, J_1, originating from direct potential exchange interaction between bands and d electrons is relatively weak and ferromagnetic. The second, J_2, arises from the hybridization between the Mn d orbitals and the band states. This "kinetic" antiferromagnetic contribution which is negligible for the conduction band becomes dominant for the valence band. Using the Schrieffer-Wolff formula for exchange via hybridization it can be expressed in a form of a classical s-d exchange hamiltonian with an exchange constant given by :

$$J_2 = - 2 \ |V_{kd}|^2 \ (\frac{1}{\mathcal{E}_k - \mathcal{E}_d} + \frac{1}{\mathcal{E}_d + U - \mathcal{E}_k})$$

where $\mathcal{E}_k$ is the energy of the band electron, $\mathcal{E}_d$ is the energy of the d electron, U is the coulomb repulsion for adding an electron to the half filled d shell and V_{kd} is the hybridization matrix element. For the valence band, as $|J_2^v| \gg J_1^v$, $\beta = J_1^v + J_2^v$ has a negative value.

A similar hybridization factor describes the superexchange between magnetic ions [40]. The comparison of increase of β and J (ion-ion exchange constant) from $Cd_{1-x}Mn_xSe$, $Cd_{1-x}Fe_xSe$ and $Cd_{1-x}Co_xSe$ [41] gives a qualitative support for a hybridization origin of β.

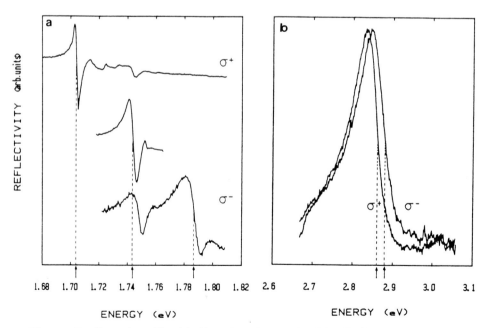

Figure 8. *Magnetoreflectivity structures in $Cd_{1-x}Mn_xTe$ at T = 4.5K for a) x = 0.09 at B = 0 T toward B = 5.5 T, b) x = 0.73 at B = 5.5T, from [31].*

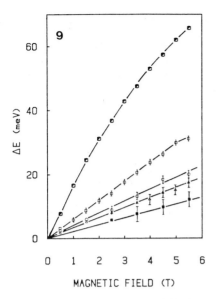
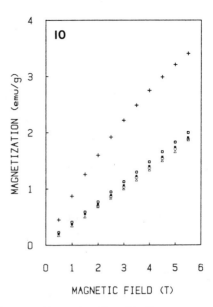

Figure 9. *Zeeman splitting of the exciton in Zn₁-xMnxTe versus magnetic field at T = 4.5K. x = 0.25 (□); x = 0.45 (□); x = 0.54 (o), x = 0.6 (Δ), x = 0.71 (■), from [17].*

Figure 10. *Magnetization of Zn₁-xMnxTe versus magnetic field at T = 4.5 K. x = 0.25 (+) ; x = 0.45 (□) ; x = 0.54 (Δ), x = 0.6 (■) ; x = 0.71 (o) from [17].*

II-4- **Deviation from mean field model for concentrated compounds**

a) Experimental evidence

Most of the studies of the Zeeman splitting of the exciton reflectance have been devoted to crystals with low Mn concentration (x < 0.3). For concentrated alloys, in reason of the alloy broadening, the magnetoreflectivity spectra exhibit only are broad component for each circular polarization. Moreover, the splitting ΔE between the σ^+ and σ^- components remains resolvable [16,17,11,31]. The figure 8 gives an example of split magnetoreflectivity structures for low and high concentrated samples. For the later, weak and strong components in each polarization are mixed and the structures are assumed to represent the strong components. The figure 9 and 10 represent the measured value of ΔE and magnetization against the field strength [17]. The comparison between the two set of curves clearly shows an important deviation from the mean field model predictions. The Figure 11 shows the drastic decrease of ΔE/x < Sz > slope when x increases. Before to question the mean field model, other possible explanations should be explored. ΔE measured when only one structure is observed for each polarization does not exactly corresponds to the shift of the strong components. Nevertheless the width of the structure is very similar to that of the zero field one. Thus the error on ΔE, assuming that the observed structure represents the strong component remains small compared to the decrease of No β depends on the position of the 3d level with respect to the valence band maximum and on the 3d - valence band hybridization matrix Vkd [32]. These values remain apparently unchanged with x [33 to 36]. Thus it is necessary to reexamine the mean field description (first order perturbation theory) of Mn-carriers exchange. In this intention, Bhattacharjee has studied the second order correction for Mn-carriers exchange Hamiltonian [37], summarized below.

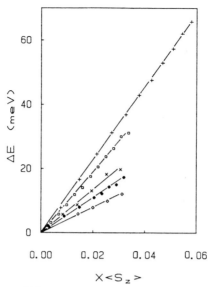

Figure 11. *Zeeman splitting of the exciton in Zn₁-xMnxTe versus the mean Mn spin value per cation, at T = 4.5K. x = 0.25 (+), x = 0.45 (□), x = 0.54 (x), x = 0.60 (◆), x = 0.71 (◇), from [17].*

b)Second order correction for exchange induced band splitting

The mean field model leads to a splitting of the conduction band into two components and the valence band in four components. The energy shift of each subband is studied using the second order perturbation calculation. This gives for the conduction bands :

$$E_{km}^{(1)} = E_{km}^{(1)} = \sum_{k=k'} \sum_{m'} \sum_{\gamma} \sum_{\gamma'} W_{\gamma} \frac{|< m\, \gamma\, k|\ Hex\ |k'\, \gamma'\, m'\ >|}{(E_{k\,m}^{(1)} + \mathcal{E}\gamma) - (E_{k'm'}^{(1)} + \mathcal{E}\gamma')}$$

where

$$E_{km}^{(1)} = Ec + \mathcal{E}(k) - No\ \alpha\ x\ < Sz >$$

is the mean field (first order) result, m = ± 1/2 denotes the eigenvalue of σz, |γ > represents the eigenstate of the Mn spin system, Wγ its statistical weight at To and $\mathcal{E}\gamma$ its energy. Using the quasistatic approximation, $\mathcal{E}\gamma = \mathcal{E}\gamma'$. The conduction band splitting, at k = 0, is then given by :

$$\Delta c = - \alpha\ No\ x\ < Sz > - \delta c$$

where

$$\delta c = (1/4) \sum_{k \neq 0} (Jok/V^2) \left(\frac{\Gamma-+}{\mathcal{E}(k) + \alpha\ No\ x\ < Sz >} - \frac{\Gamma+-}{\mathcal{E}(k) - \alpha\ No\ x\ < Sz >} \right)$$

here V is the crystal volume, Jok is the Fourier transform of the exchange interaction and :

$$\Gamma^{+-}_{-+} = \sum_{i} \sum_{j} exp[ik\ (Ri - Rj)] < Si \pm Sj^{-}_{+} >$$

are the transverse spin correlation functions.
Under the following assumptions :
   - J (r) is highly localized thus Jok ~ Joo = α
   - The Γ's do not present any structure near k = 0 and their k

178

dependence can be neglected

        - The system is in the spin glass at least for the transverse components,

the δc calculation gives :

$$\delta c = \frac{3}{4} |No\ \alpha|\ r_c\ (2\ x < S_z > + \frac{\pi}{3}\ x\ S\ (S + 1)(r_c\ |x < S_z >|)^{1/2}$$

where $r_c = |No\ (\alpha/W_c)|$, W_c being the width of the conduction band assumed parabolic.

      Analogous results are obtained for the valence band, leading to the energy difference between the strong components :

$$\Delta E = (\ No\ \beta\ (1 - \frac{3}{2}\ r_v) - No\ \alpha\ (1 - \frac{3}{2}\ r_c))\ x < S_z >$$
$$- \frac{\pi}{4}\ x\ S\ (S + 1)\ (|No\ \alpha|\ r_c^{3/2} + |No\ \beta|\ r_v^{3/2})\ \sqrt{|x < S_z >|} \tag{4}$$

The first term of (4) shows a small renormalization of exchange constants. The second which is proportional to $|x < S_z >|^{1/2}$ leads to a substantial decrease of the band splitting obtained from the first order. At large x, the curves ΔE versus $|x < S_z >|$ assume a parabolic form in the low $|x < S_z >|$ regions. Thus, instead of the slope of the curves, the decrease of ΔE with increasing x, for a given $|x < S_z >|$ value can be considered. The comparison with $Cd_{1-x}Mn_xTe$ and $Zn_{1-x}Mn_xTe$ data shows that the second order correction to the molecular field formula gives a good description of the anomalous large decrease of ΔE at large Mn concentrations [11].

III – STUDY OF MAGNETOOPTICAL EFFECTS AWAY FROM THE CENTER OF THE BRILLOUIN ZONE

      The E_1 and $E_1 + \Delta_1$ reflectivity structures corresponding to transitions in the vicinity of the L (or Λ) point of the Brillouin zone can be easily observed. Unfortunately, they are very large compared to the Zeeman splitting. So it is necessary to use a modulation polarization technique to determine the splitting. This study has been initiated by Dudziak et al [42] who performed magnetoreflectivity measurements on $Cd_{1-x}Mn_xTe$. It was followed by the work of Ginter et al [43] which studied the ratio $\Delta E^L/\Delta E^\Gamma$. Most systematic investigations has been performed by Coquillat et al in $Cd_{1-x}Mn_xTe$ [44], $Zn_{1-x}Mn_xTe$ [31] and $Hg_{1-x}Mn_xTe$ [31]. These measurements gives the opportunity to compare Mn-band carrier exchange for wide gap and small gap SMSC which have similar band structures at L point in spite of the zero gap inverted band structure of HgTe at Γ.

III-1- Measurements

      Typical experimental set up for polarization modulation magnetoreflectivity [43, 44, 31] is obtained, from figure 1, by adding a depolarizer between the light source and the sample and replacing the circular polarizer by a photoelastic modulator. The reflectivity and the polarization signal ($I^{\sigma+} - I^{\sigma-}$) are simultaneously measured by using respectively the chopper and the photoelastic modulator (circular polarization modulation) and two look in. The three signals : reference I_0, total intensity, $I^{\sigma+} + I^{\sigma-}$ and polarization $I^{\sigma+} - I^{\sigma-}$ are used to calculate the degree of polarization

$$P = (I^{\sigma+} - I^{\sigma-})/(I^{\sigma+} + I^{\sigma-})$$

and the reflectivity

$$R = (I^{\sigma^+} + I^{\sigma^-})/Io.$$

P can be written as follow [44] :

$$P = (I^{\sigma^+} - I^{\sigma^-})/(I^{\sigma^+} + I^{\sigma^-}) = \Delta R/2R$$

and the splitting Zeeman ΔE between the two σ^+ and σ^- components may be determined from :

$$R = \frac{1}{2\,R}\ \frac{dR}{dE} \qquad \Delta E = \frac{1}{2}\ \Delta E\ \frac{d(Ln\ R)}{dE}$$

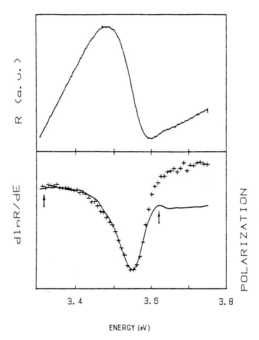

R (a. u.)

dlnR/dE

POLARIZATION

3. 4   3. 6   3. 8

ENERGY (eV)

Figure 12. *Magnetoreflectivity structure due to transitions at L point of the Brillouin zone (above) and logarithmic derivative (continuous line below) compared with polarization spectra for $Cd_{0.91}Mn_{0.09}Te$ at $T = 4.5K$ and $B = 5.5\ T$, from [44].*

An example of comparison between the calculated $d(Ln\ R)/dE$ from reflectivity and polarization is reported in Figure 12. The fit gives ΔE. The proportionality of the E_1 transition ΔE^L to ΔE^Γ visible in Figure 13 for $Zn_{1-x}Mn_xTe$ is also observed for $Cd_{1-x}Mn_xTe$. The proportionality coefficients are respectively $\Delta E^L/\Delta E^\Gamma = 1/20$ for $Cd_{1-x}Mn_xTe$ and $\Delta E^L/\Delta E^\Gamma = 1/16$ for $Zn_{1-x}Mn_xTe$. In Γ point, the splitting pattern of energy bands in narrow gap SMSC results from combination of Landau quantization and exchange effects and does not resemble that of large gap SMSC. Furthermore the available data on α and β parameters of $Hg_{1-x}Mn_xTe$ show considerable inconsistence between different authors (more than a factor 2). Therefore, in order to compare the splitting values for $Cd_{1-x}Mn_xTe$, $Zn_{1-x}Mn_xTe$ and $Hg_{1-x}Mn_xTe$, Coquillat et al [31] have chosen to plot ΔE^L as a function of $x< Sz >$ (figure 14). The splitting ΔE^L observed for $Hg_{1-x}Mn_xTe$ is three time higher than for both $Cd_{1-x}Mn_xTe$ and $Zn_{1-x}Mn_xTe$. Furthermore, the splitting for both feature (E_1 and $E_1 + \Delta_1$) determined from $Hg_{1-x}Mn_xTe$ equal in amplitude but opposite in sign.

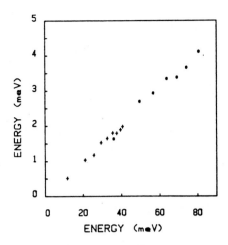

Figure 13. *Splitting ΔE^L versus ΔE^Γ in $Zn_{1-x}Mn_xTe$ at 4.5 K for x = 0.02 (+) and x = 0.17 (o), from [31].*

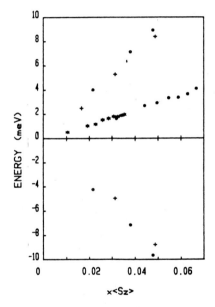

Figure 14. *Splitting of the E_1 (positive values) and $E_1 + \Delta_1$ (negative values) reflectivity structures in $Hg_{1-x}Mn_xTe$ (x = 0.05 (+) and x = 0.20 (o)), $Zn_{1-x}Mn_xTe$ (●) and $Cd_{1-x}Mn_xTe$ (*), versus the mean Mn spin value per cation, from [31].*

III-2- Theoretical analysis of magnetooptical transitions E_1 and

$E_1 + \Delta_1$

The first calculation of the splitting of E_1 and $E_1 + \Delta_1$ was proposed by Ginter et al [43] using a four orbital tight binding model and assuming α and β constants with k. This model accounts for the equal and opposite ΔE_1 and $E_1 + \Delta_1$ splitting but it predicts $\Delta E^L/\Delta E^\Gamma = 1/4$, in contrast with the experimental values 1/16 or 1/20. Recently Bhattacharjee has presented a group theoretical analysis of the Zeeman splitting of the E_1 and $E_1 + \Delta_1$ transitions [45] and a model for the k dependence of the conduction and valence bands exchange constants [46], summarized below.

a) Zeeman splitting calculation for excitonic transitions Λ (or L) point of the Brillouin zone [45]

In $Cd_{1-x}Mn_xTe$, the E_1 transition corresponds to a critical point in the Λ (near L) direction. In $Hg_{1-x}Mn_xTe$, the E_1 transition is located at L. At Λ or L point of the Brillouin zone of a zinc blende crystal, the symmetry group of the k is C_{3v}. When the spin is neglected, the first conduction band is a non degenerate Λ_1 (L_2) band and the upper valence band a doubly degenerate Λ_3 (L_3). The spin-orbit coupling splits the valence band Λ_3 in two doublets $\Lambda_{4,5}$ and Λ_6. The interband transitions $\Lambda^v_{4,5} - \Lambda^c_6$ and $\Lambda^v_6 - \Lambda^c_6$ corresponds respectively to E_1 and $E_1 + \Delta_1$ transitions. When a magnetic field is applied in an arbitrary direction (θ, φ) each zero field transition splits in four components in each circular polarization σ^+ and σ^-. In reason of they small energy spacing, the experiment corresponds to an overall splitting between σ^+ and σ^- spectra. The calculation of the average energy for each polarization gives :

$$\Delta E_1 = < E_1 >^{\sigma^+} - < E_1 >^{\sigma^+} = \frac{2 \cos^2 \theta}{1 + \cos^2 \theta} (K^c - K^v)$$

where $\qquad k^i = - \mu_B g_i B - J^i_{kk} No x < S_z >$

and $\qquad \Delta(E_1 + \Delta_1) = < E_1 + \Delta_1 >^{\sigma^-} - < E_1 + \Delta_1 >^{\sigma^+} = - \Delta E_1$

previously mentioned by Ginter et al [43]. The average of ΔE over the equivalents directions gives :

$$\Delta E_1 = ((4 - \Pi)/2) (J^v_{kk'} - J^c_{kk'}) No x < S_z >$$

where J^c_{kk} and J^v_{kk} are respectively the conduction and valence exchange parameters at the critical point k ($J^c_{oo} = \alpha$, $J^v_{oo} = \beta$).

b) variation of the exchange with k [46]

J_{kk}^c results from the direct exchange between band and d electrons. A simple four orbital empirical tight binding calculation for $Cd_{1-x}Mn_xTe$ gives $J^c_{LL} \sim 0.7 \, \alpha$.

J^v_{kk} mainly arises in the hybridization between the Mn d -orbitals and the valence band states. Neglecting the direct exchange, J^v_{kk} is calculated using the Schrieffer-Wolff formula for exchange via hybridization tacking into account the hopping integrals between the Mn d - orbitals and the nearest and next nearest neighbor anion p orbitals. It leads to $J^v_{LL}/\beta \approx 1/11$. With the know exchange parameters values :

Cd Mn Te : No α = 0.22 et, No β = - 0.88 eV [5]
Zn Mn Te : No α = 0.18 et, No β = - 1.05 eV [15]
Hg Mn Te : No α = 0.75 et, No β = - 1.5 eV [47]

The experimental slope in both $Cd_{1-x}Mn_xTe$ and $Zn_{1-x}Mn_xTe$ can be fit to

$$\Delta E / |x < S_z >| = 0.43 (J^c_{LL} - J^v_{LL}) No$$

by assuming reduction factors J^c_{LL}/α = 1/3 and J^v_{LL}/β = 1/13. The same reduction factors yeld 0.16 for the slope in $Hg_{1-x}Mn_xTe$. Its follows that the four-orbital tight binding model undestimate the variation of $\overset{\smallsmile}{J}\!kk$ whereas the k-d hybridyzation model for the valence band exchange shows an excellent agreement with the experimental reduction factors.

IV OTHER IMPORTANT MAGNETOOPTICAL EFFECTS

IV-1- Faraday rotation (FR) of SMSC

The "giant Faraday rotation" is one of the spectacular magnetooptical effects observed in SMSC. It is a direct consequence of the enhancement of the Zeeman splitting of the exciton by Mn – band carriers exchange interactions. Lot of studies have been devoted to this subject [1,3,22,48 to 55], as well to evidence and study the FR than to use its proportionality with magnetization to observe the magnetic properties of SMSC. A simple magnetotransmission set up, in Faraday geometry, in which the sample is placed between two linear polarizers is generally sufficient to measure the FR. A modulation of polarization technique may be used in the case of small angles. Typical result is reported in figure 15.

The Faraday effect is the rotation of the plane of polarization of linear polarized light propagating along the magnetic field in which the

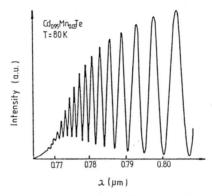

Figure 15. *Transmission spectrum through Cd₀.₉₅Mn₀.₀₅Te, in Faraday rotation geometry at T = 80K and B about 1T, from [2] (Gaj).*

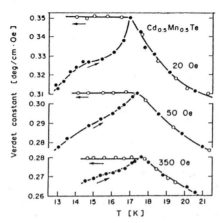

Figure 16. *Faraday rotation versus temperature in Cd₀.₉₅Mn₀.₅Te ; cooling process in H = 0, 0 - in H = 0, from [49].*

sample is placed. The rotation angle θ is then given by :

$$\theta = (E \ell / 2 \hbar c) \, (n^{-} - n^{+})$$

where ℓ is the thickness of the sample, E is the photon energy, c the speed of light in the vacuum and n^{+} and n^{-} are the refractive indexes for the two circular polarization σ^{+} and σ^{-}. Using a simple oscillator strength model and considering only the contribution to n associated with the excitonic transition, θ is given by [53] :

$$\theta = \left(\frac{\sqrt{Fo\ell}}{2 \pi c}\right) \left(\frac{y^{2}}{(1 - y^{2})^{3/2} \, Eo}\right) \Delta E$$

where Fo is a constant involving among other the oscillator strength of the excitonic transition Eo, $y = E/Eo$ and ΔE is the energy difference between the transitions observed in σ^{+} and σ^{-} polarization. As mentioned earlier, the excitonic transition Eo splits in two components for each polarization. Il is reasonable, tacking into account the respective magnitude of strong and small components to use :

$$\Delta E = Ed - Ea = No \, (\alpha - \beta) \quad x< Sz >$$

thus θ is proportional to the magnetization. The figure 16 shows an example of study of the magnetic properties of $Cd_{1-x}Mn_{x}Te$ using RF measurements.

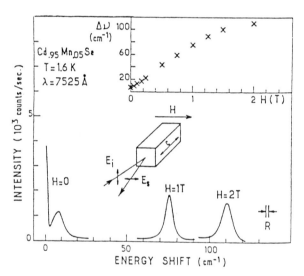

Figure 17. *Typical SFRS spectra of $Cd_{0.95}Mn_{0.5}Se$, at T = 1.6K for λ = 7525 Å. Insert : the stokes shift versus H, from [57].*

IV-2- Spin Flip Raman Scattering (SFRS)

Since the first observation of SFRS in SMSC by Nawrocki et al [57], several investigations have been devoted to this subject [57 to 69]. These studies established the nature of the SFRS lines observed in σ^{+} and σ^{-} circular polarized light in a magnetic field, associated with the Spin Flip transition of electrons bound to donors in SMSC. In such a case, the mean field model used to calculate the Zeeman splitting of conduction band can be applied. It follows :

$$\Delta E_D = No \; \alpha \; x < Sz >$$

Two important results are obtained from SFRS :
        - an accurate determination of the value of α,
        - the observation of a finite zero field splitting which is due
to Bound Magnetic Polaron.
The figure 17 shows   Typical SFRS spectra.

VI — "MAGNETOOPTICAL" EFFECTS IN ENERGY GAP OF SMSC IN ABSENCE OF APPLIED MAGNETIC FIELD

    The absorption edge of classical semiconductors usually exhibits a
blue shift when temperature decrease. In magnetic semiconductors, this
behavior is strongly affected by magnetic phase transition [70], eg : the
energy gap of EuO is reduced of about 25 % below the curie temperature.
This effect, attributed to magnetic ions-conduction band electrons
exchange is well explained by a second order calculation model [71].An
affectation of the energy gap by magnetic ions — conduction and
valence bands carriers is also observed in SMSC [72,73,16], and an
analysis, following the second order calculation used for magnetic
semiconductors, in the case of SMSC found a "magnetic" contribution to the
variation of the energy gap proportional to the temperature and magnetic
susceptibility. [74,75,76].

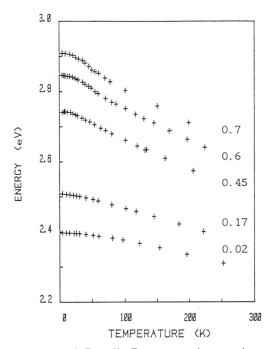

Figure 18. *Energy gap of Zn1-xMnxTe versus temperature, from [73].*

    Figure 18 shows a characteristic energy gap of SMSC versus
temperature variation. It should be noticed the "bump" at low temperatures
for the highest Mn concentrations and the strong variation of the slope of
the linear part of the curves with composition.

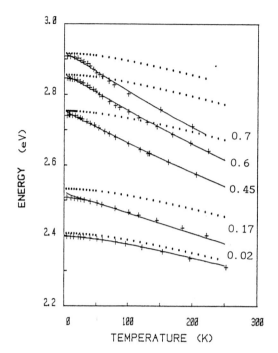

Figure 19. *Fit of the energy gap of Zn₁₋ₓMnₓTe versus temperature (lower lines). The upper lines were calculated without magnetic contribution.*

The analyze proposed by Gaj et al [75], starts from the classical exchange hamiltonian :

$$\text{Hex} = - \ (V \ / \ N) \ \sum_i J \ (\vec{r} - \vec{R}i) \ \vec{\sigma}.\vec{S}i$$

Considered as a perturbation of the band electron hamiltonian. The first order term, proportional to <Sz> is null in absence of external magnetic field. The second order correction yields the following expression for the energy band :

$$E(k,T) = Eo(k,T) + (1 \ / \ 4) \ \sum |Jq|^2 \ \Gamma q \ / \ Eo(k,T) - Eo(k + q, \ T)]$$

where

$$Jq = \int J(r) \ \exp \ (- \ i \ q \ r) \ d^3r$$

is the Fourier transform of the exchange coupling and

$$\Gamma q = N^{-1} \ \sum_j < \vec{S}o.\vec{S}j > \exp \ (i\vec{q}.\vec{R}j)$$

is the correlation function.
Introducing the following assumptions :
    - During the optical transition, the magnetic system remains frozen (static case).
    - Jq is flat near k = 0 (type III antiferromagnetic order) then

$$\Gamma q = \Gamma o \simeq N^{-1} \sum <\vec{S}o.\vec{S}j>$$

which is proportional to the product of temperature and magnetic susceptibility.
    - The non magnetic part of the energy gap variation contains a correction Ec proportional to the Mn mole fraction x and to the variation

186

of the energy gap Eg in the host crystal

$$Ec = \alpha \ x \ [Eg(0,T) - Eg(0,0)],$$

it follows :

$$Eg \ (x,T) = A \ (x) + (1 + ax) \ [Eg \ (0,T) - Eg(0,0)] - bT \ x \ (0,T) \quad (5)$$

where the values of the parameters α and β are commons for all the composition variation of the energy gap extrapolated at T =0.
The fig. 19 represents the fit of the data of The fig. 18 to (5).

REFERENCES

1) Komarov A.V., Ryabchencko S.M., Terletskii O.V., Zheru I.I., and Ivanchuk R.D., (1977) Zh. Eksp. Teor. Fiz. 73, 608 (Sov. Phys. JETP 46, 318).

2) See for example the recent review papers :
Gaj J.A., (1988) Semiconductors and Semimetals, edited by J.K. Furdyna and J. Kossut (Academic, Boston) Vol. 25, 276.
Goede O., and Heimbrodt W., (1988), Phys. Stat. Sol. (b) 146, 11.
Coquillat D., (1990) in "Diluted magnetic semiconductors", World Scientific, Singapore, to be published.

3) Gaj J.A., Galazka R.R., and Nawrocki M., (1978) Solid State Commun. 25, 193.

4) Gaj J.A., Byszewski P, Cieplak M.Z., Fishman G., Galazka R.R., Ginter J, Nawrocki M., Nguyen The Khoi, Planel R., Ranvaud R., and Twardowski A., (1978) Proceedings of 14th International Conference of Semiconductors, Edinburg, edited by B.L.H. Wilson, Inst. of Phys. Conf. Series London 1978).

5) Gaj J.A., Ginter J., and Galazka R.R., (1978) Phys. Stat. Sol. (b) 89, 655.

6) Gaj J.A., Planel R., and Fishman G., (1979) Solid State Commun. 29, 435.

7) Twardowski A., Nawrocki M., and Ginter J., (1979) Phys. Stat. Sol. (b) 96, 497.

8) Twardowski A., and Ginter J., (1982) Phys. Sta. Sol. (b) 114, 331.

9) Rebmann C., Rigaux C., Bastard G., Menant M., Triboulet R., and Giriat W., (1983) Phys. 117B&118B, 452.

10) Nguyen The Khoi, Ginter J., and Twardowski A., (1983) Phys. Stat. Sol. (b) 117, 67.

11) Lascaray J.P., Coquillat D., Deportes J., and Bhattacharjee A.K., (1988) Phys. Rev. B38, 7602.

12) Komarov A.V., Ryabchenko S.M., and Vitrikhovskii, (1978) JETP Letter 27, 413.

13) Twardowski A., Dietl T., and Demianiuk M., (1983) Solid State Commun. 48, 845.

14) Twardowski A., Von Ortenberg M., Demianiuk M., and Pauthenet R., (1984) Solid State Commun. 51, 849.

15) Barilero G., Rigaux C., Menant M., Nguyen Hy Hau, and Giriat W., (1985) Phys. Rev. B32, 5144.

16) Desjardins-Deruelle M.C., Lascaray J.P., Coquillat D., and Triboulet R., (1986), Phys. Stat. Sol. (b) 135, 227.

17) Lascaray J.P., Deruelle M.C., and Coquillat D., (1987) Phys. Rev. B35, 675.

18) Komarov. A.V., Ryabchenko S.M., and Terletskii O.V., (1980), Phys. Stat. Sol. (b) 102, 603.

19) Arciszewska M., and Nawrocki M., (1982) Proceedings of the 10th Conf. on Physics of Semiconducting Compounds, Jaszowiec, 1981, Warsaw, Poland, Polish Acad. Sci 225.

20) Arciszewska M., and Nawrocki M., (1986) J. Phys. Chem. Solids 47, 309.

21) Aggarwal R.L., Japerson S.N., Stankiewicz J., Shapira Y., Foner S., Khazai B., and Wold A., (1983) Phys. Rev. B28, 6907.

22) Aggarwal R.L., Japerson S.N., Shapira Y., Foner S., Skakibara T., Goto T., Miura N., Dwight D.,and Would A., Proc. 17th Conf. Phys. Semicond. San Francisco, (1984) 1419.

23) Gubarev S.I., (1981) Zh. Eksp. Teor. Fiz. 80, 1174, (Sov. Phys. JETP 53, 601).

24) Gubarev S.I., and Tyazholv M.G., (1986) Pis'ma Zh. Eksp. Teor. Fiz. 44, 385, (JETP Lett. 44, 494).

25) Gubarev S.I., and Tyazholv M.G., (1988) Pis'ma Zh. Eksp. Teor. Fiz. 48, 437.

26) Abramishvili, A.G., Gubarev S.I., Komarov A.V., and Ryabchenko S.M., (1984). Fiz. Tverd. Tela 26, 1095, (Sov. Phys. Solid. State. 26, 666).

27) Nawrocki M, Lascaray J.P., Coquillat D., and Demanuiuk M., (1987). Proceedings of Material Research Society, Fall meeting, edited by R.L. Aggarwal J.K., Furdyna and S. von Molnar (MRS Pittsburg, PA) Vol. 89, 65.

28) Kossut J., (1976) Phys. Stat. Sol. (b) 78, 537.

29) Bastard G., Rigaux C., Gulner Y., Mycielski J., and Mycielski A., (1978), J. Phys. 39, 87.

30) Bir G.L., and Pikus G.E., (1972) Simmetria i deformatsionnye effekty v poluprovodnikakh, Nauka, Moscow, Sections 23 and 24 (Symmetry and Strain-Induced Effects in Semiconductors, Wiley, 1974).

31) Coquillat D., Lascaray J.P., Gaj J.A., Deportes J., and Furdyna J.K., (1989) Phys. Rev. B39, 10088.

32) Bhattacharjee A.K., Fishman G., and Coqblin B., (1983) Physica 117&118B, 449.

33) Taniguchi M., Ley L., Johnson R.L., Ghijsen J., and Cardona M.,(1986) Phys. Rev. B33, 1206.

34) Taniguchi M., Fujimori M., Fujisawa M., Mori T., Souma I., and Oka Y., (1987) Solid State Commun. 62, 431.

35) Ley L., Taniguchi M., Ghijsen J., Johnson R.L., and Fujimori A., (1987) Phys. Rev. B35, 2839.

36) Chab V., Paolucci G., Prince K.C., Surman M., and Bradshaw A.M., (1988) Phys. Rev. B38, 112353.

37) Bhattacharjee A.K., (1988) Solid State Commun. 65, 275.

38) Benoit à la Guillaume C., This volume.

39) Averous M., This volume.

40) Larson B.E., Hass K.C., Ehrenreich H., and Carlsson A.E., (1988) Phys. Rev. B, 37, 4137.

41) Nawrocki M., Hamdani F., Lascaray J.P., Golacki Z., and Deportes J., (1991) Solid State Commun. 77, 111.

42) Dudziak E., Brzezinski J., and Jedral L., (1982) Proceedings of 14th International Conference on the Physics of Semicond. Compounds, Jaszowiec, Poland (Polish Academy of Science, Warsaw), 166.

43) Ginter A., Gaj J.A., and Le Si Dang., (1983) Solid State Commun. 48, 849.

44) Coquillat D., Lascaray J.P., Deruelle M.C., Gaj J.A., and Triboulet R., (1986) Solid State Commun. 59, 25.

45) Bhattacharjee A.K., (1990) Phys. Rev. B,

46) Bhattacharjee A.K., (1990) 20th, ICPS, Tessaloniki.

47) Bastard G., Gaj J.A., Planel R.,and Rigaux C., (1980) J. Phys. 41, C5, 247.

48) Kett H., Gebhardt W., Krey U., and Furdyna J.K., (1981) J. Magn. Mat. 25, 215.

49) Kierzek-Pecold E., Szymanaska W., and Galazka R.R., (1984) Solid State Commun. 50, 685.

50) Mycielski A., Rigaux C., Menant C., and Otto M., Solid State Commun. 50, 257, (1984).

51) Awschalom D.D., Halbout J.M., Von Molnar S., Siegrist T., Holtzbereg F., (1985) Phys. Rev. Letters, 55, 1128.

52) Kullendorf N., and Hök B., Appl. Phys. Letters 46, 1016, (1985).

53) Bartholomew D.U., Furdyna J.K., and Ramdas A.K., (1986) Phys. Rev. B34, 6943.

54) Bartholomew D.U., Suh E.K., Ramdas A.K., Rodriguez S., Debska U., and Furdyna J.K., (1988) Phys. Rev. B399, 5865.

55) Butler M.A., Martin S.J., and Baughman R.J., (1986) Appl. Phys. Letters, 49, 1053.

56) Eunsoon Oh., Bartholomew D.U., Ramdas A.K., Furdyna J.K., and Debska U., (1988) Phys. Rev. B38, 13183.

57) Nawrocki M., Planel R., Fishman G., and Galazka R., (1981) J. Phys. Soc. Jpn. 49, Suppl. A, 823.

58) Nawrocki M., Planel R., Fishman G., and Galazka R. (1981) Phys. Rev. Letter. 46, 735.

59) Nawrocki M., Planel R., Mollot F., and Kozielski M.J., (1984) Phys. Stat. Sol. (b) 123, 99.

60) Alov. D.L., Gubarev S.I., Timofeev V.B., and Shepel B.I., (1981) Zh. Eksper-theor, Fiz. 34, 76.

61) Alov. D.L., Gubarev S.I., Timofeev V.B., Ryabchenko S.M., and Semenov, Yu .G., (1983) Izv. Akad. Nauk SSSR, Ser. Fiz. 30, 260.

62) Alov., D.L., Gubarev S.I., and Timofeev, V.B., (1984) Zh. Eksper. Theor. Fiz. 86, 1124.

63) Peterson D.L., Petrou A., Dutta M., Ramdas A.K., and Rodriguez S;, (1982) Solid State Commun. 43, 667.

64) Shapira Y., Heiman D., and Foner S., (1982) Solid State Commun. 44, 1248.

65) Heiman D., Shapira Y., Foner S., (1983) Solid State Commun. 45, 899.

66) Heiman D., (1983) Appl. Phys. Letters, 42, 775.

67) Heiman D., Wolf P.A., Warnock J., (1983) Phys. Rev. B27, 4848.

68) Petrou A., Peterson D.L., Venugopalan S., Galazka R.R., Ramdas A.K., and Rodriguez S, (1983) Phys. Rev. B27, 3771.

69) Douglas K., Nakashima S., and Sott J.F., (1984) Phys. Rev. B.29, 5602.

70) Wachter.P., (1979) Handbook on the Physics and Chemistry of Rare Earths, North-Holland, p. 554.

71) Rys F., Helman J.S., and Baltensperger W., (1967) Phys. Kondens. Materials 6, 105.

72) Diouri J., Lascaray J.P., and El Amrani M., (1985) Phys. Rev. B31, 7995.

73) Donofrio T., Lamarche G., and Woolley J.C., (1985) J. Appl. Phys. 57 (6), 1932.

74) Bylsma R.B., Becker W.M., Kossut J., and Debska U., (1986), Phys. Rev. B33, 8207.

75) Gaj J.A., and Golnik A., (1987) Acta Phys Sol. A71, 197.

76) Gaj J.A., Golnik A., Lascaray J.P., Coquillat D., and Desjardins-Deruelle M.C., (1987) Proceedings of Material Research Society, fall Meeting, edited by R.L. Aggarwal J.K., Furdyna and S. von Molnar (MRS, Pittsburg, P.A.) Vol. 89, 59.

77) Coquillat D., (1987) Thesis, Montpellier.

II-Fe-VI SEMIMAGNETIC SEMICONDUCTORS

C. Benoit à la Guillaume

Groupe de Physique des Solides *
Universités Paris VII et Paris VI, Tour 23
2 Place Jussieu, 75251 Paris Cedex 05, France

INTRODUCTION

II-VI compounds containing iron represent a new sub-class of semimagnetic semiconductors (SMSC) [1], the most widely known being based on manganese (for a review, see ref. 2). In these alloys, the quite different behavior of Fe^{2+} comes from its 3d shell which contains one more electron than the half filled 3d shell of Mn^{2+}. The solubility of iron in II-VI compounds is quite smaller than

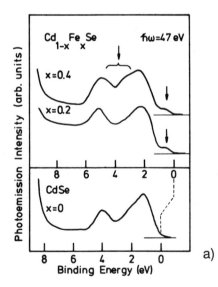
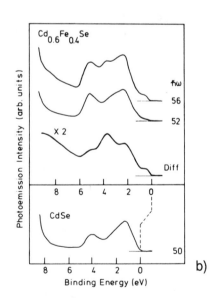

Fig. 1. a) Valence-band photoemission spectra of CdFeSe for different Fe concentrations measured at $\hbar\omega = 47$ eV. Binding energy is defined relative to the VB maximum. b) VB spectra of $Cd_{0.6}Fe_{0.4}Se$ taken just on resonance ($\hbar\omega = 56$ eV) and at antiresonance ($\hbar\omega = 52$ eV). The difference spectrum "Diff" is a measure of Fe 3d partial DOS. The spectrum of pure CdSe is also shown for comparison (ref. 3).

Semimagnetic Semiconductors and Diluted Magnetic Semiconductors
Edited by M. Averous and M. Balkanski, Plenum Press, New York, 1991

191

that of manganese: about 2 to 5 at.% in the tellurides and up to 20 at.% in the selenides. Larger concentrations can be introduced by off equilibrium methods, like molecular beam epitaxy. As a general remark, the data on Fe-based SMSC are far less complete than those on Mn-based SMSC.

ELECTRONIC STRUCTURE OF IRON IN II-VI COMPOUNDS

Experimental determination of the energy levels introduced by iron comes from photoemission. Fig 1a) shows recent experiments performed in $Cd_{1-x}Fe_xSe$ [3]; as in the case of CdMnSe [4], a resonance between the direct transition from 3d states to the vacuum continuum and the discrete $3p \rightarrow 3d$ absorption followed by a super Coster-Kronig decay gives rise to a Fano-type interference with, in the case of Fe, a resonance at 56 eV and an antiresonance at 52 eV. The Fe contributions to the density of states (DOS) is best singled out taking the difference between the two curves (see in fig. 1b) the curve "Diff"). Most of the Fe DOS falls inside the valence band (as for manganese [4]) but also a level is located about 0.5 eV above the top of valence band. Such data are explained in fig. 2 in terms of filling d orbitals, with the large jump experienced by the 6th electron (intra-ion exchange), the smaller crystal field splitting of d orbital (e_g, t_{2g}) and the strong hybridization of t_{2g} orbitals with host band states. Another model, named U, U', J according to ref. 5 (see also the lectures of K. C. Hass), taking better account of electron-electron correlations, leads to the energy scheme of fig. 3 where the $e_g\downarrow$ level lies in the forbidden gap of CdSe.

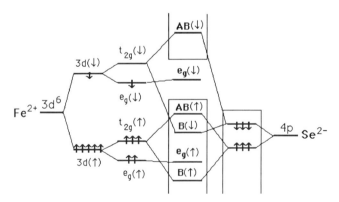

Fig. 2. Schematic energy diagram showing how the Fe-derived DOS is formed from the atomic orbitals (from ref. 3).

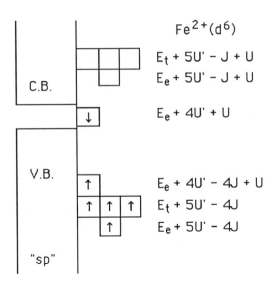

$Fe^{2+}(d^6)$

$E_t + 5U' - J + U$
$E_e + 5U' - J + U$

$E_e + 4U' + U$

$E_e + 4U' - 4J + U$
$E_t + 5U' - 4J$
$E_e + 5U' - 4J$

C.B.

V.B.

"sp"

Fig. 3. Energy scheme of Fe in CdFeSe from the U, U', J model (from K. C. Hass lectures).

Iron as a deep Donor

The existence of states in the gap is corroborated by infrared absorption and photoconductivity in the same material [6] which show that Fe acts as a deep donor, the level associated with the reaction ($Fe^{2+} \rightarrow Fe^{3+} + e$) being 1.2 eV below the bottom of the conduction band (fig. 4). Fig. 5 shows how things change in the alloys $Hg_{1-x}Cd_xSe$ as a function of x in a representation where the

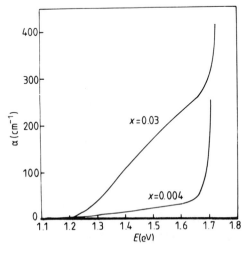

Fig. 4. Infrared absorption in $Cd_{1-x}Fe_xSe$ for x=0.004 and 0.03; the component starting at 1.2 eV correspond to $Fe^{2+} + \hbar\omega \rightarrow Fe^{3+}+e$ (ref. 6)

193

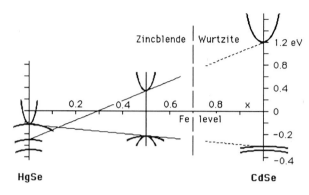

Fig. 5. Schematic evolution of Fe donor level in the alloys $Hg_{1-x}Cd_xSe$. Energy is referred with respect to the vacuum level (from ref. 6).

iron level is assumed to be invariant with respect to the vacuum level [7]. As a consequence of band offset, the Fe level crosses the bottom of conduction band when x decreases ($x \cong 0.35$) and is resonant in the conduction band of HgSe, at 0.23 eV above its bottom. Very peculiar transport properties are observed in $Hg_{1-x}Fe_xSe$ (see the lectures of T. Dietl). Experiments on inverse photoemission spectroscopy are still lacking; hence energy position where to add one more electron to Fe^{2+} is not known.

Alloying with Fe causes the usual band-gap variation. It was measured in $Cd_{1-x}Fe_xSe$ for x up to 10% (x measured by microprobe analysis) from the sharp reflectivity feature related to the lowest free exciton [8]; for $0 < x < 0.1$ at 2K the exciton energy is: $E_{FE}(x) = 1.826 + 1.15\,x$ in eV.

Fe^{2+} SPECTRUM AND MAGNETISM

The 6 electrons Fe^{2+} states are treated using crystal field perturbation :
$H = H_{cf} + H_{so}$. Fig. 6a) shows the Tanabe-Sugano diagram [9] which exhibits the effect of H_{cf} on free ion, where the transition $5_E \rightarrow 5_{T2}$ corresponding to crystal field splitting 10 D_q is observed in IR absorption around 0.35 eV and the $5_E \rightarrow 3_{T1}$ transition at 1.4 eV in CdTe [10]. The lower part of the spectrum (5_E and 5_{T2}) with L=2 and S=2 of degeneracy 5 by 5 =25 is further split by the spin-orbit coupling H_{so} and , in the case of Wurtzite type crystal, by a trigonal distortion. Fig. 6b) gives the lowest excited states of Fe^{2+} which are of crucial interest to understand its magnetism. In particular, the lowest state of A_1 symmetry is non degenerate

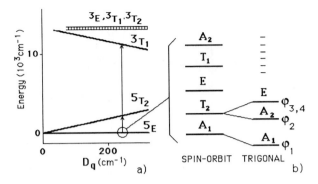

Fig. 6. a) Tanabe-Sugano energy diagram as a function of the CF parameter Dq; the position of Fe^{2+} in CdTe is indicated. b) Expanded view of the 5_E multiplet in zincblende and wurtzite type crystal.

and thus non magnetic [11]. That spectrum was determined through far-IR absoption in CdTe [10] and ZnSe [12], and also by Raman scattering in CdSe [8]. Variations of these levels in magnetic field were also measured first in CdTe [10] in low field, in ZnSe (fig. 7) [12] and in CdSe (fig. 8) [13].

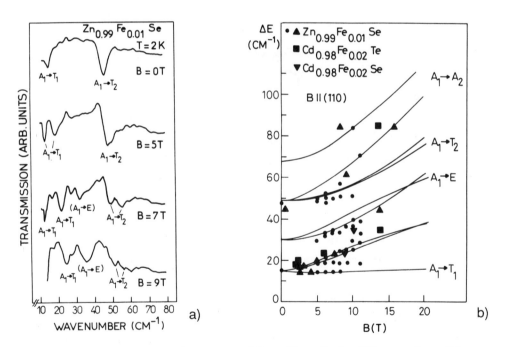

Fig. 7. a) Far-infrared transmission spectra of $Zn_{0.99}Fe_{0.01}Se$ for different values of the magnetic field B. b) Experimental resonance energies as a function of B; the lines represent results of a crystal field model for $Zn_{1-x}Fe_xSe$ (ref. 12).

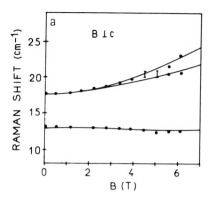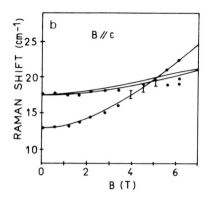

Fig. 8. Variations of the $A_1 \rightarrow A_2$ and $A_1 \rightarrow E$ transitions measured at 1.8 K by Raman
scattering as a function of the magnetic field B in $Cd_{0.999}Fe_{0.001}Se$; a) B $\perp$ c,
b) B ∥ c. The lines are theoretical (ref. 13).

Asking the question "why crystal-field model works?" despite the rather strong hybridization of some 3d orbitals with CdSe valence states, one can remark i) the model incorporate the appropriate symmetry of ion site and ii) it needs only very few fitting parameters, two for zincblende (Dq and λ) and one more for wurtzite crystal, which are determined by the above mentionned spectroscopic measurements. In fact, one of the goal is to see if the crystal field model fails to account for some experimental data.

Van Vleck Paramagnetism

In the low field limit, the ground state $|\varphi_1{}^H\rangle$ of Fe^{2+} can be obtained by the perturbation of the Zeeman term $H_z = \mu_B H.(L+2S)$ as:

$$|\varphi_1{}^H\rangle = |\varphi_1\rangle + \sum_{j \neq 1} |\varphi_j\rangle \langle \varphi_j|H_z|\varphi_1\rangle / (E_j - E_1)$$

The $|\varphi_j\rangle$'s are the states of Fe^{2+} at H=0 (j=1 to 25). Noting that $\langle\varphi_1|S|\varphi_1\rangle = 0$, the moment induced by the component H_α of the field comes from cross-terms:

$$\mu_B\langle \varphi_1{}^H|L_\alpha + 2S_\alpha|\varphi_1{}^H\rangle = \mu_B{}^2 H_\alpha \sum_{j \neq 1} |\langle \varphi_j|L_\alpha + 2S_\alpha|\varphi_1\rangle|^2 / (E_j - E_1)$$

This is Van Vleck paramagnetism, independent of temperature provided only the ground state is populated ($kT < E_2 - E_1$, where E_j is the energy of state $|\varphi_j\rangle$). Notice that the induced moment of some excited state $|\varphi_j{}^H\rangle$ may be opposite to the applied field due to sign reversal of some energy denominator; for such levels, one might speak of "Van Vleck diamagnetism". This is the reason why the susceptibility decreases so fast with temperature.

High Field Magnetization.

High field magnetization measurements where performed in $Cd_{1-x}Fe_xTe$ [14]; in this cubic material, the susceptibility, isotropic at low field [15], exhibits anisotropy above 5T, in good agreement with crystal field model (fig. 9).

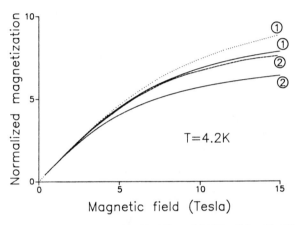

Fig. 9. Normalized magnetization M(H)/M(H=1T) at 4.2 K in $Cd_{1-x}Fe_xTe$ as a function of H, either ∥ [100] or ∥ [111] (curves 1 and 2 respectively). Solid lines are theoretical; dotted and dashed lines are experimental results for x=1-1.2 at.% (ref. 14).

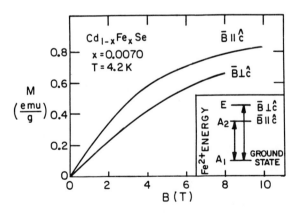

Fig. 10. Anisotropy of magnetization in $Cd_{1-x}Fe_xSe$. The inset shows what Fe^{2+} levels are coupled by the field in Van Vleck paramagnetism, depending on orientation (ref. 16).

197

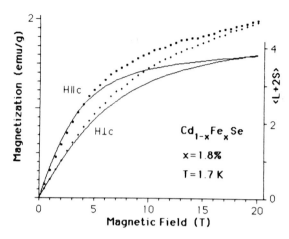

Fig. 11. High field magnetization in $Cd_{1-x}Fe_xSe$ at 1.7 K. The dots are the experiments for x=1.8 at.% (from microprobe analysis) and the continuous lines are the theoretical contributions of 1.44 at.% of isolated Fe^{2+} (ref. 17).

In $Cd_{1-x}Fe_xSe$, a wurtzite type material, the susceptibility is already anisotropic at low field [16]; this is because H_z couples the ground state A_1 to the first excited state A_2 at 13 cm^{-1} for H| |c, but to the second excited state E at 17.6 cm^{-1} for H⊥c (see inset of fig. 10). Near 20T, the anisotropy vanishes (fig. 11), in agreement with the crystal field model [17]. Notice that these measurements were performed at low x, where the contribution of Fe^{2+} pairs is small; pairs are considered at the end of the next section.

EXCHANGE EFFECTS

As in the case of usual SMSC, key parameters are the exchange integrals between Fe^{2+} ions and band carriers, noted $N_0\alpha$ at the bottom of CB and $N_0\beta$ at the top of VB [18]. They are determined using well established method [19] : comparison of magnetooptical splitting of exciton states (conveniently obtained through magneto-reflectivity) and magnetization. In the mean field approximation, the splitting reads: $\Delta E = N_0\alpha x \langle S_z \rangle$ where $\langle S_z \rangle$ is the mean ions spin value in the direction of the applied field H. On the other hand, the magnetization M_z reads:

$$M_z = \mu_B x \langle L_z + 2S_z \rangle N_{Av}/m$$

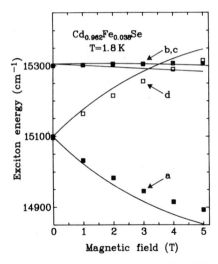

Fig. 12. Energies of exciton lines obtained in magneto-reflectivity on $Cd_{1-x}Fe_xSe$ for x=0.038 at 1.8 K for H∥c. The lines show results of calculations with $N_0\alpha$=0.23 eV and $N_0\beta$ = -1.9 eV (ref. 21).

per unit of mass, where μ_B is the Bohr magneton, N_{Av} the Avogadro number and m the molar mass of the compound. Hence, the exchange integral reads:

$$N_0\alpha=(\Delta E/M_z)(\rho\mu_B N_{Av}/2m)$$

where $\rho=\langle L_z+2S_z\rangle/\langle 2S_z\rangle$; this factor ρ can be calculated from the crystal field model; for CdFeSe, one finds ρ=1.125 or 1.135 for H respectively // or ⊥ to the c axis; it measures the small orbital contribution to M_z (the quenching of the orbital angular momentum by H_{CF} is stronger than the quenching of the spin angular momentum by H_{SO}).

Table I. Exchange integrals of Fe-based SMSC. The values in Mn-based SMSC are given for comparison.

Material	$N_0\alpha$ (eV)	$N_0\beta$ (eV)	ref.
ZnFeSe	0.22	-1.74	20
CdFeSe	0.225	-1.9	16 ; 21
	0.25	-1.53	17
ZnMnSe	0.26	-1.31	
CdMnSe	0.23	-1.26	

Avaible results for $Zn_{1-x}Fe_xSe$ [20] (see fig. 12) and $Cd_{1-x}Fe_xSe$ [17,21] are summarized in table I where the corresponding results for Mn-based compounds have been recalled for comparison : the signs are the same, positive (ferromagnetic) for the CB, negative (antiferromagnetic) for the VB. The magnitudes are also quite similar, with an increase of 20-30% of $|N_0\beta|$ in the case of Fe^{2+}. Let us recall that the negative sign of $N_0\beta$ is linked to the dominant contribution of a second order resonant term [22] like:

$$-2|V_{pd}|^2/(E_{VB}-E_0)$$

where V_{pd} is the hybridization matrix element and $E_{VB}-E_0$ the energy difference between the top of VB and the ion d orbitals. In the case of Fe^{+2}, the fact it behaves as a donor does not produce a sign reversal: this is because all d orbitals contribute to $N_0\beta$, their center of gravity being still about 3eV below the top of VB (see the curve "Diff" of fig.1b)). The increase of $|N_0\beta|$ going from Mn to Fe might result from an increase of V_{pd} and (or) a decrease of $E_{VB}-E_0$.

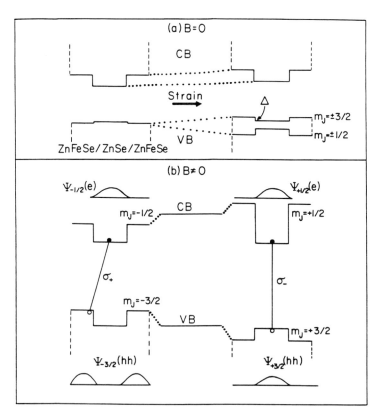

Fig. 13. ZnFeSe/ ZnSe/ ZnFeSe quantum well band diagram a) at B=0 with strain effects. b) The field B induces a splitting of the heavy hole in the barriers, making the σ_- transition of type I and the σ_+ of type II (ref. 23).

Semimagnetic Superlattices

Superlattices involving SMSC materials may exhibit new properties (see also the lectures of M. Voos). In particular, a transition type I-type II induced by a magnetic field was predicted in a quantum well of non magnetic material surrounded by SMSC barriers, provided the VB offset is small enough. This was nicely realized in ZnSe-$Zn_{1-x}Fe_xSe$ quantum well [23]; the reason why Fe works better than Mn for such purpose seems to be the nearly vanishing of VB band offset. As a result, the hole is in the well for the exciton state active in σ_- polarization, but in the SMSC barrier for the one coupled to σ_+ (fig. 13). This is proved by the strongly asymmetrical splitting of the lowest exciton state measured using modulated reflectivity (fig. 14).

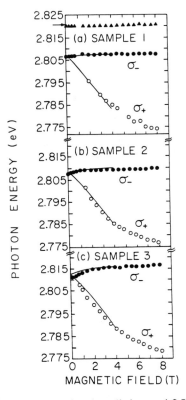

Fig. 14. Magnetic field induced excitonic splitting at 4.2 K of ZnSe quantum well between $Zn_{0.9}Fe_{0.1}Se$ barriers; well thickness is 200,150 and 100 Å respectively in a) b) and c) (ref. 23).

Exciton-Iron Inelastic Scattering

A detailed study of exciton-Fe^{2+} exchange interaction is obtained from resonant Raman scattering in CdFeSe [8]. In addition, multiple scattering involving mainly the $A_1 \rightarrow A_2$ Fe^{2+} excitation occurs when the laser energy is some meV above the A free exciton. This was explained in term of "cascade" along the polariton dispersion curve [13]; analysis of the data, including depolarization by a magnetic field, allows the determination of the time constant for exciton-Fe^{2+} inelastic scattering.

Pairs of Iron Ions

Exchange interaction between Fe^{2+} is assumed of the usual Heisenberg form $H_{ech} = -2 J \, S_i.S_j$. The energy spectrum of a pair is obtained by solving numerically the total hamiltonian $H = H_{cf} + H_{so} + H_{exc} + H_z$ in a basis restricted to the $5_E \times 5_E$ subspace [24]. A pair is still a Van Vleck system, but with a more complicated spectrum of excited states. There is no direct measurement of J_{nn} (nearest neighbour pair) at present for Fe-based SMSC. An estimated value $J_{nn}/k_B \sim 20K$ in CdFeSe and ZnFeSe was derived from high temperature susceptibility [25]. J_{nn} is needed to get contribution of pairs to the magnetization and to the low temperature specific heat [26]. Expected pair contribution to the high field magnetization seems already too small to account for the difference between experiment and the calculated contribution of isolated Fe^{2+} (see fig. 9 and 11). Level-crossing at high field has been predicted if $J > J_c$, leading to possible magnetization steps at $H > 30T$ [24].

BOUND MAGNETIC POLARONS

In SMSC, the exchange interaction of a bound carrier (donor or acceptor) with the magnetic ions under its orbit gives rise to a remarkable effect known as "bound magnetic polaron" (BMP), well studied in the case of manganese (see the lectures of P. A. Wolff). For donor BMP, where the electron-ion exchange coupling constant $\zeta = N_0 \alpha |\Psi(R)|^2$ is weak ($\Psi(R)$ is the donor envelope wavefunction), the effect is dominated by magnetic fluctuations : the most probable resultant M of N spins S has a length $[NS(S+1)]^{1/2}$ so that the electron spin can be parallel or antiparallel to M : this is the origin of zero-field spin-flip scattering, considered as a signature of the BMP effect.

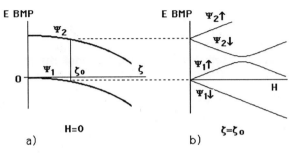

Fig. 15. a) Schematic ground state of Van Vleck BMP as a function of exchange coupling constant ζ for H=0. b) Magnetic splitting of the lowest states for a finite value of ζ.

BMP's in Iron based SMSC

In the case of iron, such fluctuations do not exist, since Fe^{2+} carries no moment and the donor spin-flip energy extrapolates to zero at H=0, as recognized in ref. 16. However, even at H=0, exchange interaction between the donor electron and the Fe^{2+} ions under its orbit induces a polarization of the ions by a mechanism analogous to Van Vleck paramagnetism: this polarization is linear in the coupling constant ζ and the energy gain goes as ζ^2 (in contrast with the case of Mn where the energy gain was linear in ζ in the regime dominated by magnetic fluctuations). This explains why, for donor, BMP effect is much weaker in SMSC with Van Vleck ions. In addition, since the Fe^{2+} polarization is induced by the electron spin, it is not possible to flip the electron spin with respect to the Fe^{2+} polarization: the BMP ground state has a degeneracy 2, so that zero-field SF energy is strictly zero for any value of ζ. Fig. 15 shows schematically how the BMP energy varies as a function of ζ and H.

A simple model. Let us show more precisely the origin of anticrossing between SF and some Fe^{2+} excitations, for H//c (a simpler case). We assume the donor wave function constant in a "box" of volume V containing N Fe^{2+} ions. The BMP ground state is noted:

$$|\Psi_1\sigma\rangle = |\sigma,\varphi_1^{~1},\ldots\ldots,\varphi_1^{~N}\rangle$$

and an excited state involving one Fe^{2+} excitation:

$$|\Psi_i^{~j}\dot{\sigma}\rangle = |\sigma',\varphi_1^{~1},\ldots,\varphi_i^{~j},\ldots,\varphi_1^{~N}\rangle$$

203

Table II. Matrix elements of S_+ and S_z between the four lowest Fe^{2+} states in CdFeSe af H=0. The S_z matrix is symmetrical. φ_3 and φ_4 were choosen in the E subspace such that $\langle\varphi_3|S_z|\varphi_4\rangle=0$. This remains true for H//c. In that case, no new non-zero off-diagonal matrix element appear; terms linear in H appear on the diagonal of S_z. The table for H$\perp$c is more complicated (from ref. 28).

S_+	φ_1	φ_2	φ_3	φ_4
φ_1	0	0	0	−1.889
φ_2	0	0	0	−1.033
φ_3	0	0	0	0
φ_4	1.889	−1.033	0	0
S_z	φ_1	φ_2	φ_3	φ_4
φ_1	0	1.515	0	0
φ_2		0	0	0
φ_3			−0.014	0
φ_4				−0.014

where φ_i^j indicates that the ion at position R^j is in the ith excited state (with H applied). The exchange perturbation is:

$$H_{ex}=(\alpha/V)\sum_j \mathbf{s}.\mathbf{S}^j .$$

Matrix elements between BMP states factorize as:

$$(\alpha/V)\,\langle\sigma'|\mathbf{s}|\sigma\rangle.\sum_j\langle\varphi_i^j|S^j|\varphi_1^j\rangle$$

With the help of Table II which gives the matrix elements of S_+, S_z for the four lowest Fe^{2+} states, we can discuss the lowest anticrossing for H∥c. Restricting our attention to a 2 by 2 matrix of interacting states, one finds two type of terms:

i) The diagonal ones, for $\sigma=\sigma'$ use the $s_z S_z$ term; one gets:

$$\alpha(N/V)\langle\varphi_1|S_z|\varphi_1\rangle=N_o\alpha x\,\langle\varphi_1|S_z|\varphi_1\rangle$$

the spin flip energy E_{SF} times $\langle\sigma|s_z|\sigma\rangle$ (since $\langle\varphi_1|S_z|\varphi_1\rangle$ is proportional to H).

ii) off-diagonal ones, for $\sigma\neq\sigma'$ use s_+S_- and s_-S_+ terms; by inspection of table II, φ_1 can be coupled only to φ_4 (not to φ_2) for H∥c. So, in the case of the pair of BMP states $|\Psi_1\uparrow\rangle$ and $|\Psi_4^j\downarrow\rangle$,the off-diagonal term is :

$\alpha/V\langle\varphi_4|S_+|\varphi_1\rangle$, but there are N equivalent $|\Psi_4^j\downarrow\rangle$ states ! One can show that, in fact, coupling takes place with the excited BMP state which looks like a k=0 Frenkel

exciton:
$$|\Psi_4^{k=0}\downarrow> = N^{-1/2}\sum_j |\Psi_4^{j}\downarrow>$$

Then, one gets a coupling term:

$$C = \alpha(N^{1/2}/V)<\varphi_4|S_+|\varphi_1> = N_o\alpha x <\varphi_4|S_+|\varphi_1> N^{-1/2}$$

Note that C vanishes for a delocalized electron (N→∞), so this is a genuine BMP effect. The 2 by 2 matrix reads:

$$\begin{vmatrix} E_1+1/2\ E_{SF} & C \\ C & E_4-1/2E_{SF} \end{vmatrix}$$

Anticrossing takes place when $E_{SF}=E_4-E_1$ where the splitting is $\Delta E=2|C|$. The displacement of the ground state at H=0 is :

$$\Delta E_p \cong C^2/(E_4-E_1)$$

but in that case, reduction to a 2 by 2 matrix is not justified.

Experiments on Van Vleck BMP's. Experimental data were obtained in CdFeSe through Raman scattering at close resonance on exciton bound to

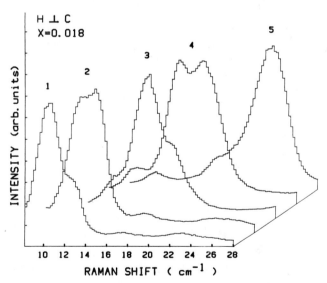

Fig. 16. Samples of resonant Raman scattering spectra at 1.8 K for several values of the field H⊥c in $Cd_{1-x}Fe_xSe$ (x=1.8 %). Curves labelled 1 to 5 correspond to field intensities 1.68, 1.9, 2.58, 2.91 and 3.81 T. Anticrossings of the SF line with the $A_1 \rightarrow A_2$ and $A_1 \rightarrow E$ Fe^{2+} excitations are observed respectively on curves 2 and 4 (ref 27).

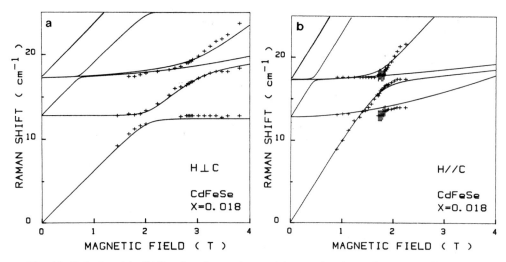

Fig. 17. Calculated (solid lines) and experimental (crosses) values of Raman shifts in $Cd_{1-x}Fe_xSe$ versus magnetic field, for x=1.8 %. a) H $\perp$ c b) H $\parallel$ c; in that case, anticrossing occurs only between SF and $A_1 \rightarrow E$ excitation (ref. 28).

neutral donors; fig. 16 shows typical Raman spectra, for x=1.8 at% and H$\perp$c [27]. ΔE at anticrossing is of the order of 3 cm^{-1} and ΔE_p about 0.2 cm^{-1} ; however, the spin polarization $\langle S \rangle$ at BMP center is not so small: about 0.06 and 0.1 in the directions respectively $\parallel$c and $\perp$c. Excellent agreement with theory is found for H$\parallel$c and H$\perp$c, as shown in fig. 17 (a more complete model incorporating one- and two- excitations BMP states is used) [28].

CONCLUSION

Let us list a few topics for future developments.

a) The possible role of Jahn-Teller effect on the 5_E mutiplet should be clarified, by more extended spectroscopic studies.

b) The antiferromagnetic coupling J_{nn} between Fe pairs should be determined, possibly through Raman scattering and the search for high field magnetization steps [24].

c) Spin glass effects which might exhibit special features in relation with the Van Vleck nature of Fe^{2+} ions.

d) Search for stronger BMP effects. An obvious candidate is the neutral acceptor A°, but preliminary attempts suggest that A° has a short lifetime in CdFeSe, probably due to electrons falling from the deep Fe donor level [29].

206

e) The study of Fe^{2+} in lower symmetry environment, which is realized in alloys like $CdTe_{1-y}Se_y$ [30].

ACKNOWLEDGMENTS

I gratefully acknowledge the contributions of my coworkers from Paris, D. Scalbert, A. Mauger, J. Cernogora and from Warsaw, J. A. Gaj, A. Mycielski and M. Nawrocki.

REFERENCES

* Unité de Recherche Associée au Centre National de la Recherche Scientifique.
1. A. Mycielski, J. Appl. Phys. **63**, 3279 (1988).
2. "Diluted Magnetic Semiconductors", edited by J. K. Furdyna and J. Kossut, Semiconductors and Semimetals Vol 25 (Academic, New York, 1988).
3. M. Taniguchi, Y. Ueda, I. Morisada, Y. Murashita, T. Ohta, I. Souma and Y. Oka, Phys. Rev. **B41**, 3069 (1990).
4. See e. g. M. Taniguchi, L. Ley, R. L. Johnson and M. Cardona, Phys. Rev. **B33**, 1206 (1986).
5. For a review, see B. H. Brendow, Advances in Physics, **26**, 651 (1977).
6. A. Mycielski, P. Dzwonkowski, B. Kowalski, B. A. Orlowski, M. Dobrowolska, M. Arciszewska, W. Dobrowolski and J. M. Baranowski, J. Phys. C: Solid State Phys. **19**, 3605 (1986).
7. M. J. Caldas, A. Fazzio and A. Zunger, Appl. Phys. Lett. **45**, 671 (1984).
8. D. Scalbert, J. Cernogora, A. Mauger, C. Benoit à la Guillaume and A. Mycielski, Solid State Commun. **68**, 1069 (1989).
9. Y. Tanabe and S. Sugano, J. Phys. Soc. Japan **9**, 766 (1954).
10. G. A. Slack, S. Roberts and J. T. Vallin, Phys. Rev. **187**, 511 (1969).
11. A discussion of the role of the parity of n in $3d^n$ ions may be found in the introduction of: M. Villeret, S. Rodriguez and E. Kartheuser, Phys. Rev. **41**, 10028 (1990).
12. M. Hausenblas, L. M. Claessen, A. Wittlin, A. Twardowski, M. von Ortenberg, W. J. M. de Jonge and P. Wyder, Solid State Commun. **72**, 253 (1989).
13. D. Scalbert, J. A. Gaj, A. Mauger, J. Cernogora, C. Benoit à la Guillaume and A. Mycielski, II-VI-89, Berlin; J. of Crystal Growth. **101**, 940 (1990).
14. C. Testelin, A. Mauger, C. Rigaux, M. Guillot and A. Mycielski, Solid State Commun. **71**, 923 (1989).

15. The linear susceptibility being a second rank tensor is necessarily isotropic in a cubic material.

16. D. A. Heiman, A. Petrou, S. H. Bloom, Y. Shapira, E. D. Isaacs and W. Giriat, Phys. Rev. Lett. **60**, 1876 (1988).

17. D. Scalbert, A. Mauger, M. Guillot, J. Cernogora, J. A. Gaj and C. Benoit à la Guillaume, submitted to Solid State Commun.

18. The Heisenberg interaction assumed for the exchange between Fe^{2+} ions and band carriers might be inaccurate in the case of VB, according to A. Blinowski and P. Kacman's seminar.

19. J. A. Gaj, R. Planel and G. Fishman, Solid State Commun. **29**, 435 (1979).

20. A. Twardowski, P. Glod, W. J. M. de Jonge and M. Demianiuk, Solid State Commun. **64**, 63 (1987).

21. A. Twardowski, K. Pakula, M. Arciszewska and A. Mycielcki, Solid State Commun. **73**, 601 (1990).

22. A. K. Bhattacharjee, G. Fishman and B. Coqblin, Physica **117B** & **118B**, 449 (1983).

23. X. Liu, A. Petrou, J. Warnock, B. T. Jonker, G. A. Prinz and J. J. Krebs, Phys. Rev. Lett. **63**, 2280 (1989).

24. H. J. M. Swagten, C. E. P. Gerrits, A. Twardowski and W. J. M. de Jonge, Phys. Rev. **B41**, 7330 (1990).

25. A. Twardowski, A. Lewicki, M. Arciszewska, W. J. M. de Jonge, H. J. M. Swagten and M. Demianiuk, Phys. Rev. **B38**, 10749 (1988).

26. A. Twardowski, H. J. M. Swagten and W. J. M. de Jonge, Phys. Rev. **B42**, 2455 (1990).

27. D. Scalbert, J. A. Gaj, A. Mauger, J. Cernogora and C. Benoit à la Guillaume, Phys. Rev. Lett. **62**, 2865 (1989).

28. D. Scalbert, J. A. Gaj, A. Mauger, J. Cernogora, M. Nawrocki and C. Benoit à la Guillaume, submitted to Phys. Rev. **B**.

29. D. Scalbert, J. Cernogora and C. Benoit à la Guillaume, unpublished.

30. A. Mycielski, W. Dobrowolski, A. Szadkowski, M. Borowiec, M. Arciszewska, C. Julien, C. Rigaux and M. Menant, ICPS 20th, Thessaloniki (August 1990).

SEMIMAGNETIC IV-VI COMPOUND SEMICONDUCTORS

G. Bauer* and H. Pascher**

* Semiconductor Physics Group, Johannes-Kepler-Universität
Linz, A-4040 Linz, Austria
** Experimentalphysik I, Universität Bayreuth
D-8580 Bayreuth, Germany

1.0 INTRODUCTION

Semimagnetic IV-VI compounds have been studied for about a decade. Nevertheless the amount of knowledge which has been accumulated is much less than compared to the situation in $A_{1-x}^{II}Mn_xB^{VI}$ or $A_{1-x}^{II}Fe_xB^{VI}$ alloys. The complications due to the many valley band structure, the fact that the band edge functions are composed of a combination of s, p, and d - like functions as well as general material properties make this class more difficult to deal with.[1] In PbTe, PbSe, PbS the group IV element has been replaced by Mn^{2+}, Eu^{2+} Gd^{2+} and Fe^{2+}. These IV-VI compounds have, for not too high concentrations of the magnetic ions, a direct narrow gap at the L-point of the Brillouin zone. A two band model is approximately valid for the electrons and holes. The mass anisotropies $K = m_l/m_t$ vary from 10 (PbTe) via 2 (PbSe) to about 1(PbS). In the pseudobinary alloys the magnetic ions cause an increase of the energy gap with composition x. Especially the systems $Pb_{1-x}Eu_xTe$[2] and $Pb_{1-x}Eu_xSe$[3] have found applications as mid infrared lasers because of the tuning of the energy gap with composition x.

The magnetic properties of the semimagnetic IV-VI compounds are determined by antiferromagnetic interactions which are weaker than those of narrow gap II-VI compounds.[4,5] The Eu-Eu exchange effects are even weaker in comparison to those of Mn-Mn, as expected.[6]

Recently a carrier concentration induced paramagnetic- ferromagnetic transition was found in $Pb_{1-x-y}Sn_xMn_yTe$ ($x \approx 0.7$, $y \approx 0.03$).[7] Thus in the magnetic phase diagram of this compound the carrier concentration appears as an additional and tunable parameter.

Apart from the study of the evolution of magnetic interactions the main interest in semimagnetic semiconductors results from the consequences of the exchange interaction between the mobile carriers in the conduction and valence bands and the localized magnetic moments.[8] Previously indications for negligibly small exchange induced effects for the electrons in $Pb_{1-x}Mn_xTe$[8-11] and $Pb_{1-x}Mn_xS$[12,13] were found. From interband magnetooptical data,[11] however, it turned out that the exchange induced corrections for the g factors are much larger for holes than for electrons. Using

Semimagnetic Semiconductors and Diluted Magnetic Semiconductors
Edited by M. Averous and M. Balkanski, Plenum Press, New York, 1991

209

direct measurements of the electron and hole g-factors by a coherent resonant Raman scattering technique in external fields, Pascher et al.[14]
established the importance of exchange induced effects in $Pb_{1-x}Mn_xTe$ and
for the first time the exchange parameters were deduced with reasonable
accuracy.

The content of this review is as follows: Band structure and mean
field theory for bulk crystals and superlattices, magnetic properties,
magnetooptical interband and intraband transitions, discussion of electron
and hole g factors, light induced magnetization in bulk crystals and
superlattices, conclusion.

2.0 BAND STRUCTURE

In contrast to the semimagnetic II-VI compounds the IV-VI materials
crystallize in the rock salt structure. In the ternary compounds with
manganese or europium the favorable situation occurs that MnSe and the
europium chalcogenides crystallize also in the NaCl structure. Only MnTe
as a volume material exists only in the hexagonal nickel arsenide structure. As a consequence it turns out to be difficult so far to grow single
phase $Pb_{1-x}Mn_xTe$ with Mn- contents above 10%. From all investigations carried out so far the Mn- and Eu- ions occupy group IV lattice sites. For
the evolution of magnetic properties with increasing magnetic ion content
it is important to note that each ion is surrounded by six nearest neighbors, whereas in the zincblende structure this number is four.

2.1 Band Structure of Nonmagnetic Host Crystals

For the lead compounds PbTe, PbSe and PbS apart from the lowest lying
conduction and highest valence level which form the minimum gap two further conduction and two valence bands have to be taken into account. These

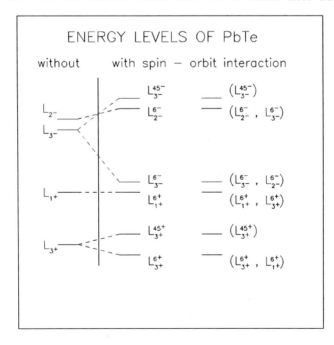

Fig.1. Group theoretical classification of the energy levels clos to the
fundamental energy gap in the notation of Bernick and Kleinman[15]
and Mitchell and Wallis.[19]

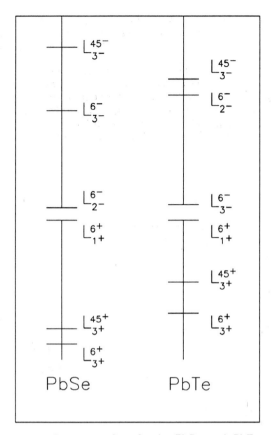

Fig.2. Sequence of energy levels in PbSe and PbTe after Ref.15.

six levels are within an energy spread of about 3 - 5 eV, whereas all other levels are about 8 eV removed. Empirical pseudopotential calculations [15-18] have established the ordering of those six levels. In Fig.1 for PbTe the group theoretical classification without and with spin orbit interaction is given. The spin orbit interaction is huge, it not only removes the degeneracy of the L_3 levels but the originating energy gaps are of the order of the gaps between L_1 and L_3 levels. The notation shown in parentheses indicates explicitly that the levels forming the minimum gap consist of a linear combination of L_{3-}^{6-}, L_{2-}^{6-} and L_{1+}^{6+}, L_{3+}^{6+}.

The symmetry of the band edge functions for the conduction band is given by: [19]

$$L_6^-(L_2^-)_\alpha = -\sin\Theta^- \; Z \uparrow - \cos\Theta^- \; X_+ \downarrow$$

$$L_6^-(L_2^-)_\beta = \sin\Theta^- \; Z \downarrow - \cos\Theta^- \; X_- \uparrow \tag{1}$$

and for the valence band:

$$L_6^+(L_1^+)_\alpha = i \cos\Theta^+ \; R \uparrow + \sin\Theta^+ \; S_+ \downarrow$$

$$L_6^+(L_1^+)_\beta = i \cos\Theta^+ \; R \downarrow + \sin\Theta^+ \; S_- \uparrow \tag{1a}$$

The relative mixing of the states by spin orbit coupling is determined by the parameters Θ^+, Θ^-. [15] R, Z, X_{+-} and S_{+-} denote functions which are derived from functions with s,p,d symmetry, respectively and are explicitely given in Ref.19. The system of coordinates is chosen such that $x \parallel [1\bar{1}2]$, $y \parallel [1\bar{1}0]$ and $z \parallel [111]$.

The masses and their anisotropies are determined by the energy gaps and the values of the momentum matrix elements. Whereas in PbTe the contribution of the L_3^{6-} dominates the lowest conduction level, for PbSe it is the L_2^{6-} (Fig.2). Due to spin orbit mixing a coupling of both the longitudinal and transverse momentum matrix elements is possible for a single pair of levels which explains the large differences in the mass anisotropy.

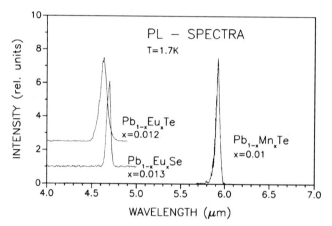

Fig.3. Photoluminescence (PL) spectra of some semimagnetic IV-VI compounds. The onset of the Pl spectra of the host materials would be at 6.5μm (PbTe) and 8.5μm (PbSe.

2.2 Energy Gaps of Semimagnetic IV-VI Compounds

The increase of the energy gaps with increasing content of magnetic ions is shown in the photoluminescence spectra, reproduced in Fig.3. Taking into account that the gap of PbTe corresponds to 6.5μm and that of PbSe to 8.5μm, the figure demonstrates that incorporation of Eu shifts the gaps by a larger amount than Mn. From the energy gaps which can be found out from the low energy side of the lines the x- values can be calculated in a very easy and accurate way by the following equations, valid for He-temperatures and small concentrations x:[11,20,2]

$$Pb_{1-x}Mn_xTe: \quad E_g(x)/meV = 190 + 2510 \cdot x \quad (2)$$

$$Pb_{1-x}Eu_xSe: \quad E_g(x)/meV = 114 + 11370 \cdot x \quad (3)$$

$$Pb_{1-x}Eu_xTe: \quad E_g(x)/meV = 190 + 6000 \cdot x \quad (4)$$

A collection of data on Mn- compounds shows Fig.4a, for Eu- compounds Fig.4b.

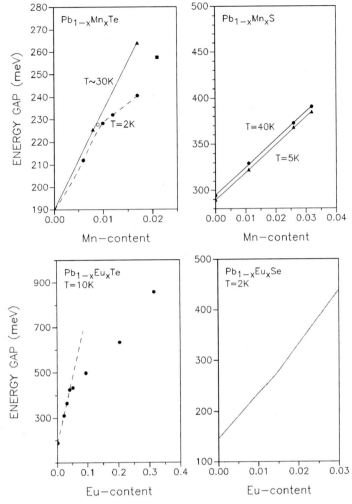

Fig.4. Energy gap versus composition. a) (upper half) Mn- compounds
b) (lower half) Eu- compounds.

2.3 Calculation of Landau States: Mean Field Theory for IV-VI Compounds

2.31 Homogeneous Crystals

The $\vec{k} \cdot \vec{p}$ band model for the lead compounds has been derived by several authors. In the following we use the notation of Adler et al.[21] who treat exactly the interaction of the upper most valence and the lowest conduction band level whereas the interaction with two more distant conduction and valence levels is treated in k^2- approximation. The 4x4 matrix Hamiltonian for the calculation of the Landau states resulting from this procedure is given by:[21,22]

$$H = \begin{bmatrix} h_v & h_{vc} \\ h_{cv} & h_c \end{bmatrix} \qquad (5)$$

with

$$h_c = \left(E_c + \frac{k_3^2}{2m_l^-} + \frac{k_1^2 + k_2^2}{2m_t^-} \right) I_2 + \left(g_1^- \sigma_z B_3 + g_t^- (\sigma_x B_1 + \sigma_y B_2) \right) \mu_B/2 \qquad (6)$$

$$h_v = \left(E_v - \frac{k_3^2}{2m_l^+} - \frac{k_1^2 + k_2^2}{2m_t^+} \right) I_2 - \left(g_1^+ \sigma_z B_3 + g_t^+ (\sigma_x B_1 + \sigma_y B_2) \right) \mu_B/2 \qquad (7)$$

$$h_{cv} = P_\parallel k_3 \sigma_z + P_\perp (k_1 \sigma_x + k_2 \sigma_y) \qquad (8)$$

$$h_{vc} = h_{cv}^\dagger \qquad (9)$$

I_2 is the 2x2 unit matrix, σ are the Pauli spin matrices (σ_z diagonal), μ_B is the Bohr magneton and the indices 1,2,3 refer to the valley axes system with axis 3 parallel to the valley main axis. $P_\parallel$ and $P\perp$ are the momentum matrix elements for the two band interaction perpendicular and parallel to <111>, $m_{l,t}^{+-}$ are the far band contributions to the conduction band (−) and valence band (+), parallel (l) and perpendicular (t) <111>, $g_{1,t}^{+-}$ the same for the g-factors.

For semimagnetic semiconductors with rather small contents x of the paramagnetic ions we assume that the sequence of energy levels is not altered with respect to the host crystal. In the spirit of the mean field approximation[23] (MFA) an exchange Hamiltonian is added to the 4x4 matrix Hamiltonian.[9,10,11,14] It describes the interaction between the electron spin $\vec{\sigma}$ and spin $\vec{S}$ the paramagnetic ions is given by:[23]

$$H_{exch} = J(|\vec{r}-\vec{R}|) \; \vec{\sigma} \cdot \vec{S} \qquad (10)$$

J is the exchange integral.

The $\vec{k} \cdot \vec{p}$ Hamiltonian is finally obtained for S=5/2:

$$H_{exch} = 0.5 \times S_o B_{5/2} \left[\frac{5\mu_B |\vec{B}|}{k (T+T_o)} \right]$$

$$\times \begin{bmatrix} A\sigma_z \alpha_3 + a_1 (\sigma_x \alpha_1 + \sigma_y \alpha_2) & 0 \\ 0 & B\sigma_z \alpha_3 - b_1 (\sigma_x \alpha_1 + \sigma_y \alpha_2) \end{bmatrix} \qquad (11)$$

(For the Eu- compounds S=5/2 has to be replaced by S=7/2).
$\alpha_{1,2,3}$ are the direction cosines of $\vec{B}$ in the valley axes system, $B_{5/2}$ is the modified Brillouin function describing the paramagnetic orientation of the magnetic moments,[8] S_o and T_o are adjustable parameters, determined from magnetization data, A, a_1, B, b_1 are the four exchange parameters for the valence and conduction bands, respectively.[11] This 4x4 exchange Hamiltonian contains off diagonal terms which cause a strong mixing of Landau states with different spin and Landau quantum numbers. In order to calculate the energies of the 0^+ and 0^- Landau states of the conduction and valence bands a 12x12 secular determinant $\tilde{\mathcal{H}} = \tilde{\mathcal{X}} + \tilde{\mathcal{Y}}$ is used where $\tilde{\mathcal{X}}$ has three 4x4 blocks according to Eq.(5) along the main diagonal and $\tilde{\mathcal{Y}}$ is given by Eq.(11).[24] A, a_1, B, b_1 depend on the exchange integrals and the spin orbit mixing parameters Θ^+, Θ^-:

214

$$A = a_1 - a_2 = \alpha \cdot \cos^2\Theta^+ - \delta \cdot \sin^2\Theta^+$$

$$B = b_1 - b_2 = \beta_{\parallel} \cdot \sin^2\Theta^- - \beta_{\perp} \cdot \cos^2\Theta^- \qquad (12)$$

Due to the symmetry properties of the wave functions at the L-point of the Brillouin zone instead of two exchange integrals as in zincblende semiconductors now four are needed where the first two are for the holes and the second two for the electrons:[9,10,11]

$$\alpha = (R|J|R)/\Omega, \qquad \delta = (S_{\pm}|J|S_{\pm})/\Omega,$$

$$\beta_{\parallel} = (X_{\pm}|J|X_{\pm})/\Omega, \qquad \beta_{\perp} = (Z|J|Z)/\Omega \qquad (13)$$

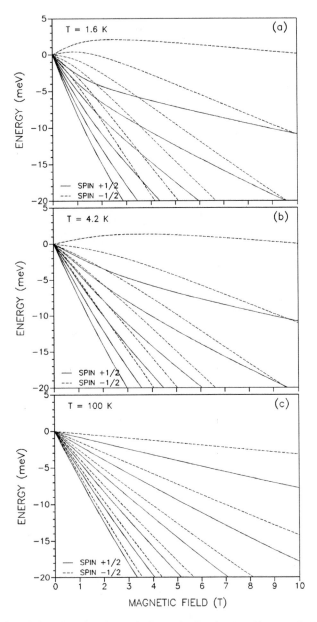

Fig.5. Valence band Landau levels of $Pb_{1-x}Mn_xTe$ (x=0.01) for 3 temperatures

Using the band parameters of Ref.14 and the exchange parameters of Table I in Figs.5a,b,c Landau levels for the conduction and valence band are shown calculated for three temperatures, $\Phi = 90°$ for $Pb_{1-x}Mn_xTe$ (x=0.01). At T=1.6 K the exchange induced corrections alter drastically the magnetic field dependence of the Landau states causing even a positive slope of the 0^- valence band level at small magnetic fields. For higher temperatures the influence of the exchange interaction gradually decreases and finally the magnetic field dependence of the Landau states resembles nearly the diamagnetic host material. In $Pb_{1-x}Eu_xSe$ the exchange induced effects are weaker but of different sign and even a crossing of the n^- and n^+ levels occurs in the valence band.

2.32 Superlattices

The electronic properties of superlattices composed of either III-V or IV-VI semiconductors which have a direct gap at the Γ or L point of the Brillouin zone, respectively are readily described by the envelope function formalism.[25]

For a superlattice the envelope function Schrödinger equation Hf=Ef has position dependent material parameters. Mainly the band edge energies are varying and the exchange interaction vanishes in the PbTe layers. The differential operators have of course to be written in a hermitized form.[25]

For Voigt geometry with $\vec{B}$ parallel to the layers we take the coordinate system with axis z parallel to the growth direction and x parallel to B. The vector potential in Landau gauge is oriented along y. The envelope functions f_b (b=1,..,4) are of the form:

$$f_b = \exp(ik_Bx) \exp(ik_yy) Z_b(z-k_y\ell^2) \qquad (14)$$

where the momentum k_B and the center coordinate $z_M=k_y\ell^2$ are good quantum numbers.[25] The functions Z_b are obtained as solutions of the four coupled differential equations Hf=Ef, where the substitution

$$\begin{bmatrix} k_1 \\ k_2 \\ k_3 \end{bmatrix} = T \begin{bmatrix} k_B \\ k_y - z/\ell^2 \\ -i\partial/\partial z \end{bmatrix} \qquad (15)$$

has been performed. T is the transformation matrix from the x,y,z coordinate system to the valley system.

3.0 MAGNETIC PROPERTIES

Magnetic properties of dilute magnetic lead compounds were studied by low field susceptibility, high field magnetization and electron paramagnetic resonance experiments. Susceptibility measurements were performed on $Pb_{1-x}Mn_xS$,[5] $Pb_{1-x}Mn_xSe$,[27] $Pb_{1-x}Mn_xTe$,[27] $Pb_{1-x}Eu_xTe$,[6] as well as $Pb_{1-x}Gd_xTe$.[27] For these materials the high temperature extrapolation yields small negative values implying weak antiferromagnetic interaction. The nearest neighbor exchange interaction 2J/k varies from about 3 K($Pb_{1-x}Mn_xS$) to 0.2 K ($Pb_{1-x}Eu_xTe$) and thus is much smaller than in the narrow gap II-VI semimagnetic semiconductors.

High field magnetization studies were carried out on the same materials as mentioned above.[5,6,11,28-30] In all cases it turned out that a modified Brillouin function (see Eq.(11)) is well suited for the description of the data at temperatures above 2 K. Due to the rather small values of the exchange coupling constants J_i (i numbering the shells of nearest neighbors) so far no steps in magnetization as a function of field were observed.

3.1 Free Carrier Induced Ferromagnetism

One of the most fascinating magnetic properties of the semimagnetic lead salt alloys found so far, has been described by Story et al.[7] and has since been quantitatively explained by several authors.[31-33] In $Pb_{1-x-y}Sn_xMn_yTe$ (for $x\approx0.7$, $y\approx0.03-0.08$) a ferromagnetic ordering has been observed above a critical hole concentration of about 3×10^{20} cm^{-3}. The transition temperature (0-5K) increases first with increasing hole concentration. The ferromagnetic phase has been unambiguously identified by magnetization, magnetic susceptibility and specific heat measurements and experiments under hydrostatic pressure. The interesting feature is the following: for such large hole concentrations a second set of valence levels (Σ-band) becomes populated apart form the L-levels.[31] The position of the Σ levels relatively to the L-levels strongly depends on chemical composition and hydrostatic pressure. The interaction which mediates this ferromagnetic transition is the Ruderman-Kittel-Kasuya-Yosida (RKKY) interaction which was originally developed for parabolic bands. Swagten et al.[31] have extended this RKKY interaction to the two valence band situation (L-Σ).

The Curie-Weiss parameter is calculated by the expression:

$$\theta = \frac{2S(S+1)x}{3k} \sum_{i=1}^{2} \sum_{j\geq1} J_i^{eff}(R_j) \tag{16}$$

R_j: distance between two lattice points in the NaCl lattice
$S = 5/2$ for Mn^{2+}
J_i: exchange constant for L or Σ band
a_0: lattice parameter

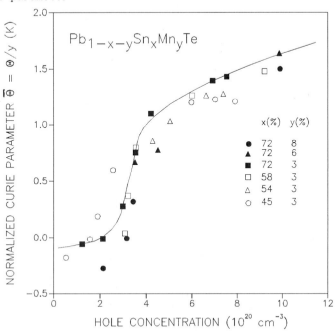

Fig.6.: Normalized paramagnetic Curie- Weiss parameter $\bar{\theta}=\theta/y$ for $Pb_{1-x-y}Sn_xMn_yTe$ versus free hole concentration. After Ref.32

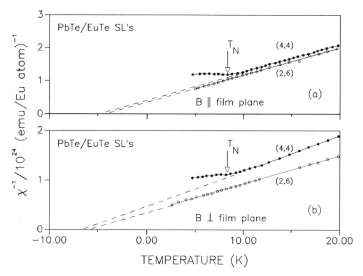

Fig.7. Magnetic properties of short- period PbTe/EuTe superlattices:
inverse susceptibility vs temperature for B perpendicular (bottom)
and parallel (top) to the layers. The (4/4) sample exhibits an
antiferromagnetic phase transition whereas the (2/6) sample remains paramagnetic. After Ref.34.

J^{eff} is calculated for intraband RKKY interaction between the magne
tic moments, and a mean field approximation for the spin system is used.
Using for J_i = 0.3 eV, for the mean free path λ values between 7 and 13 Å
corresponding to the observed mobilities of about 20 cm^2/ Vs) it was possible to derive a critical hole concentration of about p = 2.4 x 10^{20} cm^{-3}
with the energy gaps E_g=0.15eV and $\Delta(E_L - E_\Sigma)$ = 0.6eV. It turned out that
even the experimentally observed step like increase of the Curie-Weiss-
parameter θ (Eq.8) for hole concentrations p > p_{crit} (θ>0) and the further
dependence of θ on p is correctly described by the model of Swagten et al.
(Ref.31). A refinement of this model with the correct description of the
compositional evolution of the band structure in $Pb_{1-x-y}Sn_xMn_yTe$ describes
quantitatively the carrier concentration dependence of its magnetic properties as shown in Fig.6 (Story et al.[32]).

3.2 Magnetic Properties of Layered IV-VI Diluted Magnetic Semiconductors

The study of magnetic properties in quasi two dimensional systems is
of particular interest both for the aspects of magnetism as well as of
fundamental importance for an understanding of the electronic properties.
The first experiment of this kind was a susceptibility study on PbTe/ EuTe
short period superlattices.[34] PbTe is diamagnetic, EuTe paramagnetic and
below 9.58 K antiferromagnetic. A systematic study on MBE grown material[2]
was undertaken using three samples consisting of $(EuTe)_m/(PbTe)_n$ layer
structures with (m,n) = (1,3); (2,6) and (4,4) and periods between 100 and
400. Susceptibility measurements were performed using a SQUID magnetometer
and applying external fields both parallel and perpendicular to the
layers. Results on the (4/4) and (2/6) samples are shown in Fig.7. $\chi^{-1}(T)$
follows a Curie-Weiss-law with anisotropic extrapolated Curie-Weiss-
parameters. For the (4/4) sample the antiferromagnetic phase transition is
indeed observed at a Néel temperature of T = 8.5 K, only slightly below
the bulk value for EuTe. No phase transition was observed for the (2/6)
and (1/3) samples where the Eu atoms do not have all of their 6 nearest

neighbors in contrast to the (4/4) sample. The anisotropy of θ was explained by involving effects of strain ($\varepsilon \approx 10^{-2}$) on the nearest and next nearest neighbor interactions. The disappearance of the antiferromagnetic properties in the ultra thin (two and one monolayer EuTe) samples is at least consistent with the tendency and the predictions of the Mermin-Wagner-theorem.[8]

4.0 MAGNETOOPTICAL INTERBAND TRANSITIONS

Magnetooptical interband transitions in $Pb_{1-x}Eu_xSe$ ($x \leq 0.01$) and $Pb_{1-x}Mn_xTe$ ($x \leq 0.02$) require radiation wavelengths between 4.5 and 7.5 μm. Coherent radiation in the desired range was obtained either from a CO-laser or by frequency doubling of the radiation of a Q-switched CO_2-laser in a phase matched Te-crystal.

The transmission spectra were observed in Faraday configuration $\vec{B} \| \vec{k} \| [111]$ with circularly polarized σ^+ and σ^- radiation and in Voigt configuration $\vec{B} \perp \vec{k}$, $\vec{B} \| [1\bar{1}0]$, $\vec{E} \| \vec{B}$.

Figure 8a shows the interband transmission of a $Pb_{1-x}Mn_xTe$ sample with x=0.006 as a function of the magnetic field for the two circular polarizations in Faraday configuration, Fig.8b the same for $Pb_{1-x}Eu_xSe$, x=0.008. It is evident that the resonant magnetic fields are different for the two polarizations in contrast to the PbTe and PbSe case, respectively. Whereas in $Pb_{1-x}Mn_xTe$ the σ^+ resonances occur at higher magnetic fields than σ^-, in $Pb_{1-x}Eu_xSe$ it is vice versa. From this fact one can conclude that the spin splitting of the valence band is diminished in $Pb_{1-x}Eu_xSe$ by the exchange interaction of free carriers with localized spins and increased in $Pb_{1-x}Mn_xTe$. From a comparison of Faraday- and Voigt data it follows that this interaction mainly affects holes in the valence band.

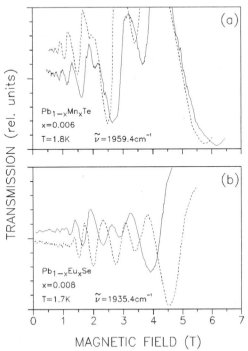

Fig.8: Interband magnetotransmission: a) $Pb_{1-x}Mn_xTe$, b) $Pb_{1-x}Eu_xSe$ in Faraday configuration.
Full curves: σ^+ and broken curves σ^- circular polarization.

From a large number of such recordings taken with various infrared laser frequencies fan charts for $Pb_{1-x}Mn_xTe$ were obtained as shown in Fig.9a for $Pb_{1-x}Mn_x$ and in Fig.9b for $Pb_{1-x}Eu_xSe$ in Faraday configuration. The full lines correspond to calculated transitions based on the equations given in Sec.2. with the selection rules $\Delta n=0$, $\Delta s=\pm 1$ (Faraday geometry). We would like to point out that the mean field theory of Sec.2. is capable of explaining all of the experimental data.

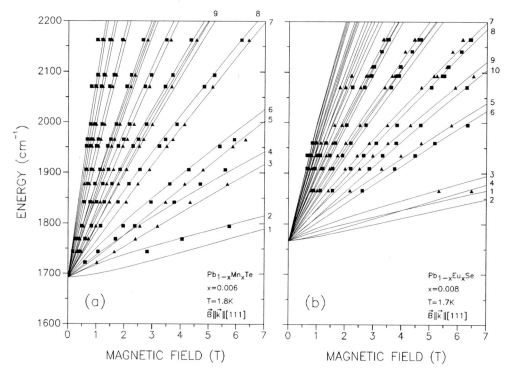

Fig.9: Fan chart for interband magnetooptical transitions in Faraday geometry: a) PbMnTe; b) PbEuSe ■, ▲ experimental data (σ^-, σ^+); full lines: calculated data; identification:
1: $0^- \rightarrow 0^+ (\Phi=70.53°)$; 2: $0^+ \rightarrow 0^- (\Phi=70.53°)$; 3: $0^- \rightarrow 0^+ (\Phi=0°)$;
4: $0^+ \rightarrow 0^- (\Phi=0°)$; 5: $1^- \rightarrow 1^+ (\Phi=70.53°)$; 6: $1^+ \rightarrow 1^- (\Phi=70.53°)$;
7: $2^- \rightarrow 2^+ (\Phi=70.53°)$; 8: $2^+ \rightarrow 2^- (\Phi=70.53°)$; 9: $1^- \rightarrow 1^+ (\Phi=0°)$;
10: $1^+ \rightarrow 1^- (\Phi=0°)$;

In Voigt geometry one relatively strong series of interband transitions is observed which does not obey the selection rules $\Delta n=0$, $\Delta s=0$. It turns out that a transition $0^+(vb) \rightarrow 0^-(cb)$, i.e. $\Delta s=-1$ is observed, within the $\Phi=35,26°$ valleys. The allowed transitions $0^+(vb) \rightarrow 0^+(cb)$ for $\Phi=35.26°$ also appears, whereas $0^-(vb) \rightarrow 0^-(cb)$ is blocked due to the position of the Fermi energy. Experiments were performed on several $Pb_{1-x}Mn_xTe$ samples with x values up to x=1.7%. The band parameters from the fits are given in Ref.14, the exchange parameters in Table I.

5.0 COHERENT RAMAN SCATTERING

In nonmagnetic semiconductors the cyclotron resonance is infrared active, whereas the spin resonance of the free carriers is Raman active. However, Raman spectra of narrow gap semiconductors are very difficult to be obtained by conventional Raman methods since visible light is strongly absorbed by interband transitions. Spontaneous Raman scattering in the infrared is hardly observable due to the poor sensitivity of infrared detectors and due to the frequency dependence of scattering cross sections.

With the coherent Raman methods which are based on optical four- wave mixing the scattered light leaves the sample as a collimated beam, by far easier to be detected than the spontaneously scattered light which is emitted into the whole solid angle. This makes coherent Raman scattering applicable to narrow gap semiconductors [35] and allows an observation of spin flip transitions of electrons and holes. If the sample is held in Voigt configuration in a magnetic field and two laser beams with frequencies ω_L and ω_S are collinearly focused onto the sample a resonant enhancement of the scattered intensity at $\omega_{AS} = 2\omega_L - \omega_S$ is observed if $g^* \mu_B B = \hbar(\omega_L - \omega_S)$. Observations of these resonances yield highly precise data on the effective g factors which are of special importance with semimagnetic semiconductors.

5.1 Experimental Results and Discussion

The experiments were performed with two simultaneously Q-switched CO_2 lasers. The peak powers were attenuated as much as possible to avoid power broadening of the lines and saturation of the spin resonances. For $\hbar\omega_L \approx E_g$ powers of the order of 50mW are sufficient, for $\hbar\omega_L \approx E_g/2$ 500W to 2kW are necessary. The pulse lengths are about 100ns. Therefore the energy per pulse is low enough to avoid heating of the crystal lattice. For collinear adjustment of the laser beams the optimum sample thickness is equal to the coherence length $L_c = \pi / |\Delta \vec{k}|$. With thin epitaxial films $L << L_c$ and the phase factor can be approximated by 1. Due to the proportionality $I_{AS} \propto L^2$ in very thin films the signals are difficult to observe without taking advantage of the band gap resonance $\hbar\omega_L \approx E_g$. In this case using a double monochromator and a sensitive MCT photovoltaic detector a sample thickness of at least $3\mu m$ is preferable.

In $Pb_{1-x}Mn_xTe$ as well as $Pb_{1-x}Eu_xSe$ with $x \leq 0.012$ there are CO_2 laser lines with $\hbar\omega_L > E_g/2$. Using these lines the laser radiation produces a considerable number of minority carriers by two photon absorption. Therefore electron and hole resonances can be observed in one sample independent of its doping.

In Figs. 10a-f results on the spin splittings of valence and conduction bands as obtained by CARS are compared for $Pb_{1-x}Mn_xTe$ and $Pb_{1-x}Eu_xSe$. The full lines are calculated by the model outlined in Sec. 2. The plots show that the spin splitting of the valence band is increased in $Pb_{1-x}Mn_xTe$ and is diminished in $Pb_{1-x}Eu_xSe$, with respect to the non magnetic host materials. This demonstrates that the exchange interaction between free carriers and magnetic ions is ferromagnetic in $Pb_{1-x}Mn_xTe$ and antiferromagnetic in $Pb_{1-x}Eu_xSe$. The anisotropy both of the effective masses and the effective g-factors is much smaller in $Pb_{1-x}Eu_xSe$ than in $Pb_{1-x}Mn_xTe$ as it is for PbSe [36] compared to PbTe. [37,38] From an inspection of the data further follows:
(i) the contributions to the spin flip transition energies induced by the exchange interaction are much larger for the valence band than for the conduction band
(ii) for fields in excess of about 2 T, the difference between the spin flip energies for electrons and holes for the oblique $\Phi = 35.26°$ valleys as

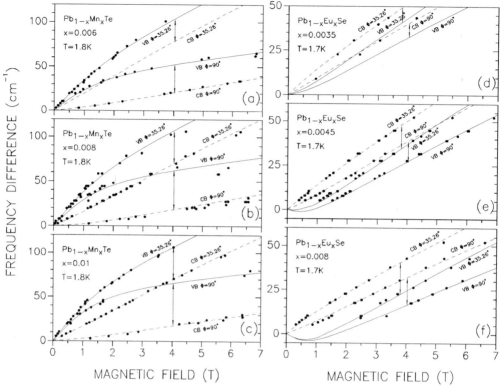

Fig.10. Results of CARS measurements (frequency differences) vs magnetic
field for x=0.006 (a), x=0.008 (b) and x=0.01 (c) for $\vec{B}\parallel[110]$,
T=1.8K, PbMnTe. Experimental data: ●, calculated data for holes:
full lines for $\Phi=35.26°$ and $\Phi=90°$ valleys; electrons: broken
lines for $\Phi=35.26°$ and $\Phi=90°$ valleys.
PbEuSe: x=0.0035 (d), x=0.0045 (e), x=0.008 (f); T=1.7 K.

well as for the $\Phi=90°$ valleys changes only slightly with field (arrows in
Figs.10a-f) for fields above 2T. This energy difference is the same within
experimental error which is observed for interband magnetooptical transi-
tions in $\vec{B}\parallel\vec{k}$ -geometry if energy differences for the σ^+ and σ^- polarized
radiation $(0^-(\text{vb}) \rightarrow 0^+(\text{cb}), 0^+(\text{vb}) \rightarrow 0^-(\text{cb}))$ are derived from the experimental data (see Fig.9).

The temperature dependence of these transition energies is quite dramatic, as shown in Figs.11a-f for experiments performed up to 12K for a
sample with a Mn-content of x=0.006 and an Eu-content of x=0.008. The temperature dependence is particularly pronounced for the hole spin flip
transitions. With increasing lattice temperature the sequence of spin flip
transition energies becomes more and more PbTe-(PbSe-) like apart from the
fact that the absolute values of the g-factors are smaller due to the larger gap.

The calculated transition energies are presented by the full lines in
the figures using the model outlined in Sec.2. Actually, the four exchange
parameters listed in Table I as well as T_0 were determined from the fits
shown in the figures, whereas the two band parameters $P_\perp$, $P_\parallel$ and the energy gap were obtained from the magnetooptical interband transition energies. The far band parameters were assumed to be identical to those of
PbTe and PbSe, respectively. The excellent agreement between experimental
results and calculated transition energies is obvious.

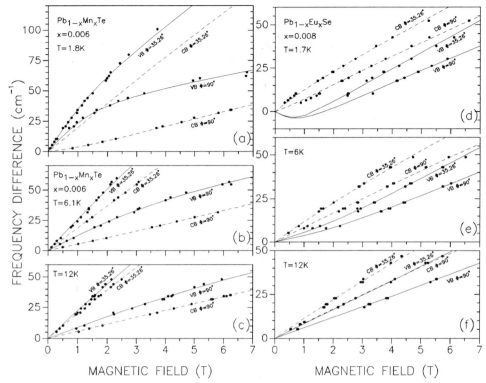

Fig. 11. Results of CARS measurements (frequency differences) vs magnetic
field for T=1.8K (a), T=6.1K (b) and T=12K (c) for B̄∥[1̄10],
x=0.006, PbMnTe. Experimental data: ●, calculated data for holes:
full lines for Φ=35.26° and Φ=90° valleys; electrons: broken
lines for Φ=35.26° and Φ=90° valleys.
PbEuSe x=0.008: T=1.7K (d), T=6.0K (e); T=12.K (f).

5.2 Effective Electron and Hole g Factors

   The spin resonant four wave mixing experiments provide an ideal means
for the study of the magnetic field dependence of the spin splitting between the 0^--0^+ states both in the conduction and in the valence bands. In
Fig. 12a experimentally derived g-factors of $Pb_{1-x}Mn_xTe$ (x=0.01) are plotted as a function of magnetic field together with calculated data based on
Sec. 2, Ref. 14 and Table I. Figure 12b shows the analogous plot for
$Pb_{1-x}Eu_xSe$ (x=0.008). For an orientation of the magnetic field B//[1̄10]
four g factors are experimentally accessible, two for the conduction and
two for the valence band. An explicit expression for the longitudinal effective g factors can be derived using Eqs. 10-12:

$$\text{VB:} \quad g_{v,1}^{eff} = - \frac{4\,P_{\perp}^2}{m_o E_g} + g_1^+ + x\,A\,\frac{\langle S_z \rangle}{\mu_B B} \qquad (17)$$

$$\text{CB:} \quad g_{c,1}^{eff} = + \frac{4\,P_{\perp}^2}{m_o E_g} + g_1^- + x\,B\,\frac{\langle S_z \rangle}{\mu_B B} \qquad (18)$$

$\langle S_z \rangle$ denotes the average spin per magnetic ion site and is connected with
the magnetization M by:

$$M = -\,x\,N_o\,g\,\mu_B\,\langle S_z \rangle \qquad (19)$$

g is the g factor for the magnetic ions and N_o the number of cations per unit volume.

   With increasing magnetic field the absolute values of the g-factors (i) decrease in the valence band of $Pb_{1-x}Mn_xTe$ for both kinds of valleys. The relative change is larger for the $\Phi=90^o$ valleys than for the $\Phi=35.26^o$ valleys In $Pb_{1-x}Eu_xSe$ the valence band g factors increase. If x is large enough, at small magnetic fields the ordering of the spin levels is inverted; g goes through zero.
(ii) In the conduction bands the deviation from the case without exchange interaction is small for both materials. There is a small increase in the conduction band of $Pb_{1-x}Mn_xTe$ for both kinds of valleys, whereas in $Pb_{1-x}Eu_xSe$ the g-factor for the $\Phi=35^o$ valleys decreases and for $\Phi=90^o$ increases.

   The mean field approach yields an excellent overall agreement for all g factors involved. In order to demonstrate the effect of the exchange interaction on the magnetic field dependence of the g factors we note that nonparabolicity decreases the absolute values of the g factors with increasing field for electrons and holes for all types of valleys. In Figs.12a,b for comparison also theoretical values of the magnetic field dependence of the g factors are plotted without exchange interaction for $0^- \rightarrow 0^+$ transitions. With increasing field the electron g factors in $Pb_{1-x}Mn_xTe$ approach the values of the host material from <u>below</u> whereas the

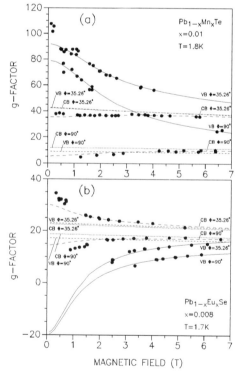

Fig.12: ● Magnetic field dependence of g factors derived from CARS; full lines: calculated for holes, dashed lines for electrons; dotted lines: calculated without exchange interaction. a) $Pb_{1-x}Mn_xTe$, x=0.01, T=1.8K. b) $Pb_{1-x}Eu_xSe$, x=0.008.

actual hole g-factors approach the corresponding g factors without taking into account the exchange interaction from <u>above</u>. In $Pb_{1-x}Eu_xSe$ it is vice versa (apart from $\Phi=90^o$, CB).

5.3 Coherent Raman Spectroscopy of PbTe/$Pb_{1-x}Mn_xTe$ Superlattices

$PbTe/Pb_{1-x}Mn_xTe$ is a quantum well (QW) structure with the non magnetic PbTe being the well and the semimagnetic component $Pb_{1-x}Mn_xTe$ being the barrier. In such materials a study of the exchange interaction between the free carriers, confined in the non magnetic components and the paramagnetic ions within the barriers yields information on the penetration of the wave function into the barrier. From this knowledge a determination of the band offset should be possible. Photoluminescence experiments show for the $PbTe/Pb_{1-x}Mn_xTe$ superlattices an efficiency comparable or even higher than in bulk PbTe and PbMnTe, respectively. This establishes the present structures to be type I superlattices.[39]

In Fig.13 experimental recordings of the CARS intensity versus magnetic field are compared for bulk PbTe and for a n-type superlattice with 50 periods of 4 nm $Pb_{1-x}Mn_xTe$ (x=0.021) and 7 nm PbTe. As expected from the theory the effective g- factors of the superlattice are reduced with respect to bulk PbTe. The resonant magnetic fields for the 35° valley (≈1.2T) as well as for $90^o(\approx3.5$T) are shifted to higher fields in the superlattice.

By the method outlined in Sec.2.32 the Landau level energies as a function of a magnetic field parallel to the layers have been calculated for varying band offsets. The resulting conduction band spin splitting ([111] valley, B=3.5T) is plotted in Fig.14 versus the band offset for the following parameters: 4 nm $Pb_{1-x}Mn_xTe$ (x=0.021, E_g=257meV), 7 nm PbTe. For such small layer thicknesses the magnetic wave function averages over several periods and thus there is almost no dependence on the center coordinate. The experimental data and the calculated dependence of the spin splitting versus offset indicates that almost the entire offset occurs in the conduction band.

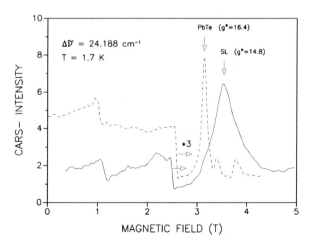

Fig.13. CARS- Intensity versus magnetic field. — — — PbTe
———— PbTe/$Pb_{1-x}Mn_xTe$ SL, 4nm $Pb_{1-x}Mn_xTe$ (x= 0.021), 7nm PbTe.

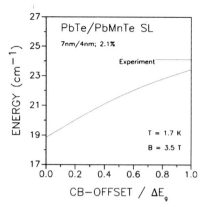

Fig.14. Conduction band spin splitting, valley oriented perpendicular to $\vec{B}$
versus band offset; SL, 4nm $Pb_{1-x}Mn_xTe$ (x=2.1%), 7nm PbTe, T=1.7K.

6.0 FAR INFRARED SPECTROSCOPY

Magnetooptical intraband transitions in Faraday geometry were investigated in a number of n- and p-type samples to get information on the
effective masses and their anisotropy by cyclotron resonance absorption.
Investigations on n- and p-$Pb_{1-x}Mn_xTe$ are presented in Refs.10,40,41,42.
For $\vec{B} \| [111]$ two resonant frequencies associated with cyclotron transitions
in the [111] (ω_{c1}, $\Phi=0°$) and in the three obliquely oriented <111>
($\Phi=70.53°$) valleys are observed. In samples with high mobility and consequently long relaxation times with FIR laser spectroscopy both $n^- \to (n+1)^-$
, $n^+ \to (n+1)^+$ transitions can be observed. The identification of the resonance frequencies is generally based on oscillator fits using a model dielectric function [22,43] since for laser energies $\hbar\omega_L$ below about 14 meV,
$\hbar\omega_{TO} < \hbar\omega_L < \hbar\omega_{LO}$ where ω_{TO} and ω_{LO} denote the TO and LO optic mode phonon
frequencies.

In Voigt geometry, $\vec{E} \| \vec{B}$, experiments were performed again on n- and
p-type samples with $\vec{B} \| [1\bar{1}0]$, i.e. $\Phi=35.26°$ and $\Phi=90°$. For the PbTe band-
structure just the oblique valley cyclotron resonance ($\Phi=35.26°$) is allowed whereas no cyclotron transition is possible for the two $\Phi=90°$ valleys. In Fig.15 FIR transmission experiments on n-PbTe and n-$Pb_{1-x}Mn_xTe$
are compared with each other. A strong second resonance is observed at
fields higher than the cyclotron resonance transition. As already shown in
Refs.10,40, the second resonance exhibits a strong temperature dependence
and vanishes at higher temperatures. Zawadzki [10,44] has calculated magnetooptical intraband selection rules based on the exchange contribution
(Eq.11), treating the nondiagonal exchange terms as a perturbation. The
nondiagonal terms mix the conduction spin states as well as the valence
spin states and enhance additional combined spin flip resonances as well
as pure spin flip resonances. In PbTe like materials without magnetic ions
such resonances are very weak since they result from far band interactions
only. The observed transitions for several FIR laser lines are summarized
in Fig.15 together with calculated data (based on the parameters of Ref.14
and Table I). From these data the additional transition seems to be a spin
flip resonance in the oblique ($\Phi=35.26°$) valleys.

From the transmission data alone, due to the dielectric anomalies associated with both the CR and SF transitions, no direct conclusion on the oscillator strengths of both transitions can be drawn. In addition the data were obtained on thin biaxially strained epitaxial samples, where the tensile strain at low temperatures enhances the SF transition as well.

Similar FIR results in Voigt geometry $\vec{E}\|\vec{B}\|[1\bar{1}0]$ for p-type $Pb_{1-x}Mn_xTe$ (x = 0.01) also exhibit two resonances but are qualitatively different. Again a cyclotron resonance of the holes of the oblique valley ($0^- \to 1^-$, $\Phi=35.26°$) is observed together with an additional resonance which is is likely to be a spin flip resonance of the $\Phi=90°$ valleys.[45]

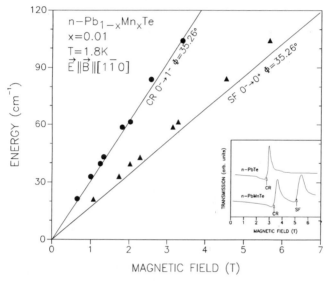

Fig.15. Experimental (●, ▲) and calculated resonance positions in Voigt geometry ($\vec{E}\|\vec{B}\|[1\bar{1}0]$)] for n-$Pb_{1-x}Mn_xTe$ together with identification: CR: cyclotron resonance, SF: spin resonance.
Inset: Far infrared transmission in Voigt geometry, $\lambda=118.8\mu m$. In PbTe just the oblique valley resonance ($\Phi=35.26°$) is observed, in $Pb_{1-x}Mn_xTe$ additional spin flip resonance occurs.

Cyclotron resonance experiments on bulk $Pb_{1-x}Mn_xTe$ using strip- line techniques were performed by Gorska et al.[41] Due to the rather moderate mobilities the resonances are broadened considerably and only limited informations can be extracted.

Further experimental investigations, especially on the temperature dependence of the observed resonances in Voigt geometry in p-type $Pb_{1-x}Mn_xTe$ as well as in $Pb_{1-x}Eu_xSe$ are necessary. In addition calculations on the oscillator strengths of the transitions involving spin flips are required.

7.0 EXCHANGE PARAMETERS

Neither the band ordering nor the energy separation for the two higher conduction and two lower valence levels at the L-point of the Brillouin zone changes considerably when paramagnetic ions (Eu, Mn) are introduced into the host lattice, for small x-values (x<0.02). The two band and the far band parameters for $Pb_{1-x}Mn_xTe$ and $Pb_{1-x}Eu_xSe$ were taken from fits to the magnetooptical interband transitions (Refs.37,36,4).

The exchange induced coefficients A, a_1, B, b_1 and the parameter T_o in the modified Brillouin function (see Eq.11) are obtained from least square fits to the experimentally observed $0^- \to 0^+$ transitions, i. e. from the CARS experiments which determine these spin splittings most accurately. Since in these experiments for $\vec{B} \| [1\bar{1}0]$ two kinds of valleys, (i.e. $\Phi=35.26°$ and $\Phi=90°$) yield two g-factors both for the conduction as well as for the valence band all band and exchange parameters can be determined unambiguously. S_o (Eq.11) is taken from fits to the experimental magnetization data for $Pb_{1-x}Mn_xTe$ which yield also values for T_o. The latter are in agreement with those obtained from the CARS data. For $Pb_{1-x}Eu_xSe$ no magnetization data are available so far. We assumed S_o to be 7/2, T_o derived from the CARS data turned out to be OK.

Using Eqs.12,13 the actual exchange constants α, δ (cb) and $\beta_\|$, $\beta_\perp$ (vb) are derived and given in Table II as a function of composition and temperature. In order to calculate these values the spin orbit mixing parameters $\Theta^{\pm}$ as given by Bernick and Kleinman are used.[15] The large absolute value of δ for $Pb_{1-x}Mn_xTe$ is most probably an artifact caused by the uncertainty of $\sin\Theta^+$ (Ref.15) and a small change of this parameter drastically affects the value of δ. Unfortunately, even the general uncertainty in the spin orbit mixing parameters $\Theta^{\pm}$ is too large to derive reliable values for α, δ (vb) and $\beta_\|$, $\beta_\perp$ (cb). Nevertheless Zasavitskii et al. (Ref.46) have tried to obtain exchange integrals from luminescence data.

Table I. Exchange parameters

Mn-content	T(K)	A(meV)	a_1 (meV)	B(meV)	b_1 (meV)
x=0.010	1.8	−182±15	−288±15	−33±10	27±5
x=0.008	1.8	−192±15	−315±15	−66±10	55±5
x=0.006	1.8	−225±15	−314±15	−-	50±5
x=0.006	3.5	−142±15	−279±15	−41±10	50±5
x=0.006	4.4	−124±15	−279±15	−51±10	51±5
x=0.006	12.0	−51±15	−288±15	−-	59±5
Eu-content					
x=0.045	1.7	102±10	96±10	21±5	24±5
x=0.008	1.7	78±10	65±10	19±5	8±5
x=0.008	6.0	77±10	59±10	25±5	3±5
x=0.008	12.0	70±10	56±10	31±5	7±5

Table II. Exchange integrals

Mn-content	T(K)	α(meV)	δ(meV)	$\beta_{\parallel}$(meV)	$\beta_{\perp}$(meV)
x=0.010	1.8	-305	-2460	109	80
x=0.006	1.8	-305	-1700	225	--
x=0.006	3.5	-305	-3290	225	120
x=0.006	4.4	-305	-4040	225	140
x=0.006	12.0	-305	-5900	225	140
Eu-content					
x=0.045	1.7	98	-144	73	4
x=0.008	1.7	61	-433	18	-19
x=0.008	6.0	61	-408	18	-28
x=0.008	12.0	61	-240	18	-37

For PbSe the spin orbit mixing parameters $\Theta^{\pm}$ as given by Bernick and Kleinman are more accurate than for PbTe since the g- factors and masses deduced in Ref.15 agree much better with experimental data for PbSe.[38] We conclude that thus in $Pb_{1-x}Eu_xSe$ also α, δ (vb) and $\beta_{\parallel}$, $\beta_{\perp}$ (cb) are more reliable than in $Pb_{1-x}Mn_xTe$.

The exchange induced effects in the valence band are larger than in the conduction band and therefore we restrict the following discussion to these, more precisely determined parameters. For $Pb_{1-x}Mn_xTe$ the parameter $|A|$ decreases with increasing temperature whereas for $Pb_{1-x}Eu_xSe$ it stays approximately constant in the same temperature interval from 1.7 to 12 K. In addition with increasing x in both materials, $|A|$ and $|a_1|$ decrease. The variation of exchange parameters is a hint for the necessity for further improvements in the theoretical description of semimagnetic IV-VI compounds in external fields. It demonstrates that the temperature dependence of the hole g-factors does not reflect the temperature dependence of the macroscopic magnetization. In contrast to the semimagnetic II-VI compounds,[8] the physical origin of the s,p,d- d or s,p,d- f exchange constants in the IV-VI compounds has not been analyzed by ab initio calculations. Since in the IV-VI's the minimum gap is at the L- point instead of Γ and furthermore already the band edge functions contain admixtures of d-like functions, the problem of the exchange interaction with the localized moments is more delicate than in the II-VI's.

8.0 LIGHT INDUCED MAGNETIZATION

8.1 Bulk $Pb_{1-x}Mn_xTe$

Light induced magnetization is a method by which spinpolarized carriers are created through interband excitation using circularly polarized light. The spinpolarized carriers interact via exchange interaction with the Mn-ions and induce a magnetic moment which is detected by sensitive Squid methods. It was first demonstrated by Krenn et al.[47] who found a magnetization signal in $Hg_{1-x}Mn_xTe$ which depended characteristically on the degree of polarization of the excitation radiation. Later on Awschalom et al.[48] extended this method to the observation of time resolved phenomena in the light induced magnetization of $Cd_{1-x}Mn_xTe$.

Pumping with circularly polarized radiation produces an initial occupation difference of the degenerate Kramers pair α,β. The amount of this occupation difference depends on the symmetry properties of the carrier wavefunctions involved. In PbTe for light propagation along [111] direction the four L-valleys are not equivalent and consequently the initial polarization P_i

$$P_i = \frac{n_\alpha - n_\beta}{n_\alpha + n_\beta} \tag{20}$$

is either 1 (for the valley oriented parallel to [111]) or~ 0.56 (for the three obliquely oriented valleys (Kaufmann et al.[49], P. Vogl,[50] unpublished).

In somewhat more general terms the initial occupation difference of the degenerate Kramers pair α,β is expressed by the density matrix

$$\rho^{in}_{\alpha\beta}(k) = \sum_i < \vec{k},\alpha|\pi|i><i|\pi^+|\vec{k}, \beta>\delta(E_k-E_i\pm\hbar\omega) \tag{21}$$

Due to scattering process there will be a stationary $\rho(\vec{k})$ which is determined by a rate equation. Thus the circularly polarized light produces a stationary "spin density" $(n_\alpha-n_\beta)$ oriented in the direction of the light propagation. In the presence of the exchange interaction the Mn spins experience a molecular field pointed along the z direction[51] which is e.g. for electrons in the [111] valley given by:

$$B^* = - \frac{n_\alpha-n_\beta}{2g\mu_B} (b_1- b_2) \tag{22}$$

(where $g = 2$ is the Lande factor of the Mn spins, $(b_1- b_2)$ see Eq. 12).

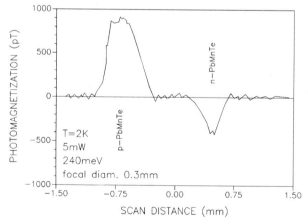

Fig. 16. Photomagnetization of p- and n- $Pb_{1-x}Mn_xTe$ mounted in the same pick up coil. Sign of signal shows the sign of the exchange interaction, the amplitude measures the magnitude.

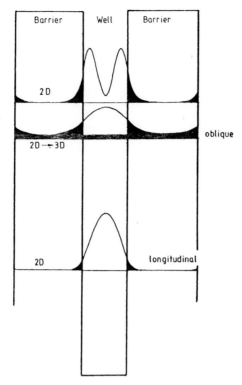

Fig.17. Wavefunctions of electrons in PbTe/Pb$_{1-x}$Mn$_x$Te SL structure
(5.8nm/28.8nm, 20 periods, x=0.027). Lowest and upper most level:
i=1,2 subbands of longitudinal valley; second level: i=1, oblique
forming a miniband due to wavefunctions overlap.

B* induces a magnetization M which is written in the form

$$M = \chi(T) \cdot B^* \tag{23}$$

where $\chi(T)$ describes the temperature dependent static magnetic susceptibility of the Mn-ions. Provided that n_α and n_β are temperature independent,
the T-dependence of $\chi(T)$ determines that of M(T).

From Eq.(19) it is immediately apparent that due to the difference of
the total exchange interaction for electrons and holes, in n and p-type
materials the sign of the light induced magnetization is different. Experimentally this is indeed the case as shown in Fig.16.One n-type and one
p-type Pb$_{1-x}$Mn$_x$Te sample are mounted adjacent to each other within the
pick up coil of the Squid and the circularly polarized laser beam is scan-
ned across their surface. The two samples were epitaxial films of
Pb$_{1-x}$Mn$_x$Te with x-values of 1.2% and 1.29 % respectively and electron and
hole concentrations of 2.4 · 10^{17} cm^{-3}). Not only the photo magnetization
signal changes sign but the absolute amount of the induced magnetization
is much higher for the holes than for the electrons (both samples are degenerate at T = 2 K and thus after illumination either the nonequilibrium
holes or nonequilibrium electrons determine the sign of the PM signal according to Eq.22. It is evident from this result that the method of the
light induced magnetization is capable to determine directly the sign of
the exchange interaction in an experiment where no externally applied magnetic field is necessary.

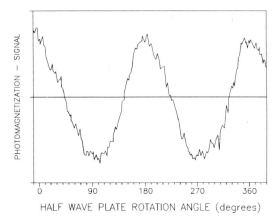

PHOTOMAGNETIZATION – SIGNAL

HALF WAVE PLATE ROTATION ANGLE (degrees)

Fig.18. Photomagnetization signal as a function of state of polarization
of exciting light for the sample of Fig.17.

8.2 Photomagnetization in Quantum Wells

This method has recently been extended [52] to a study of
$Pb_{1-x}Mn_xTe/PbTe$ quantum well structures,[53,54] where a periodic sequence of
diamagnetic wells (PbTe) is embedded in a sequence of paramagnetic
($Pb_{1-x}Mn_xTe$) barriers. If the excitation energy of the laser is lower than
the bandgap of $Pb_{1-x}Mn_xTe$, which forms the barrier, "spin" polarized carriers are created within the PbTe wells only. However due to the leakage
of the wave function of the carriers confined within the diamagnetic wells
into the paramagnetic barriers, the Mn ions can be polarized as a consequence of the contact spin-spin exchange interaction as explained in
Fig.17 and experimentally demonstrated in Fig.18. In this experiment on a
n- $Pb_{1-x}Mn_xTe/PbTe$ multi quantum well sample the Squid signal is shown as
a function of the rotation angle of a half wave plate which is inserted
between the circular polarizer and the sample. The means the polarization
is switched from right hand to linear to left hand and so on.

These experiments were carried out recently, for samples with different well and barrier width.[53] In Fig.19 experimental data are shown together with a photoluminescence spectrum on a n- type SL-structure. Whereas
luminescence probes the distance between the closest lying confined states
in the conduction and valence band, the PM spectra resemble rather an excitation experiment. The excitation of carriers will only occur for photon
energies sufficiently high to excite carriers from the valence band subbands into allowed states above the Fermi level within the conduction
band.

In order to explain the shift quantitatively, Kriechbaum (in Ref.53)
has performed envelope function calculations for the determination of the
SL band structure (also see 3.22). For the sample used in Fig.19, as shown
in Fig.20, for the i = 1 subband of the oblique valleys already miniband
formation occurs. This subband is below the Fermi energy for the given
carrier concentration. Therefore photoexcited spinpolarized carriers interact within this miniband effectively with the Mn-ions in the barriers,
Since these carriers are no longer localized within the PbTe wells. It was
found that for thicker $Pb_{1-x}Mn_xTe$ barriers where the subbands do not expe-

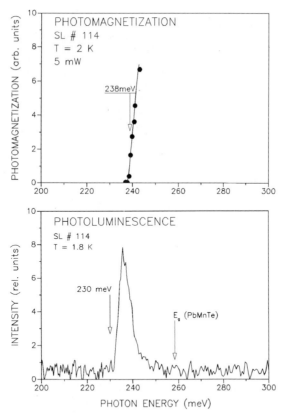

Fig. 19. Comparison of photomagnetization and photoluminescence of
SL # 114 (sample parameters see Fig.17). Note that the onset of
PM occurs at higher energies than of PL.

rience any broadening and the carriers are confined in separate quantum
wells the amplitude of the light induced magnetization signal is much
smaller.

The amount of wave function penetration associated with the longitudinal and oblique valleys and the two ladders of electric subbands are
shown in Fig.17. Since the mass along z ([111])-direction is smaller for
the oblique in comparison with the longitudinal valleys ($m_z = m_l$) a 2D behavior can be seen even for the higher lying i=2 longitudinal subband as
compared to the miniband associated with the oblique i = 1 level.

As with any interband excitation experiment it turns out to be difficult to locate the offsets.[54] However, from the fact that clearly at type
I band offset occurs, the sign of the PM signal and its magnitude as well
as its evolution with the width of the $Pb_{1-x}Mn_xTe$ layer indicates that
there is definitely no flat conduction band across the interface
$Pb_{1-x}Mn_xTe \rightarrow PbTe$. This is in agreement with the CARS results presented in
Sec.5.3.

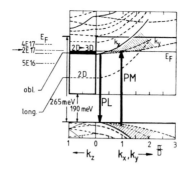

Fig.20. EFA bandstructure calculation of SL # 114 (parameters see Fig.17). Dispersion of levels is shown along growth (z-) direction (period D) and in the plane of the layers (k_x, k_y). Electrons occupy states up to the i=1 oblique valley subband. Bandoffset: ΔE_v=0. Because of much smaller density of states PL in longitudinal valley is not observed.

9.0   CONCLUSION

In the semimagnetic IV-VI compounds four exchange parameters determine the electron and hole - ion (s, p, d, - d (f)) interaction instead of two parameters as common in the II-VI's. This fact is due to the symmetry properties of the carrier wave functions at the L point of the Brillouin zone. The position of the Mn^{2+}- $3d^5$ and the Eu^{2+}- $4f^7$ levels with respect to the band structure of the IV-VI host crystal has not yet been established experimentally.

The magnetic properties are determined by an antiferromagnetic exchange in all of the Mn and Eu compounds mediated by a superexchange mechanism. The magnitude of the exchange parameters J/k varies from about 3 K ($Pb_{1-x}Mn_xS$) to about 0.24 K ($Pb_{1-x}Eu_xTe$) for materials with comparatively small magnetic ion content x<0.04. The exchange parameters are by about a factor 5 - 50 smaller than those found in narrow gap II-VI materials. In $Pb_{1-x-y}Sn_xMn_yTe$ with hole concentrations of the order of 10^{20} cm^{-3} carrier concentration dependent ferromagnetic ordering of the magnetic ions was observed which is caused by RKKY interaction. This is a property, peculiar to semimagnetic IV-VI compounds.

Precise experimental determinations of electron and hole g factors in $Pb_{1-x}Mn_xTe$ and $Pb_{1-x}Eu_xSe$ have established their magnetic field and temperature dependence. A mean field picture of the carrier- ion exchange yields the overall features apart from the fact that some of the exchange parameters turn out to be temperature dependent. The absolute value of the exchange induced corrections to the g factors is much larger for holes than for electrons in both $Pb_{1-x}Mn_xTe$ and $Pb_{1-x}Eu_xSe$. However, the sign of these corrections is different for holes in $Pb_{1-x}Mn_xTe$ in comparison to $Pb_{1-x}Eu_xSe$.

New developments are the study of semimagnetic IV-VI compound superlattices, both by light induced magnetization as well as by CARS spectroscopy. The new aspects are connected with wave function penetration of carriers confined in diamagnetic wells into the paramagnetic barriers and the related consequences for the contact exchange interaction on electronic properties.

ACKNOWLEDGEMENTS

　　Experimental results were obtained by P.Röthlein, K.Kaltenegger, H.Krenn, E.J.Fantner, S.Gerken, I.Roschger, G.Meyer and M.von Ortenberg. Samples were grown by L.Palmetshofer, H.Clemens, N.Frank and M.Tacke. For helpful discussions on the theoretical aspects we thank W.Zawadzki, M.Kriechbaum and E.Bangert. The work was supported by the Deutsche Forschungsgemeinschaft (Bonn) and Fonds zur Förderung der wissenschaftlichen Forschung (Vienna).

REFERENCES

1.　G. Bauer, Mat. Res. Soc. Proc. 89: 107 (1987).
　　(eds. R. L. Aggarwal, J. K. Furdyna and S. von Molnar).
2.　D. L. Partin, IEEE J.Quantum Electronics QE 24: 1716 (1988).
3.　P. Norton and M. Tacke, J. Crystal Growth 81: 405 (1987).
4.　J. R. Anderson and M. Gorska, Solid State Commun. 52: 601 (1984).
5.　G. Karczewski, M. von Ortenberg, Z. Wilamowski, W. Dobrowolski, and J. Niedwodniczanska-Zawadzka, Solid State Commun. 55: 249 (1985).
6.　G. Braunstein, G. Dresselhaus, J. Heremans, and D. L. Partin, Phys. Rev. B35: 1969 (1987).
7.　T. Story, R. R. Galazka, R. B. Frankel, P. A. Wolff, Phys. Rev. Lett. 56: 777 (1986)
8.　For a recent general review on diluted magnetic II-VI compounds see: J. K. Furdyna, J. Appl. Phys. 64: R29 (1988) and references cited therein.
9.　J. Niedwodniczanska-Zawadzka, J. Kossut, A. Sandauer, W. Dobrowolski, Lecture Notes in: Physics 133: 245 (1980). (Springer, Berlin, Heidelberg and New York).
10.　J. Niedwodniczanska-Zawadzka, J. G. Elsinger, L. Palmetshofer, A. Lopez-Otero, E. J. Fantner, G. Bauer, W. Zawadzki, Physica B+C 117 and 118B: 458 (1983).
11.　H. Pascher, E. J. Fantner, G. Bauer, W. Zawadzki, M. v.Ortenberg, Solid State Commun. 48: 461 (1983).
12.　G. Karczewski and L. Kowalczyk, Solid State Commun. 48: 653 (1983).
13.　G. Karczewski and M. von Ortenberg, in: "Proc. 17th Int. Conf. on the Physics of Semiconductors", San Francisco 1984, J.D. Chadi and W. A. Harrison, eds., Springer, New York (1985), p. 1435.
14.　H. Pascher, P. Röthlein, G. Bauer, L. Palmetshofer, Phys. Rev. B36: 9395, (1987).
15.　R. L. Bernick and L. Kleinman, Solid State Comm. 8: 569 (1970).
16.　G. Martinez, M. Schlüter and M. C. Cohen, Phys. Rev. B11: 651 (1975); B11: 660 (1975).
17.　A. Jedrzejcak, D. Guillot and G. Martinez, Phys. Rev. B17: 829 (1978).
18.　B. Rennex, R. Glasser, F. A. Frazier and J. Kinoshita, J. Phys.C 16: 3105 (1983).
19.　D. L. Mitchell and R. F. Wallis, Phys. Rev. 151: 581 (1966).
20.　R. Roseman, A. Katzir, P. Norton, K.-H. Bachem, H. M. Preier, IEEE J. Quant. El. QE-23: 94 (1987).
21.　M. S. Adler, C. R. Hewes and S. D. Senturia, Phys. Rev. B7: 186 (1973).
22.　G. Bauer in: "Narrow Gap Semiconductors, Physics and Applications", Vol.133 of Lecture Notes in Physics, W. Zawadzki, ed., Springer Berlin (1980).
23.　R. R. Galazka and J. Kossut, in: Ref. 22.

24. G. Bauer and H. Pascher _in:_ "Diluted Magnetic Semiconductors", M. Jain, ed. World Scientific, Singapore (1990).
25. M. Kriechbaum _in:_ "Springer Series in Solid State Sciences" 67: 120 (1986), G. Bauer, F. Kuchar, H. Heinrich, eds.
26. M. Kriechbaum, P. Kocevar, H. Pascher and G. Bauer, _IEEE J. Quant. El._ 24: 1727 (1988).
27. M. Gorska and J. R. Anderson, _Phys. Rev._ B38: 9120, (1988).
28. H. Pascher, P. Röthlein, G. Bauer, M. von Ortenberg, _Phys. Rev._ B40: 10469 (1989).
29. J. R. Anderson, G. Kido, Y. Nishina, M. Gorska, L. Kowalczyk, Z. Golacki, _Phys. Rev._ B41: 1014 (1990).
30. S. Bruno, J. P. Lascaray, M. Averous, J. M. Broto, J. C. Ousset, J. F. Dumas, _Phys. Rev._ B35: 2068 (1987).
31. H. J. M. Swagten, W. J. M. de Jonge, R. R. Galazka, P. Warmenbol, J. T. Devreese, _Phys. Rev._ B37: 9907 (1988).
32. T. Story, G. Karczewski, L Swierkowski, M. Gorska, R. R. Galazka, _Semicond. Sci. Technol._ 5 : S138 (1990).
33. W. J. M. de Jonge, H. J. M. Swagten. S. J. E. Eltink, N. M. J. Stoffels, _Semicond. Sci. Technol._ 5: S131 (1990).
34. J. Heremans and D. L. Partin, _Phys. Rev._ B37: 6311 (1988).
35. C. K. N. Patel, R. E. Slusher, P. A. Fleury, _Phys. Rev. Lett._ 17: 1011 (1966).
36. H. Pascher, G. Bauer, R. Grisar, _Phys. Rev._ B38: 3383 (1988)
37. H. Pascher, _Appl. Phys._ B34: 107 (1984).
38. H. Pascher, _Semicond. Sci. Technol._ 5: S141 (1990).
39. H. Pascher, P. Röthlein, M. Kriechbaum, N. Frank, G. Bauer, _Superlattices and Microstructures_, in print.
40. G. Bauer, _in:_ "Physics of Semiconducting Compounds" R. R. Galazka, ed. Ossolineum, Warszawa (1983) p. 62.
41. M. Gorska, T. Wojtowicz, W. Knap, _Solid State Commun._ 51: 115 (1984)
42. M. von Ortenberg, G. Bauer and G. Elsinger, _in:_ "Proc. Int. Conference Millimeter Waves", Marseille (1985),
43. H. Burkhard, G. Bauer and W. Zawadzki, _Phys. Rev._ B19: 5149 (1979).
44. W. Zawadzki, to be published.
45. H. Pascher, P. Röthlein, I. Roschger, G. Bauer, _in:_ "Proc. 19th Int. Conf. on the Physics of Semiconductros", W. Zawadzki, ed. Institute of Physics, Polish Academy of Sciences (1988), p.1535
46. I. I. Zasavitskii, L. Kowalczyk, B. N. Matsonashvili, A. V. Sazonov, _Fiz. Tekh. Poluprovodn._ 22: 2188 (1988) [_Sov. Phys.- Semicond._ 22: 1338 (1988)].
47. H. Krenn, W. Zawadzki, G. Bauer, _Phys. Rev. Lett._ 55: 1510, (1985).
48. D. D. Awschalom, J. Warnock and S. von Molnar, _Phys. Rev. Lett._ 58: 812 (1987).
49. B. Kaufmann, R. J. Nicholas, G. Bauer, P. Vogl, _Physica_ 117B & 118B: 505 (1983).
50. P. Vogl, unpublished.
51. H. Krenn, K. Kaltenegger, T. Dietl, J. Spalek, G. Bauer, _Phys. Rev._ B39: 10918 (1989).
52. D. D. Awschalom, J. Warnock, J. M. Hong, L. L. Chang, M. B. Ketchen, W. J. Gallagher, _Phys. Rev. Lett._ 62: 199 (1989).
53. H. Krenn, K. Kaltenegger, N. Frank, G. Bauer, H. Pascher, M. Kriechbaum, _in:_ "Proc. 20th Int. Conf. Phys. Semicond.", E. Anastassakis, ed., World Scientific, Singapore (1990).
54. H. Krenn, _in:_ "Springer Series in Solid State Sciences 93", F. Kuchar, H. Heinrich, G. Bauer, eds., Springer, Heidelberg (1990).

SEMIMAGNETIC SEMICONDUCTOR SUPERLATTICES

Michel Voos

Laboratoire de Physique de la Matière Condensée de l'Ecole
Normale Supérieure
24 rue Lhomond, 75231 Paris Cedex 05, France

ABSTRACT

At first, we describe in details the situation encountered in type III superlattices (SL) like HgZnTe-CdTe because this will be helpful to understand the behavior of HgMnTe-CdTe SL's. Far infrared magneto-absorption experiments done in these semimagnetic semiconductor superlattices (SMSL) are then presented, and the data are analyzed from band structure calculations which include magnetic field effects. Besides, the temperature dependence of the electron cyclotron resonance evidences the exchange interaction between the localized Mn d-electrons and the conduction band electrons in the SMSL. In wide-gap CdTe-CdMnTe SMSL's, magneto-pholuminescence and photoluminescence excitation spectroscopy experiments are described, and the results are compared to calculations of the level energies and exciton binding energies, including field-induced negative band offsets. From these studies, a type I to type II transition is evidenced at rather low magnetic fields, and the valence band offset is found to be of the order of 15 or 20% of the difference between the bandgaps of the host materials.

Only some aspects of SMSL's are presented here and, to conclude, we wish to emphasize that II-VI compound SL's involving Hg and/or Mn are quite interesting, at least from the point of view of basic physics, because they offer in principle a larger variety of investigations than well-known systems like GaAs-AlGaAs heterostructures for instance.

INTRODUCTION

There are essentially two kinds of semimagnetic superlattices (SMSL), depending on the characteristics of the involved materials. For CdTe-CdMnTe SMSL's for instance, these materials are wide-gap semiconductors, and they correspond to type I heterostructures like GaAs-AlGaAs superlattices (SL).

II-VI superlattices (SL) formed with CdTe and a zero-gap mercury based compound belong to another category, namely Type III heterostructures because of the unique inverted band structure of the zero-gap material. They can be either semiconducting or semimetallic[1], depending on the layer thicknesses and, as shown here, their band structure can be determined from far-infrared magneto-absorption experiments.

At first, we present the situation in HgZnTe-CdTe type III SL's because this will facilitate the understanding of HgMnTe-CdTe SMSL's. For that purpose, we report cyclotron resonance measurements on high electron mobility HgZnTe-CdTe SL's grown by molecular beam epitaxy in the (100) direction. The SL electron cyclotron resonance is measured as a function of θ, which is the angle between the direction of the magnetic field

Semimagnetic Semiconductors and Diluted Magnetic Semiconductors
Edited by M. Averous and M. Balkanski, Plenum Press, New York, 1991

237

and the SL axis. Comparison of the results with band structure calculations reveals the semimetallic nature of these heterostructures at low temperature and support a large valence band offset value ($\geq$ 300 meV) between $Hg_{1-x}Zn_xTe$ (x < 0.1) and CdTe. Experiments performed in a rather wide temperature range (2 - 200 K) demonstrate the existence of a semimetal-semiconductor transition induced by temperature in agreement with band structure calculations.

Then we describe far-infrared magneto-absorption experiments performed on a $Hg_{0.96}Mn_{0.04}Te$-CdTe SL for temperatures ranging from 1.5 to 10 K. The temperature dependence of the electron cyclotron resonance evidences the exchange interaction between the localized Mn d-electrons and the conduction band electrons in this semimagnetic SL. The results are interpreted from band structure calculations which include magnetization effects.

Finally, in the case of wide gap CdTe-CdMnTe SMSL's, it is shown from magneto-photoluminescence and photoluminescence excitation spectroscopy that a type I to type II transition occurs as a function of the magnetic field. From these investigations, it results that the valence band offset at each interface is necessarily small.

BAND STRUCTURE OF HgZnTe-CdTe SUPERLATTICES

We present here results obtained on two SL's (S_1 and S_2) grown on a (100) GaAs substrate with a 2 μm CdTe buffer[2] and consisting of 100 periods of $Hg_{1-x}Zn_xTe$-CdTe, the CdTe layers containing 15% HgTe[3]. For both samples, the layer thicknesses and the Zn composition x, as well as the electron mobility μ and concentration at 25 K are listed in Table I. The samples were prepared intentionally with thin CdTe barriers to study the carrier transport along the SL axis and were n type in the temperature range investigated (2 - 300 K), with Hall mobilities larger than 2 x 10^5 cm^2/Vs at low temperature.

We discuss first the SL band structure and Landau levels. All the calculations are performed using a 6 x 6 envelope function hamiltonian[4], taking for CdTe the band parameters given in reference 4. For HgZnTe, we obtained the Γ_6-Γ_8 energy separation using x = 0.12 for the semimetal-semiconductor transition at 2 K[5], the other parameters being identical to those of HgTe. Figure 1 gives the calculated band structure of S_1 in the plane of the layers (k_x) and along the SL growth axis (k_z) with T = 2 K. Reported experimental and theoretical values of the valence band offset Λ between CdTe and HgTe lie between 0 and 500 meV[4,6,7] and, in the case of sample S_1, we give the calculated band structure for Λ = 40 meV(a) , Λ = 300 meV (b) and Λ = 450 meV (c). The zero of energy is taken at the CdTe valence-band edge. E_1, LH_1 (HH_1, HH_2), respectively, denote the first light particles (heavy hole) bands for the motion along the growth axis z. For Λ = 40 meV (fig.1a), the SL is a semiconductor, E_1 being the conduction band and HH_1 the highest valence band. For large valence band offset, the E_1-HH_1 separation decreases until finally E_1 and HH_1 meet and then cross in the k_z direction for $k_z = k_c$. For Λ = 300 meV (fig.1b), HH_1 becomes the conduction band and the superlattice is semimetallic with a nearly zero gap and k_c = 0.1 π/d. For Λ = 450 meV (fig.1c), E_1 is at about 16 meV below HH_1 and k_c = 0.3 π/d. The most important change in the band structure when increasing Λ is the k_z dispersion of the SL conduction band. For Λ = 40 meV, this band is nearly isotropic around $\mathbf{k}$ = 0. However, for Λ = 300 meV and 450 meV, it is dispersionless for 0 < k_z < k_c corresponding to a strong mass anisotropy.

Table I. Parameters of the HgZnTe-CdTe superlattices used here. d_1 and d_2 are the HgZnTe and CdTe layer thicknesses, respectively, x is the Zn content; n and μ are measured at 25 K.

Sample	d_1(Å)	d_2(Å)	x	μ(cm^2/Vs)	n(cm^{-3})
S_1	105	20	0.053	2.1x10^5	3.7x10^{15}
S_2	80	12	0.053	2.1x10^5	2.6x10^{15}

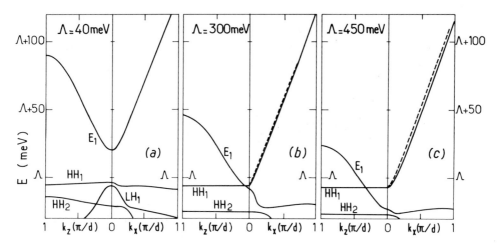

Fig. 1. Calculated band structure of S_1 for T = 2 K in the plane of the layers (k_x) and along the growth axis (k_z) for Λ = 40 meV (a), Λ = 300 meV (b) and Λ = 450 meV (c). The dashed lines in figures 1b and 1c give the in-plane dispersion relations for $k_z = k_c$.

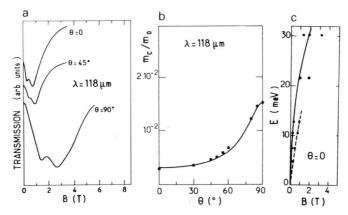

Fig. 2. a) Typical transmission spectra of S_1 for λ = 118 µm and T = 2 K.
b) Cyclotron mass versus θ for λ = 118 µm and T = 2 K. The dots correspond to the experimental points and the solid line to the calculated values in the ellipsoidal approximation described in the text.
c) Energy of the observed transmission minima versus B (dots). The solid and dashed lines give the calculated SL cyclotron resonances.

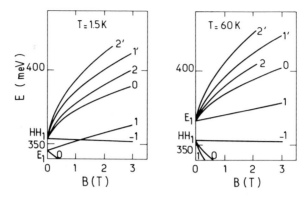

Fig. 3. Calculated Landau level energies in S_1 for $\Lambda = 360$ meV and $\theta = 0$ in the semimetallic (left panel) and semiconducting situations (right panel).

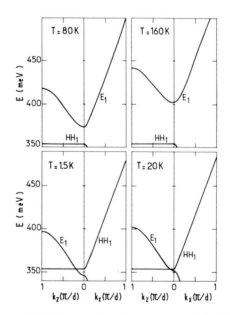

Fig. 4. Calculated band structure of S_1 for different temperatures and $\Lambda = 360$ meV.

In any case the in-plane mass m_x is extremely small near $k_x = 0$ ($m_x \approx 2.10^{-3} m_0$), as a result of the small HH_1-E_1 separation. The dashed line in figure 1b and 1c gives the in-plane dispersion relation calculated for $k_z = k_c$. One can see that m_x depends on the growth-direction wavevector because $m_x \propto |HH_1$-$E_1|$ which depends on k_z. Figure 1 shows that the measurement of the conduction band anisotropy[8] can yield the nature (semiconductor or semimetallic) of the SL. Similar investigations done for S_2, demonstrate that S_2 undergoes a semimetal-semiconductor transition for $\Lambda \sim 250$ meV at 2 K.

We have done[9] magnetoabsorption experiments at 2 K, using a molecular gas laser ($\lambda = 41$ - 255 µm). The transmission signal was measured at fixed infrared photon energies E, while B could be varied up to 12 T. Figure 2a shows typical transmission spectra of S_1 at $\lambda = 118$ µm for $\theta = 0$ (Faraday geometry), 45° and 90° (Voigt geometry). For $\theta = 0$, two well-developed resonance lines appear at low magnetic field (< 1 T). These two lines are shifted towards higher magnetic fields when θ is increased. The energy positions of the transmission minima as a function of B are given in figure 2c. At $\theta = 0$, the two observed transitions extrapolate to $E \approx 0$ at $B = 0$ and are therefore attributed to intra-conduction band transitions, considering the n-type nature of the sample. To interpret these data, we have calculated [4] the SL Landau levels energy at $\theta = 0$ which are presented in figure 3 for $\Lambda = 360$ meV. The solid line in figure 2c is the calculated energy of the - 1 $\rightarrow$ 0 transition. The agreement with the experimental data is quite satisfying, so that the low magnetic field transition can be reasonably attributed to the SL electronic cyclotron resonance. The dashed line in figure 2c corresponds to the calculated energy of the 0 $\rightarrow$ 1' transition, which is also allowed in the Faraday geometry. The fact that both - 1 $\rightarrow$ 0 and 0 $\rightarrow$ 1' transitions are observed at the same magnetic field arises from the existence of an electron accumulation layer near the interface between the SL and the CdTe buffer layer. Such an electron accumulation results from charge transfer from CdTe towards the SL and is usually observed in type III heterostructures grown in the (100) direction in similar conditions[9], for instance in HgCdTe-CdTe heterojunctions[10].

In the Voigt geometry ($\theta = 90°$), the cyclotron resonance line occurs near 1 T for $\lambda = 118$ µm corresponding to a cyclotron orbit radius of 250 Å, so that electrons should tunnel through several interfaces. In figure 2b, we have shown the measured cyclotron mass m_c versus θ for $\lambda = 118$ µm. We assume an ellipsoidal dispersion relation for the conduction band near k = 0 so that the cyclotron mass is given[8] at low magnetic field by $m_c^2 = (m_x^2 m_z)/(m_z \cos^2\theta + m_x \sin^2\theta)$ where m_x is the in-plane mass at the photon energy and m_z the mass along the growth axis. The solid line in figure 2b, which is the calculated cyclotron mass using $m_z = 0.065\ m_0$, is in very good agreement with the experimental data. The measured mass-anisotropy ratio m_z/m_x is ~ 25 for $\lambda = 118$ µm. Comparison to the band structure calculations shown in figure 1 clearly supports the semimetallic nature of the superlattice. Indeed calculations presented in the semiconductor configuration for $\Lambda = 40$ meV (fig.1a) lead to an anisotropy ratio of ≈ 1 at $\lambda = 118$ µm, while the anisotropy is found to be ≥ 10 for $\Lambda \geq 300$ meV. Similar experiments done in sample S_2 give a mass-anisotropy ratio ~ 6 at $\lambda = 118$ µm, again consistent with the calculated band structure for $\Lambda \geq 300$ meV. It is noteworthy that we have not observed any plasma shift of the cyclotron resonance for $\theta \neq 0$ in the photon energy range used in our experiments ($\hbar\omega \geq 5$ meV) [9].

Thus, these results show clearly that S_1 and S_2 are semimetallic at 2 K and that the electron mass anisotropy is large, so that the valence band offset is necessarily large, in agreement with previous XPS experiments[7].

We investigate now the temperature dependence of the SL band structure and cyclotron resonance. Figure 4 gives the calculated band structure of S_1 for different temperatures between 1.5 and 160 K using $\Lambda = 360$ meV and a linear temperature variation of 0.57 meV K^{-1} for the Γ_8-Γ_6 energy separation in $Hg_{1-x}Zn_xTe$ with low Zn content[11]. S_1 is semimetallic up to T = 25 K and becomes semiconductor for T $\geq$ 25 K, the SL band gap (E_1 - HH_1) increasing with temperature. From figure 4 it is clear that, when T is raised, a decrease of m_z, the mass along the SL axis, occurs near 25 K and that the in-plane mass m_x increases for T > 25 K, as a result of the increasing E_1 - HH_1 separation.

We have done magneto optical experiments between 1.5 K and 200 K for $\theta = 0$ and $\theta = 90°$, and typical transmission spectra obtained for $\lambda = 118$ µm in S_1 are given in figure

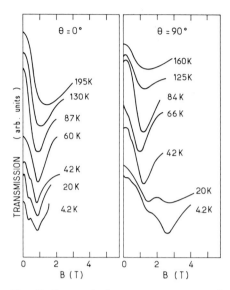

Fig. 5. Transmission spectra measured
in S_1 at $\lambda = 118$ μm for different
temperatures.

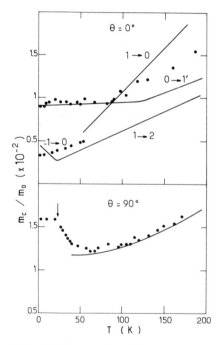

Fig. 6. Measured cyclotron masses
(dots) in the Faraday (upper
panel) and the Voigt (lower
panel) geometry. The solid
lines correspond to the
calculations discussed in the
text.

5. For $\theta = 0$, two resonance lines appear at low temperature and shift towards higher magnetic fields with increasing temperature, the lower field minimum vanishing progressively. At $\theta = 90°$, only the lower field transition observed at low temperature has an energy which extrapolates to $E = 0$ at $B = 0$ and is therefore attributed to the SL electron cyclotron resonance. The transmission minimum shifts at first to low fields when T is raised up to 60 K, and then to higher fields when the temperature is further increased. The cyclotron masses corresponding to the different transmission minima observed for $\theta = 0$ and $\theta = 90°$ at $\lambda = 118$ μm are plotted in figure 6 as a function of temperature. A drop of ~ 25% can be seen around 25 K in the curve $m_c(T)$ at $\theta = 90°$ shown in figure 6 (lower panel), in agreement with the calculated semimetal- semiconductor transition for $\Lambda \sim 350$ meV which is indicated by the arrow. The solid line in figure 6 (upper panel) corresponds to the calculated cyclotron masses for the main intraband transitions using the Landau level calculations shown in figure 3. At low temperature, the two resonances observed for $\theta = 0$ are attributed to the $-1 \rightarrow 0$ and $0 \rightarrow 1'$ transitions which are two ground interband transitions in the semimetallic configuration (fig. 3). For higher temperatures, the two minima correspond to the $1 \rightarrow 2$ and $0 \rightarrow 1'$ transitions, as expected in the semiconducting configuration (fig. 3). For T ≥ 100 K, the thermal energy kT becomes comparable to the photon energy and one observes only a single broad minimum which corresponds to the mixing of the different intraband transitions. The agreement between the calculated masses and experiments obtained for $\Lambda = 360$ meV is quite satisfying. To interpret the data at $\theta = 90°$, we assume an ellipsoidal dispersion relation for the conduction band near $k = 0$, so that the cyclotron mass is thus given by $m_c = (m_x m_z)^{1/2}$. The solid line in figure 6 (lower panel) gives the theoretical variation $m_c(T)$ in the semiconducting situation. In these calculations, the mass m_z at the photon energy in the conduction band is obtained for each temperature from the band structure calculations presented in figure 4, while the mass m_x is simply deduced from the cyclotron mass at $\theta = 0$, i.e. from the energy of the $0 \rightarrow 1'$ transition which corresponds to the dominant minimum observed in the spectra. Finally the drop around 25 K, as well as the monotonic increase of m_c observed for T > 50 K, are well interpreted from the band structure calculations with $\Lambda = 360$ meV. Analogous experiments were done on sample S_2 and a drop in the curve $m_c(T)$ at $\theta = 90°$ is observed around 40 K, which is also in agreement with band structure calculations which $\Lambda \sim 350$ meV.

It is concluded that our investigations provide an accurate determination of the SL conduction band dispersion and show that the observed temperature dependence of the effective masses m_x and m_z is due to the temperature-induced semimetal-semiconductor transition which is typical of type III heterostructures.

HgMnTe-CdTe SEMIMAGNETIC SUPERLATTICES : EXCHANGE INTERACTION

Many studies have been directed toward semimagnetic SL's (SMSL) in search of low-dimensionality magnetic effects[12-15]. The $Hg_{1-x}Mn_xTe$-CdTe type III system is unique in the sense that the two-dimensional electron confinement occurs in the semimagnetic layers, and this system should be particularly interesting to probe the exchange interaction between the localized Mn magnetic moments and the SL conduction band electrons. In HgMnTe-CdTe SL's, a strong effect of the exchange interaction on the Landau level energy is expected which can be evidenced from the temperature dependence of the electron cyclotron resonance.

We describe here investigations done in a $Hg_{0.96}Mn_{0.04}Te$-CdTe SL grown by molecular beam epitaxy on a (100) GaAs substrate with a 2 μm CdTe (111) buffer layer[16]. The SL consists of 100 periods of 168 Å thick $Hg_{0.96}Mn_{0.04}Te$ layers interspaced by 22 Å thick CdTe barriers. It is n type with an electron density $n = 6 \times 10^{16}$ cm^{-3} and a mobility $\mu = 2.7 \times 10^4$ cm^2/Vs at 25 K.

The SMSL band structure and Landau level calculations are done using a 6 x 6 envelope function hamiltonian[4,17]. The Γ_6-Γ_8 energy separation is taken to be - 130 meV in $Hg_{0.96}Mn_{0.04}Te$ at low temperature. For the $Hg_{0.96}Mn_{0.04}Te$ layers, the theoretical model is modified to take into account the exchange interactions between localized d electrons bound to the Mn^{2+} ions and the Γ_6 and Γ_8 s- and p-band electrons. The s-d and p-d interactions are introduced in the molecular field approximation through two additional parameters, r and A, where $r = \alpha/\beta$ is the ratio between the Γ_6 and Γ_8 exchange integrals, α and β,

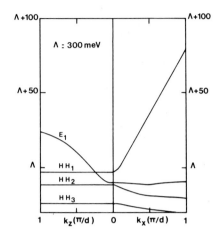

Fig. 7. Calculated band structure (at 2 K) along k_x and k_z for the $Hg_{0.96}Mn_{0.04}Te$-CdTe superlattice investigated here.

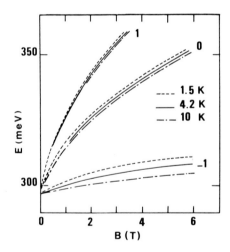

Fig. 8. Calculated conduction Landau levels for several temperatures and $\Lambda = 300$ meV.

respectively[18,19]. Here A is the normalized magnetization defined by $A = (1/6)\beta x N_0 <S_z>$, where N_0 is the number of unit cells per unit volume of the crystal, x is the Mn composition (x = 0.04), and $<S_z>$ is the thermal average of the spin operator along the direction of the applied magnetic field B. Extensive magneto-optical data[18,19] and direct magnetization measurements[20,21], obtained on bulk alloys with dilute Mn concentration, have shown that $<S_z>$ is well described by a modified spin - 5/2 Brillouin function, where the temperature T is replaced by an empirical effective temperature $(T + T_0)$. T_0 is found[21] to be ~ 5 K for x = 0.04. The exchange integral[19] $N_0\beta$ is known to be of the order of 1 eV and r ~ - 1. The magnitude of A which satisfies the modified Brillouin function for x = 0.04 is given in Table II for different magnetic fields and temperatures between 1.5 and 10 K. The resulting band structure along k_x and k_z, calculated for zero magnetic field using a valence band offset Λ = 300 meV, is shown in figure 7 and one can note that the SMSL is semimetallic. Figure 8 shows the SMSL conduction Landau levels calculated for T = 1.5, 4.2 and 10 K using the bulk $Hg_{0.96}Mn_{0.04}Te$ magnetization. It is noteworthy that as T decreases, the magnetization |A| is increased (see Table II) and the energy of all the conduction levels is increased.

Far infrared magneto-absorption experiments were done between 1.5 and 10 K. Figure 9 shows transmission spectra obtained at λ = 41 μm for different temperatures. The main absorption minimum corresponds to the - 1 $\rightarrow$ 0 transition (figure 8) which is the ground electron cyclotron resonance. For each temperature, the calculated field position of the line is indicated by the arrows in figure 9, and one can see that the agreement between experiments and Landau level calculations is quite good. The most striking point is that the - 1 $\rightarrow$ 0 transition shifts to lower B with increasing temperature, the variation being ΔB ~ - 0.6 T between 1.5 and 10 K. There are two contributions to this temperature-dependent shift : the temperature dependence of the $Hg_{0.96}Mn_{0.04}Te$ energy gap and the temperature dependent Mn magnetization. In bulk HgTe, as T increases from 1.5 K to 10 K, the magnitude of the energy gap decreases by less than 3 meV. The effect in $Hg_{0.96}Mn_{0.04}Te$ is expected to be similar. Such a small shift in energy gap is calculated to have only a slight effect on the cyclotron resonance transition, at most - 0.05 T. The observed shift of the - 1 $\rightarrow$ 0 transition is an order of magnitude larger than would imply the temperature dependence of the energy gap alone, and the Mn magnetization can account for the enhanced temperature dependence of the cyclotron resonance line. As T is increased, the Mn magnetization decreases dramatically at lower magnetic fields (see Table II). This has a direct effect on the Landau levels in the superlattice as shown in Figure 8. In particular, the n = - 1 level is depressed more than the n = 0 level, increasing the cyclotron resonance energy as temperature is increased. In bulk semimagnetic semiconductors, the exchange interaction has been observed from magneto-absorption primarily in the spin and combined (cyclotron plus spin) resonances. The observation of the exchange interaction in the cyclotron resonance data corresponds to the increased band mixing in a superlattice[17].

To conclude, we have observed cyclotron resonance in a $Hg_{0.56}Mn_{0.04}Te$-CdTe superlattice which is in good agreement with band structure calculations achieved in the envelope function approximation using a valence band offset Λ ~ 300 meV. The temperature dependence of the cyclotron resonance transitions evidences the effect of the exchange interaction in this semimagnetic semiconductor superlattice. The magnetization observed in the superlattice is consistent with the bulk $Hg_{0.96}Mn_{0.04}Te$ magnetization, as a result of the thick $Hg_{0.96}Mn_{0.04}Te$ layers and thin CdTe barriers in the investigated superlattice.

Table II . Magnetization A in meV/T at different magnetic fields and temperatures for bulk $Hg_{0.96}Mn_{0.04}Te$.

T(K) \ B(T)	0.5	3	5	7	10
1.5	- 2.3	- 1.3	- 1.0	- 0.8	- 0.6
4.2	- 1.4	- 1.1	- 0.9	- 0.7	- 0.6
10.0	- 0.5	- 0.5	- 0.45	- 0.45	- 0.4

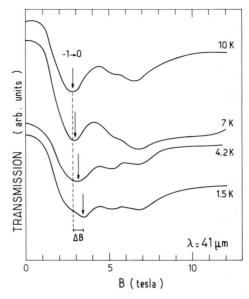

Fig. 9 : Far-infrared magnetotransmission spectra at $\lambda = 41 \ \mu m$ for temperatures ranging from 1.5 to 10 K. The arrows give the calculated cyclotron resonance positions.

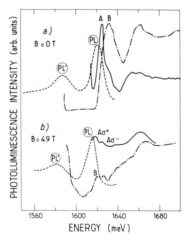

Fig. 10. Photoluminescence spectrum (dashed line), photoluminescence excitation spectra of the (PL) line (solid line) and of the (PL') line (dashed-dotted line) for the $CdTe/Cd_{0.93}Mn_{0.07}Te$ superlattice investigated here at (a) B = 0 T and (b) B = 4.9 Teslas.

CdTe-CdMnTe SMSL : TYPE I TO TYPE II TRANSITION

The recent growth by molecular beam epitaxy of semimagnetic semiconductor compounds has made possible the study of optical[22] or magnetical[23] properties of superlattices such as CdTe/CdMnTe. One of the key differences with the GaAs/GaAlAs system is the large magnetic field variation of the CdMnTe bandgap, and thus of the CdTe/CdMnTe conduction-band and valence-band offsets. These effects arise from the exchange interaction between the magnetic ions and the carriers[24,25]. Associated with the possible small value of the valence band offset Λ[22], a magnetic tuning of Λ might lead to a type I → type II transition at a moderate magnetic field in the case of low Mn concentration in the barrier. This transition has been first observed very recently in another system, ZnSe-ZnFeSe[26], and it is due to the very low value of Λ and the lattice mismatch.

We describe[27] here some evidences for a type I → type II magnetic field-induced transition for one spin component of the holes in a <111> grown 86 Å - 86 Å CdTe/Cd$_{0.93}$Mn$_{0.07}$Te 25 periods superlattice. This superlattice was grown on top of a <100> GaAs substrate after a 1500 Å thick CdTe buffer layer. This is supported by three observations : a well-marked bump in the field dependence of the fundamental σ^+ E$_1$ - HH$_1$ excitonic transition; a break in the variation of the photoluminescence efficiency; and a strong red shift of the onset of the photoluminescence excitation spectrum of a line associated with impurities located in the barrier. These three features take place around the same field value ≈ 2 T. We point out that the variation of the binding energy of the exciton in this low Λ (and even negative Λ) system is essential for the comparison with experiment.

Figure 10a shows the luminescence spectrum (dashed line) of the SMSL at zero magnetic field. Two lines, (PL) and (PL'), are observed which peak at ≈ 1622 and 1585 meV respectively. The photoluminescence excitation spectrum of the (PL) line (solid line in fig. 10a) shows a well marked A peak at ≈ 1624 meV flanked by a weak shoulder around 1635 meV. The onset of the excitation spectrum of the (PL') line (dashed-dotted line in fig. 10a) is located near 1635 meV (line B), while the rest of the spectrum is essentially the same as that of the (PL) line. Complementary reflectivity and piezomodulation experiments, whose details are not given here, have shown that the in-plane stress in the system for CdTe is ε_{CdTe} ≈ - 0,2%, which leads to a heavy-hole - light hole splitting of 12 meV for the CdTe buffer, the light-hole lying at higher energy. For an elastic accomodation of the CdMnTe barrier, we find a stress value ε_{CdMnTe} ≈ - 0,09% for the barrier[28]. A calculation of the energy sublevels in the framework of the three band envelope function approximation[29] has been performed using the following parameters : m_e = 0.096 m_0, m_h = 0.6 m_0, spin-orbit splitting Δ = 1 eV, Eg(Cd$_{1-x}$Mn$_x$Te) = 1610 + 1564 x (meV) and a variable valence band offset. This calculation, complemented by that of the exciton binding energy (see below), shows that the A line is associated with the excitonic HH$_1$ - E$_1$ 1s transition energy while the B line might be related to the excitonic E$_1$ - LH$_1$ transition. Note that comparison between theory and experiment does not provide a precise measurement of Λ, as usually in superlattice systems. The (PL) line corresponds to weakly bound excitons whereas the low energy (PL') line is likely to be associated with acceptor-like defects (as the sample is p-type) of the CdMnTe barrier. Besides, the onset of its excitation spectrum corresponds to the onset of rather delocalized light-hole states (the LH$_1$ miniband width is 7 meV for an offset Λ ~ 15 meV). This promotes a trapping of the photocreated light-holes by the defects of the barriers, in opposition with the heavy holes involved in the A line and corresponding to a wavefunction almost not penetrating in the barrier, at least at zero magnetic field and Λ large enough.

The sample was inserted in a superconducting magnet providing magnetic fields up to 5 Teslas and the temperature was 1.7 K. Laser ingoing and photoluminescence outgoing light beams were collected by a multi-fiber-optic guide. Fig. 10b shows the photoluminescence and photoluminescence excitation spectra obtained at B = 4.9 Teslas. The (PL) and (PL') lines exhibit a red shift whereas the X$_{E1-HH1}$ excitonic A line is divided into two A$_{\sigma^+}$ and A$_{\sigma^-}$ lines in the standard way[25]. The onset B of the excitation spectrum of the (PL') line is strongly red shifted and appears almost at the same energy as the A$_{\sigma^+}$ line, in striking contrast with the zero field situation. The experimental magnetic field dependence of the peak energies are plotted in fig. 11 and correspond to crosses (PL), and open circles (B). A plateau in the red shift is observed in the (PL) variation near 2 T. In the same field range, the B onset energy displays a rather fast decrease. Though it is not presented here for the sake of

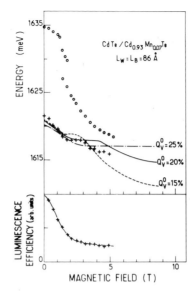

Fig. 11.Upper panel : experimental
variation of the PL (+) and B(o)
lines vs. magnetic field. The
other curves give the theoretical
dependence of the energy of the
PL line for different Q_v^0. Lower
panel : experimental variation
with B of the PL line efficiency.

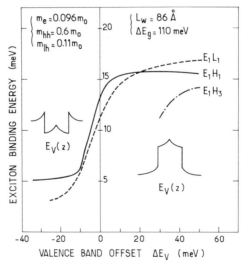

Fig. 12. Calculated exciton binding energies
as a function of the valence band
offset ($\Delta E_v = \Lambda$).

simplicity, the $A_{\sigma-}$ line follows the usual variation towards high energy with magnetic field and without any accident in the observed shift. Finally, the measurement of the relative radiative yield ρ of the (PL) line (fig. 11) indicates that ρ decreases by a factor of two up to 2-2.5 Teslas and remains almost constant at larger magnetic fields.

The results of the calculation of the exciton binding energy, which is necessary for a comparison between theory and experiments, are plotted in fig. 12 as a function of the unknown Λ offset. The details of the calculation, whose assumptions are close to those of Chang et al.[30], will be published elsewhere[31]. Briefly, a variational approach, in the diagonal approximation of the excitonic Hamiltonian[29], has been used under the assumption that, due to the large conduction band offset, the longitudinal electron motion is forced by the quantum well potential (E_1 state), while the longitudinal hole motion takes place in an effective one-dimensional potential which is the sum of the superlattice potential and of the coulombic potential averaged over the probability densities of electron and reduced in-plane motion. The potential seen by the hole is shown in fig. 12 both for a positive (right panel) and negative (left panel) Λ. Taking a 1S-like exponential wavefunction for the in-plane exciton motion, the energy difference between the transition energies with and without coulombic interaction provides the exciton binding energy E_x both for positive and negative offsets. A fast decrease of E_x from ~ 15 meV down to ~ 5 meV is found in an offset range of 10 meV around the vanishing offset.

The large exchange mechanisms in the CdMnTe layer modify the actual valence and conduction band offsets, Λ (B, m_J) and Λ_c (B, m_J), respectively, in the presence of magnetic field. We took the bulk parameters reported in reference 24 for the calculation of the conduction, light hole and heavy hole giant Zeeman splittings. For each value of B, that is say for each value of Λ (B, m_J) and Λ_c (B, m_J), the calculations of the band-to-band transitions and of the exciton binding energies have been performed. The diamagnetic shift of the exciton was neglected with respect to the giant Zeeman splitting. In our calculations, we took into account the stress induced by the lattice mismatch. The strained values of the zero-magnetic field offsets are deduced from the strain-free values by $\Lambda^{(HH)} = \Lambda^0 + 2.2$ meV and $\Lambda_c = \Lambda_c{}^0 - 2.6$ meV for heavy-holes and electrons, respectively[27].

The theoretical variation of the $X_{E1HH1}(\sigma^+)$ transition energies are plotted in fig. 11 for three values of Λ^0, i.e. 0.15 $\Delta E_G{}^0$ (dashed line), 0.2 $\Delta E_G{}^0$ (solid line) and 0.25 $\Delta E_G{}^0$ (dashed-dotted line) with $\Delta E_G{}^0 = 110$ meV. The presence of a bump or a plateau in the field variation of the fundamental transition is striking and strongly resembles that found experimentally. This bump occurs near the type I $\rightarrow$ type II transition and is related to the fast decrease of the exciton binding energy. This decrease is faster than the decrease of the band to band energy and therefore induces a small increase, or at least a plateau, in the variation of the transition energy versus magnetic field. The type I $\rightarrow$ type II transition occurs theoretically at B = 1.2, 1.9 and 2.5 Teslas for $\Lambda^0/\Delta E_G{}^0 = 0.15$, 0.2 and 0.25, respectively.

Two other experimental results corroborate this manifestation of a magnetic field-induced type I $\rightarrow$ type II transition. With decrasing Λ, the heavy-hole wavefunction penetrates more and more in the CdMnTe barrier. Thus the overlap with possible defects of the ternary alloy is larger, leading to an increase of possible non-radiative channel efficiency (and also a decrease of the radiative rate). When the offset becomes negative enough, the coulombic interaction keeps the hole near the interface, leading to a constant binding energy of the exciton (fig. 12) and to a constant interaction with the defects of the barrier. The variation of the radiative efficiency induced by the variation of the magnetic field becomes weaker, as observed experimentally. Finally, the strong red shift in the (B) onset of the excitation spectrum of the (PL') line is easily explained if one considers that this line is related to trapping of holes localized in the CdMnTe barrier. At low field, the lowest lying hole state delocalized in the barrier is the light hole state. We found theoretically a probability of presence in the barrier for the LH$_1$ (HH$_1$) in the range of 35% (10%) at B = 0. Above the type I $\rightarrow$ type II transition, the heavy hole state becomes localized in the barrier (50% at vanishing offset) and the (B) line corresponds to the $X_{E1HH1}(\sigma^+)$ line, leading to a decrease of the energy of the observed optical transition, in agreement with the data of figure 11.

To summarize, we have obtained strong evidences for a magnetic field-induced type I $\rightarrow$ type II transition in a semimagnetic CdTe/Cd$_{0.93}$Mn$_{0.07}$Te superlattice. This effect is due to the giant Zeeman splitting of the heavy hole CdMnTe band edge. The calculations demonstrate that the rapid variation of the exciton binding energy with the vicinity of this transition is an essential feature for its understanding. This transition occurs for a value of the magnetic field close to 2 Teslas, which is theoretically sensitive to the zero-field and strain-free valence band offset Λ^0. From our investigations, we get a value of the relative strain-free valence band offset $Q_v^0 = \Lambda^0/\Delta E_G^0$ of the order of 15-20%, a result which is somewhat larger than those reported up to now[22].

CONCLUSION

We have presented here only some features of semimagnetic semiconductor superlattices. It is believed that, at least from the point of view of basic physics, II-VI compound SL's involving Hg and/or Mn are quite interesting and attractive because they present properties which are not encountered in well-known systems like GaAs-AlGaAs for instance, so that they offer a larger variety of investigations.

ACKNOWLEDGEMENTS

The author would like to thank G. Bastard, J.M. Berroir, C. Delalande, E. Deleporte, Y. Guldner, J. Manassès and J.P. Vieren from the LPMC in Paris and B. Gil from the GES in Montpellier for their contribution to the work described here. He wishes also to thank L.L. Chang and J.M. Hong from IBM who provided the CdTe-CdMnTe samples, and J.P. Faurie from the University of Illinois at Chicago for the supply of the Hg-based superlattices used in the studies reported here.

REFERENCES

1. N.F. Johnson, P.M. Hui and H. Ehrenreich, Phys. Rev. Lett. 61:1993 (1988).
2. X. Chu, S. Sivananthan and J.P. Faurie, Superlattices and Microstructures, 4:175 (1988).
3. J. Reno, R. Sporken, Y.J. Kim, C. Hsu and J.P. Faurie, Appl. Phys. Lett. 51:1545 (1987)..
4. J.M. Berroir, Y. Guldner, J.P. Vieren, M. Voos and J.P. Faurie, Phys. Rev. B34 :891 (1986).
5. B. Toulouse, R. Granger, S. Rolland and R. Triboulet, J. Phys. 48:247 (1987).
6. Y. Guldner, G. Bastard, J.P. Vieren, M. Voos, J.P. Faurie and A. Million, Phys. Rev. Lett. 51:907 (1983).
7. S.P. Kowalczyk, J.T. Cheung, E.A. Kraut and R.W. Grant, Phys. Rev. Lett. 56:1605 (1986); Tran Minh Duc, C. Hsu and J.P. Faurie, ibid. 58:1127 (1987).
8. N.W. Ashcroft and N.D. Mermin, "Solid State Physics", Holt, Rinehart and Winston, New York (1976).
9. J.M. Berroir, Y. Guldner, J.P. Vieren, M. Voos, X. Chu and J.P. Faurie, Phys. Rev. Lett. 62:2024 (1989).
10. Y. Guldner, G.S. Boebinger, J.P. Vieren, M. Voos and J.P. Faurie, Phys. Rev. B36:2958 (1987).
11. M.H. Weiler, in : "Semiconductors and Semimetals", R.K. Willardson and A.C. Beer, eds., Academic, New York (1981), vol. 16, p. 119.
12. R.N. Bicknell, R.W. Yanka, N.C. Giles-Taylor, E.L. Buckland, and J.F. Schetzina, Appl. Phys. Lett. 45:92 (1984).
13. L.A. Kolodziejski, T.C. Bonsett, R.L. Gunshor, S. Datta, R.B. Bylsma, W.M. Becker and N. Otsuka, Appl. Phys. Lett. 45:440 (1984).
14. K.A. Harris, S. Hwang, Y. Lansari, J.W. Cook Jr. and J.F. Schetzina, Appl. Phys. Lett. 49:713 (1986).
15. D.D. Awschalom, J.M. Hong, L.L. Chang and G. Grinstein, Phys. Rev. Lett. 59:1733 (1987).
16. X. Chu, S. Sivananthan and J.P. Faurie, Appl. Phys. Lett. 50:597 (1987).
17. G.S. Boebinger, Y. Guldner, J.M. Berroir, M. Voos, J.P. Vieren and J.P. Faurie, Phys. Rev. B36:7930 (1987).

18. G. Bastard, C. Rigaux, Y. Guldner, J. Mycielski and A. Mycielski, J. Phys. (Paris) 39:87 (1978).
19. G. Bastard, C. Rigaux, Y. Guldner, A. Mycielski, J.K. Furdyna and D.P. Mullin, Phys. Rev. B24:1961 (1978).
20. W. Dobrowolski, M. von Ortenberg, A.M. Sandauer, R.R. Galazka, A. Mycielski and R. Pauthenet, in : "Physics of Narrow Gap Semiconductors", Vol.152 of Lecture Notes in Physics, E. Gornik, ed., Springer, Heidelberg (1982), p.302.
21. J.R. Anderson, M. Gorska, L.J. Azevedo and E.L. Venturini, Phys. Rev. B33:4706 (1986).
22. A.V. Nurmikko, R.L. Gunshor and L.A. Kolodziejski, IEEE J. Quant. Electro. QE 22:1785 (1986).
23. D.D. Awschalom, J. Warnock, J.M. Hong, L.L. Chang, M.B. Ketchen and W.J. Gallagher, Phys. Rev. Lett. 62:2309 (1989).
24. J.A. Gaj, R. Planel and G. Fishman, Solid State Commun. 29:435 (1979).
25. J. Wornock, A. Petrou, R.N. Bicknell, N.C. Giles-Taylor, O.K. Blanks and J.F. Schetznina, Phys. Rev. B 32:816 (1985).
26 X. Liu, A. Petrou, J. Warnock, B.T. Jonker, G.A. Prinz and J.J. Kretz, Phys. Rev. Lett. 63:2280 (1989).
27. E. Deleporte, J.M. Berroir, C. Delalande, G. Bastard, J.M. Hong and L.L. Chang, to be published.
28. For the values of the parameters, see "Semiconductors and semimetals", Vol. 25, Diluted Magnetic Semiconductors, J.K. Furdyna and J. Kossut, eds; Academic Press, New York (1988).
29. G. Bastard, "Wave mechanics applied to semiconductor heterostructures", Les Editions de Physique, Les Ulis (1988).
30. S.K. Chang, A.V. Nurmiko, J.W. Liu, L.A. Kolodziezski and R.L. Gunshor, Phys. Rev. B 37:1191 (1988).
31. E. Deleporte, B. Gil, J.M. Berroir, C. Delalande, J.M. Hong and L.L. Chang, to be published.

MAGNETISM OF Fe-BASED DILUTED MAGNETIC SEMICONDUCTORS

A.Twardowski

Institute   of   Experimental   Physics,   Warsaw
University, Hoza 69, 00-681 Warsaw, Poland

INTRODUCTION

General   properties   of   Fe-based   Diluted   Magnetic
Semiconductors (DMS)[1,2] have been discussed in the   preceding
Chapter.   In   this   Chapter   we will focus   on   the   magnetic
properties of these materials.

Substitutional Fe^{++} ions in II-VI compounds have d^6
configuration   resulting   in nonzero spin momentum   (S=2)   as
well as nonzero orbital momentum (L=2). In this case free ion
ground   term ^5D is split by tetrahedral crystal   field   into
orbital doublet ^5E and orbital triplet ^5T separated by energy
gap of the order of 4000K.   These terms are split furthermore
by spin-orbit interaction (which is effective since L≠0).   In
particular ^5E term,   which determines magnetic properties   of
the ion,   is split into a singlet A_1 , a triplet T_1 , a doublet
E,   a triplet T_2 and a singlet A_2 ,   all of them separated by
energy gaps of the order of 20K[3,4]. Since the ground state of
Fe^{++} ion   is   a   singlet A_1 ,   there is   no   permanent   magnetic
moment associated with Fe ion.   Magnetic moment of Fe^{++} ions
can   be only induced by external magnetic field,   leading   to
typical Van Vleck-type paramagnetism[2].

This   situation   is substantially   different   than   that
existing   for   Mn-based DMS. We recall that Mn^{++} (d^5) ions
possess   only spin momentum (S=5/2,   L=0).   Consequently   the
ground   state   is   magnetically active sextet $^6A_1$ yielding
permanent   magnetic   moment of Mn ions (so   called   Brillouin
type paramagnetism).

The   difference   between Mn- and Fe-based   DMS   is   well
manifested   in magnetic properties.   We notice in   particular
crucial difference between Mn and Fe ions in contribution   to
magnetic   specific   heat in the absence   of   magnetic   field:
excited states of Fe ion (T_1 ,   E, T_2 and A_2) are close enough
to   the   ground state (A_1) to be thermally populated   in   low
temperatures and then producing excess specific heat.   On the
other   hand in the case of Mn ion there are no close   excited
states (first one is at about 25000K above the ground   state)
which means that noninteracting Mn ions do not contribute   to
the specific heat.   Substantial differences in   magnetization
and magnetic susceptibility are also expected for Mn- and Fe-
type DMS.

Semimagnetic Semiconductors and Diluted Magnetic Semiconductors
Edited by M. Averous and M. Balkanski, Plenum Press, New York, 1991

253

A survey of experimental data concerning typical magnetic properties of Fe-based DMS is given in Sec.II. In this review our interest is confined to bulk materials, which were mostly studied so far. Low dimensional magnetism of Fe-DMS is not considered here. In Sec.III we discuss Hamiltonian describing Fe ion system, together with two approximative methods of partial solution of the problem: High Temperature Expansion series for susceptibility and Extended Nearest Neighbour Pair Approximation. From that point of view we consider in some detail specific heat and high field magnetization. Finally we speculate on the nature of interaction coupling Fe ions. We conclude by pointing open problems of magnetism of Fe-type DMS.

EXPERIMENTAL DATA

Available Fe-based DMS crystals are mostly II-VI selenides (HgFeSe, HgCdFeSe, CdFeSe, ZnFeSe) and sulfides (ZnFeS, CdFeS), which were synthesised as bulk materials with Fe concentration up to 26% . On the other hand II-VI tellurides grown so far contain much less iron (typically 1-3%).

In this Section we present representative data of magnetic specific heat (C_m), magnetization (M_n) and susceptibility (χ). Some related subjects as FIR spectroscopy of Fe ion energetical structure or s,p-d interaction are not discussed here (see preceding Chapter).

Specific heat

Magnetic contribution to the specific heat (C_m) is obtained as a difference of total heat capacity of the crystal (C_p) and heat capacity of nonmagnetic lattice (C_{latt}):

(1) $C_m = C_p (A_{1-x} Fe_x B) - C_{latt}$

Usually pure respective II-VI compound is used to obtain lattice specific heat $C_{latt} = C_p (A_{1-x} Fe_x B, x=0)$. This procedure is well justified in the case of DMS for which nonmagnetic cations (A) have similar mass and size to those of Fe ions (for instance ZnFeS and ZnFeSe). On the other hand substitution of heavy Hg (or Cd) ions by relatively light Fe ions changes vibrational lattice modes considerably and consequently, the lattice specific heat of the mixed crystal differs significantly from that of pure AB system. In this case $A_{1-x} Fe_x B$ "nonmagnetic" lattice must be simulated. Some simulation procedures were considered in Ref.25.

Typical experimental data of C_m are shown in Fig.1 for ZnFeS for several Fe concentrations. These data display general features of all the Fe-based DMS:
- at low temperatures (T<5K) C_m increases exponentially with increasing temperature, which is typical for Shottky-type anomaly (i.e. ground state separated by the energy gap from the excited states).
- no structure is observed at very low temperatures (0.3<T<1K) in contrast to Mn-type DMS [5,6].
- magnetic specific heat calculated per one Fe ion (C_m/x) decreases with increasing Fe concentration x, which indicates

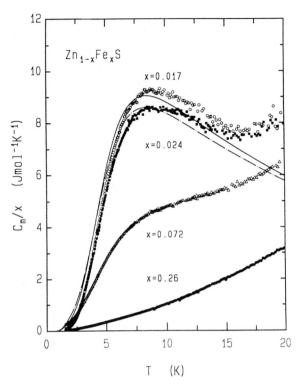

Fig.1 Magnetic specific heat of $Zn_{1-x}Fe_xS$ at B=0. The lines represent ENNPA calculations[5,6,7] for x=0.017 and 0.024 ($Dq=315.75cm^{-1}$, $\lambda=-94.8cm^{-1}$ and $J_{NN}-22K$).

interaction between Fe ions (otherwise C_m should scale with x).
- C_m is practically magnetic field independent[7,8,9] suggesting singlet ground state of Fe ions (multiplet ground state would be split in magnetic field resulting in series of closely lying levels which could be thermally populated, yielding large increase of C_m, as was the case of Mn-DMS [5,6,10]).

These observations are in agreement with energy pattern of Fe ion discussed in Introduction and could be expected for very diluted systems (x<0.01) where the major contribution originates from isolated (noninteracting) Fe ions.

Magnetization

Representative magnetization data are shown in Fig.2 for CdFeSe (see also Figs.8, 10 of the preceding Chapter). In general magnetization reveals typical Van Vleck-type behaviour:
- relatively strong variation with magnetic field. Saturation effects are observed only for rather strong fields (B>20T), in contrast with Mn-type DMS [11].
- temperature independent magnetization at low temperatures[12].

Magnetization calculated per one Fe ion (M/x) decreases with increasing x, indicating antiferromagnetic (AF) interaction between Fe ions (increasing number of Fe ions is "frozen" in clusters, which results in decreasing

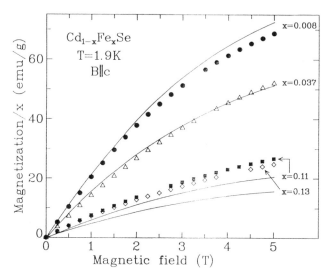

Fig.2 Magnetization of $Cd_{0.992}Fe_{0.008}Se$ at T=1.9K for B parallel ($\Theta=0^{\circ}$) and perpendicular ($\Theta=90^{\circ}$) to the crystal c axis. The solid lines show calculated (ENNPA) magnetization (after Ref.29)

magnetization).

Pronounced anisotropy of magnetization is observed for cubic crystals (CdFeTe[13]) as well as for hexagonal crystals (CdFeSe[14,15,16]). These effects are well predicted by single ion model, since energy level pattern depends strongly on orientation of magnetic field relatively crystal axis for strong magnetic fields[17]. For hexagonal crystals magnetization anisotropy occurs also at low magnetic fields due to field mixing of the ground state with different excited states for field parallel and perpendicular to the crystal hexagonal axis[15,16].

Susceptibility

In Fig.3 we show typical low field susceptibility data for ZnFeSe. In the high temperature range (100–300K) susceptibility of ZnFeSe, CdFeSe, HgCdFeSe and HgFeSe reveal typical Curie–Weiss behaviour with negative Curie–Weiss temperatures (Θ), again suggesting AF interaction between Fe ions (this suggestion is further supported by decreasing susceptibility per one Fe ion with increasing Fe concentration).

At lower temperatures (T<70K) pronounced downward bending of χ^{-1} is observed (Fig.3). This bending changes its slope as T–>0 (upward bending), contrary the situation found for Mn–type DMS[5,6]. Following the energetical structure of Fe ions described in Introduction, one could expect temperature independent susceptibility for T<4–5K (which is typical for Van Vleck–type paramagnetism). This is however not the case: all the data collected so far for II–VI selenides and ZnFeS reveal gradual increase of susceptibility with T–>0 [18,19,20,7]. This particular behaviour was ascribed to the presence of permanent magnetic moments in the studied

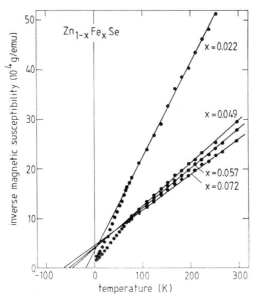

Fig.3 Inverse magnetic susceptibility of $Zn_{1-x}Fe_xSe$ as a function of temperature. Straight lines show Curie-Weiss law. After Ref.19.

crystals. Such a moments would contribute to the susceptibility in a conventional Curie-type way, masking true Van Vleck-type susceptibility of Fe^{++} ions. Permanent magnetic moments can originate from residual paramagnetic impurities of DMS crystals. In fact such impurities (mainly Mn^{++}) were detected by EPR in ZnFeSe[21] and ZnFeS[9]. In the case of HgFeSe, except of residual impurities one is dealing with Fe^{+++} ions (d^5 configuration, equivalent to Mn^{++} from magnetic point of view), existing in this material as a result of selfionization of Fe^{++} ions[1]. Although the number of paramagnetic ions in Fe-DMS is rather small ($10^{17}-10^{18}$ cm^{-3}) it can provide dominant contribution to susceptibility in very low temperatures[21,20]. We should stress however that the available data are not sufficient to derive pertinent conclusions concerning low temperature susceptibility. In particular one can not exclude, at the present stage, that non-Van Vleck susceptibility reflects complicated energetical structure of large Fe ion clusters (larger than isolated ions and pairs discussed in the next Section).

Spectacular feature of Mn-DMS is existence of spin-glass phase depending on Mn concentration and temperature range[5,6,10]. There are some experimental suggestions that similar phase transition occurs also for Fe-DMS. In particular for the crystals with high Fe content (x>0.1) maxima in a.c. susceptibility were observed at temperatures depending on Fe concentration[18,7,20]. Since no anomalies were detected in specific heat at the corresponding temperatures, these maxima were ascribed to spin-glass phase transition[18,7]. We should stress however, that this conjecture should be approved by more extensive study (for instance field-cooled and zero-field-cooled d.c. susceptibility data are necessary[22]. We point out that complex investigation of possible spinglass-paramagnetic phase transition requires crystals with high Fe concentrations. One can expect problems in studying crystals with x<0.1 due to unclear situation with paramagnetic impurities discussed above.

Concluding the review of experimental data we should stress similar magnetic behaviour of all Fe-type DMS studied so far irrespective of different host lattices (ZnFeS, ZnFeSe, CdFeSe, CdHgFeSe, CdFeTe and HgFeSe - see preceding Section). Magnetic properties of the diluted compounds are in general agreement with the energetical structure of Fe^{++} ions discussed in the Introduction. Interaction between Fe ions is evident and is considered in the next Section.

d-d INTERACTION BETWEEN Fe IONS

The full Hamiltonian describing the system of (interacting) Fe ions can be written in the following form:

(2) $H= \sum_i H_i + \sum\sum_{i\,j} H_{i\,j}$

where H_i is isolated ion Hamiltonian and $H_{i\,j}$ describes interaction between ions i and j (summation runs over all Fe ions present in the crystal, i≠j). It was discussed in the preceding Chapter that isolated Fe^{++} ion Hamiltonian can be approximated by:

(3) $H_i = (H_{cf} + H_{mo} + H_B)_i$

where: H_{cf} describes crystal field (cubic or hexagonal[23]),
         $H_{mo} = \lambda \mathbf{LS}$ is spin-orbit term,
         $H_B = \mu_B (\mathbf{L} + g\mathbf{S})\mathbf{B}$ is Zeeman term

and $\mathbf{L}=(L_x, L_y, L_z)$ and $\mathbf{S}=(S_x, S_y, S_z)$ are orbital momentum and spin momentum operators, respectively. We notice that Hamiltonian (3) could be augmented by some other terms (in particular term describing Jahn-Teller (JT) coupling with the crystal lattice), but the available experimental data do not provide clear evidence of the additional terms. In fact most of the data (especially FIR data) can be reasonably well described by simple Hamiltonian (3) (e.g. FIR spectra of Fe^{++} in ZnS or ZnSe). Only energy structure of ^5E term of CdTe:Fe can not be described by the energy spectrum resulting from (3)[4]. Additional absorption lines observed in CdTe:Fe were ascribed to JT coupling[24]. Another suggestion of JT effect arose from analysis of specific heat of CdFeSe[25]. However recent analysis suggests that anomaly large specific heat of CdFeSe is rather due to too simple simulation of "nonmagnetic" lattice specific heat C_{latt} (neglecting local phonon modes) than to JT coupling[26]. Precise FIR spectroscopy experiments are necessary to establish exact form of single ion Hamiltonian.

Hamiltonian describing interaction between Fe ions ($H_{i\,j}$) has to be revised since one is dealing with interaction between field-induced magnetic moments, for which both $L,S\neq0$. General form of Hamiltonian for such a situation is complicated and contains many unknown parameters[27]. However in the case of Fe-DMS magnetic properties are determined mainly by ^5E term (excited ^5T term is at about 4000K above ^5E term and can be hardly thermally populated even at room temperatures), which has orbital momentum largely quenched. Actually it was shown that orbital momentum contributes to total magnetic moment of Fe ion an order of magnitude less than spin momentum does[28,29]. Therefore it seems reasonable to start with isotropic Heisenberg-type exchange as the first order approximation of H_{ij}:

(4) $H_{ij} \approx -2J\ \mathbf{S_i S_j}$

where J is exchange constant. Another words this is the simplest approximation in which we assume that Heisenberg exchange is dominant and the other terms in Hamiltonian (for instance anisotropic Dzialoszynski-Morija term) can be neglected. Practical motivation for Hamiltonian (4) is that it requires only one adjustable parameter (J).

We notice that even such crude approximation yields Hamiltonian (2) which can not be solved. Therefore the "real" system of Fe ions was considered in two approximative ways.

First one is High Temperature Expansion Series (HTE) for magnetic susceptibility. This conventional procedure is based on assumption that for $T \to \infty$ magnetic moments at different lattice sites can be regarded as completely independent. It was shown for Fe-DMS[19] that HTE results in Curie-Weiss type susceptibility with Curie-Weiss temperature Θ given by:

$$(5)\quad \Theta(x) = \frac{\langle M_z^2\rangle \langle E\rangle - \langle M_z^2 E\rangle}{\langle M_z^2\rangle k_B} + \frac{2\langle M_z S_z\rangle^2}{\langle M_z^2\rangle} x \sum_p z_p (J_p/k_B)$$

where $\langle .. \rangle$ are quantum averages of appropriate operators calculated for isolated Fe ion wavefunctions at $T \to \infty$, J_p is exchange integral for p-th coordination sphere and z_p is coordination number. Using experimental values of Θ and calculated averages $\langle .. \rangle^1$ one can estimate nearest neighbour interaction parameter (neglecting long ranged interaction)[19]. This was done for several materials and data are collected in Tab.1 together with the data for corresponding Mn-DMS.

We notice that coupling between Fe ions is considerably stronger than between Mn ions. We return to this point below. It is worthwhile to stress that high temperature susceptibility is up to now the only way to estimate exchange constants.

Second approximative treatment of Fe-DMS crystal is so called Extended Nearest Neighbour Pair Approximation (ENNPA)[35,10]. We recall that in this model the whole magnetic

_____

[1] Originally[19] averages $\langle .. \rangle$ were calculated for 10 states of ^5E term (and neglecting ^5T term) i.e. assuming that experimental temperature is infinite from the point of view of ^5E term but is still neglegible concerning ^5E-^5T energy separation. This seems to be reasonable approximation for experimental temperature range (T<300K) but in fact results in at about 10% overestimation of exchange constant. Such an error is comparable with experimental accuracy of Θ.

system is factorized into pairs coupled by (principally) long range exchange interaction. The statistical weights of different pair configurations are given by a random distribution of magnetic ions. We stress that ENNPA is

Table 1
Nearest neighbour exchange integrals J_{NN} (K)

material	Y=Fe	Y=Mn
HgYSe	−18[25]	−11[30]
		−6 [31]
HgCdYSe	−18[25]	
CdYSe	−19[25]	−8.7[32]
		−7.9[33]
ZnYSe	−22[19]	−13[34]

limited to rather diluted systems (x<0.05) because clusters of ions larger than pairs are not considered. For more details of this model we refer to Ref.10. The basic element of ENNPA is solution of a pair problem. Since Fe-Fe pair is the simplest interacting ion system we consider this case in some detail.

Under assumptions discussed so far Fe-Fe pair Hamiltonian has the following form:

(6) $H = H_1 + H_2 - 2J\mathbf{S}_1 \mathbf{S}_2$

where H_1, H_2 are given by (3).

The basis pair wavefunctions $\Psi_{i,j}$ used for evaluating Hamiltonian matrix were constructed as products of single Fe ion eigenstates Φ_i [3,6]:

(8) $\Psi_{i,j} = \Phi_i(1) \; \Phi_j(2)$

where indices 1,2 indicate ions in the pair and i,j denote number of the ion eigenstates (i,j=1...10, since basis was limited to $^5E \otimes {}^5E$ subspace[3,6]). Obviously basis (8) diagonalizes H_1 and H_2. One notice that neglecting offdiagonal elements of Hamiltonian (7) the pair ground state is a singlet since $E_{ground} = E_{11} = E_1(1) + E_1(2) = 2E_1$ is given by only one state $\Psi_{11} = \Phi_1(1) \; \Phi_1(2)$. On the other hand the first excited state which has energy $E_{12} = E_1(1) + E_2(2) = E_2(1) + E_1(2) = E_1 + E_2$ is a sextet ($\Psi_{12} = \Phi_1(1) \; \Phi_2(2)$, and Φ_2 is triplet T_1). Therefore only exchange interaction can rearange pair levels and possibly brought excited multiplets below singlet Ψ_{11}. However detailed calculations show[3,6,7] that due to interaction between Ψ_{11} and excited states, Ψ_{11} is only downward shifted instead of being crossed by the excited states. Example results of numerical diagonalization of the full Hamiltonian matrix (7) are shown in Fig.4. We notice that Heisenberg-type interaction essentially does not change Fe energy structure, in the sense that the ground state is still a singlet one. However the first excited state approaches the ground state when J increases. Consequently

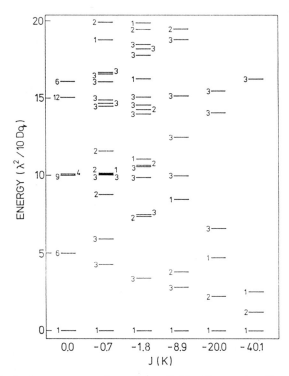

Fig.4 Low lying energy levels of Fe–Fe pair for example J
values (Dq=293cm⁻¹, λ=–85cm⁻¹). After Ref.39

Fe–Fe pair should behave qualitatively in a similar way as an
isolated ion does. This is exemplified in Fig.1 where results
of ENNPA calculations are shown.

The essential difference between an isolated Fe ion and
Fe–Fe pair is found in the presence of strong magnetic field.
In the case of the isolated ion the ground state is never
crossed by any of the excited states which results in gradual
increase of magnetization with increasing magnetic field. On
the other hand for Fe–Fe pair one finds crossing of the
ground state with the excited states at high magnetic fields
(Fig.5). Such a level crossing means rapid changement of the
level energy variation with magnetic field, yielding step-
like structure of magnetization (Fig.5). Similar step-like
structures in high field magnetization were observed for Mn–
DMS (see Chapter by Y.Shapira in this book). In contrast to
the situation for Mn–DMS, in the present case level crossings
(and steps in magnetization) occur only if the exchange
interaction is strong enough. Weak Fe–Fe interaction produces
no steps in magnetization (the treshold value of J is of the
order of 12–15K[37]).

We mention that the pair energy structure is strongly
anisotropic (even in cubic structure) as was the case for an
isolated Fe ion (Fig.6).

The first step in magnetization is predicted for
magnetic field B≈30–35T (Fig.6). However preliminary
experimental data[2] show no steps up to 40T. Experimental
confirmation of this fact (including extension of the
magnetic field range) would indicate existence of additional
interaction between the ground and the excited states which
prohibits level crossing. This would be strong suggestion for

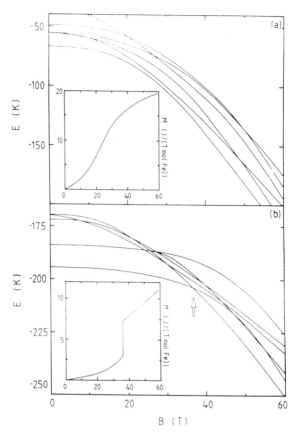

Fig.5 Low lying energy levels and magnetization at T=0 (see insets) of the Fe-Fe pair for B along the crystallographic (111) axis: a). J=-10K, b). J=-22K. The arrow indicates the position of the magnetization step. After Ref.37

development more extensive pair Hamiltonian.

One of the conclusions resulting from Fe-Fe pair calculations discussed above is relative irrelevance of the long range interaction between Fe ions. Assuming that interaction between Fe ions falls down by an order of magnitude between nearest neighbours (NN) and next nearest neighbours (NNN), as was the case for Mn ions[10], one finds that weakly coupled NNN Fe-Fe pair has energetical structure very similar to the structure of an isolated ion. Actually the difference between specific heat or magnetization of isolated Fe ion or NNN Fe-Fe pair is so tiny that NNN pairs can be reasonably approximated by the isolated ions[22]. Therefore in the spirit of ENNPA one can express any

_____

[22]We stress that this situation is essentially different in the case of Mn-DMS. The magnetic properties of an isolated Mn ion are determined by single energy level producing no excess specific heat. On the other hand Mn-Mn pair, even weakly coupled has series of levels which contribute to the specific heat (even in the absence of magnetic field).

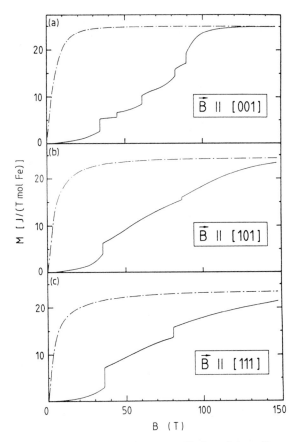

Fig.6 Calculated magnetization at T=0 of an Fe^{++} single ion (broken line) and an Fe-Fe pair (solid line, J=-22K) for B along three crystallographic axes: a). (001), b). (101), c). (111). After Ref.37.

thermodynamic quantity A(B,T,x) (per one ion) in the following way:

(9) $A(B,T,x) = A_s P_s + \frac{1}{2} A_p P_p$

where indices s and p denote single ion and NN pair, respectively. $P_{s,p}$ is probability of finding considered ion having no magnetic nearest neighbours (s) or being coupled in NN pair (p). For random distribution of magnetic ions we have for fcc or hcp lattices: $P_s = (1-x)^{12}$ and $P_p = 12x(1-x)^{18}$. Usually we assume that all the ions which are not coupled in NN pair are isolated ions and then $P_p = 1 - P_s$.

In particular this procedure can be applied to the specific heat yielding:

(10) $C_m(x) = C_m{}^s P_s + \frac{1}{2} C_m{}^p P_p$

Low temperature specific heat is mostly determined by the energy gap between the ground and the first excited states. This energy gap is of the order of 20K for isolated ion (as evaluated by FIR spectroscopy[4,38,39,40,41]) and 10K for NN pair (as calculated for $J_{NN} \approx -20K$, cf. Tab.1). Therefore one

263

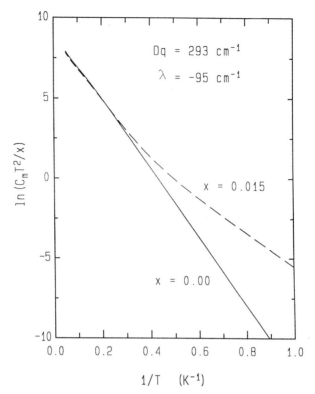

Fig.7 Calculated magnetic specific heat per mole of Fe ions (C_m/x) plotted as $\ln(C_m T^2/x)$ versus $1/T$ ($Dq=293cm^{-1}$, $\lambda=-95cm^{-1}$, $J_{NN}=-22K$). The line for $x=0.00$ corresponds to the case of noninteracting Fe ions. After Ref.25.

can expect that total C_m of the actual crystal is fixed by two different energy gaps (which contributions are weighted by P_m and P_p). These two energy gaps can be well exhibited by appropriate choice of coordinates used for displaying C_m.

We recall that for simple Shottky-type anomaly (i.e. singlet ground state separated by energy gap Δ from the excited multiplet, with g-folded degeneracy) one has:

(11) $\quad C_m(T) = g(\Delta/k_B T)^2 \exp(-\Delta/k_B T)$

Therefore plotting $\ln(C_m T^2/x)$ versus $1/T$ one should obtain a straight line with the slope $-\Delta/k_B$. Similar linear behaviour is found for C_m calculated on the base of the real (five-level) single Fe ion energy pattern[25] (Fig.7). Isolated ion energy gap (A_1-T_1) inserted in the model is recovered with accuracy better than 3%. Moreover including pairs into calculations (ENNPA) one finds another linear part of the plot at very low temperatures (Fig.7), which corresponds to the pair energy gap. Since this gap is much smaller than single ion gap, it dominates low temperature specific heat. At higher temperatures single Fe ions, with larger energy gap start to contribute to C_m. For low Fe concentrations $P_m \gg P_p$ and single ion contribution is dominant at $T>3-4K$.

The experimental data for various compounds indeed show linear part at temperatures $T<10K$ (Figs.8 and 9) yielding single ions energy gaps well comparable with the values

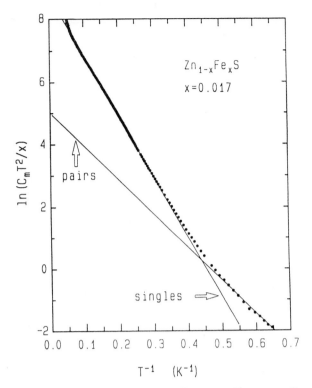

Fig.8 Magnetic specific heat of $Zn_{0.983}Fe_{0.017}S$ plotted as $\ln(C_mT^2/x)$ versus $1/T$ indicating the contribution of isolated ions and pairs.

Table 2
Energy gap (in K) for isolated ion and Fe—Fe pair

		ZnFeS	ZnFeSe	CdFeSe	HgFeSe
isolated Fe ion:	C_m	20	21	19	16
	FIR	21[38]	23[39]	19[42,43]	—
Fe—Fe pair:	C_m	<10	<13	<9	<10
	HTE		10.5	13	

obtained from FIR spectroscopy (Tab.2).
The pair contribution is well visible for T<2.5K. Although the experimental temperature range is too limited to observe distinct linear part of the plot one can estimate upper limit for the pair energy gaps, which are given in Tab.2. For comparison we also inserted pair energy gap calculated using exchange integral J_{NN} estimated from high temperature susceptibility (Tab.2,HTE). We notice strong reduction of the pair energy gap, as expected, although the value obtained for CdFeSe from specific heat is somehow smaller than the value resulting from HTE. The data for still lower temperatures (T<1.5K) are necessary to draw pertinent conclusions.
It is worthwhile to mention that the presented analysis of the specific heat provides the first experimental evidence of the reduction of the pair energy gap (no FIR spectroscopy

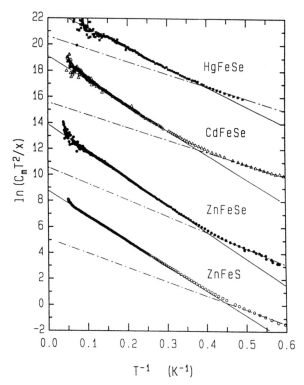

Fig.9 Specific heat of ZnFeS (x=0.017), ZnFeSe (x=0.015), CdFeSe (x=0.008) and HgFeSe (x=0.015) plotted as $\ln(C_m T^2/x)$ versus 1/T (plots are shifted over 5 units in vertical scale).

data for Fe-Fe pair were reported so far). It seems that low temperature specific heat may serve as a useful means for energy structure determination (even in the case that FIR spectroscopy can not be used because of selection rules).

d-d interaction nature

First of all we notice that exchange interaction coupling Fe ions is of the same sign (AF) and same order of magnitude as reported for Mn-DMS (Tab.1). For Mn-DMS it was rather well established (by both experimental and theoretical study) that *superexchange* (SE)[44] is dominant interaction coupling Mn ions[45,46]. It is known that SE strength depends strongly on probability of hopping of electrons between d and p orbitals (i.e. hybridization of d and p orbitals)[44,45]. In the case of tetrahedrally coordinated lattice sites only t-type orbitals hybridize strongly with p-type orbitals. Hybridization of e-type orbitals is very weak (see Fig.2 in the preceding Chapter). Therefore only electrons occupying t-type orbitals (t_+ or t_-) are supposed to provide major contribution to SE coupling. We notice that for both Mn^{++} and Fe^{++} ions the same number of t-type orbitals is occupied: (Mn^{++}: $d^5 = e_+^2 t_+^3 e_-^0 t_-^0$ and Fe^{++}: $d^6 = e_+^2 t_+^3 e_-^1 t_-^0$) and thus SE should be similarly effective in both cases.

The other mechanisms of interaction between Fe ions seem to be much less important, similarly as in the case of Mn-DMS:

−conventional *dipole-dipole interaction* between magnetic moments may be effective only at milikelvin temperatures and thus apparently can not be responsible for effects observed above 1K,

−*RKKY interaction* [47], which is mediated by free carriers can be also ruled out since concentration of free carriers in Fe−DMS never exceed $10^{19} cm^{-3}$, which is one−two orders of magnitude too low for RKKY to be effective,

−in *Blombergen-Rowland* mechanism (BR)[48,49] d electrons are exchanged via conduction and valence bands and thus energy gap between these bands is important. In general one may expect this interaction being strong only for narrow gap materials (like HgFeSe). In contrast we observe similar strength of interaction irrespective to the energy gap (Tab.1: ZnSe, CdSe − wide gap materials, HgCdSe − narrow gap and HgSe − zerogap material). However one should remember that for BR interaction conduction and valence band states far from Brillouin zone center may be very important and in such a case the energy separation between bands for II−VI compound are comparable. Anyway no clear evidence of importance of BR interaction was reported so far neither for Fe−DMS nor for Mn−DMS.

The problem of the interaction nature could be enlightened by the information about the interaction range. However the available data for Fe−DMS are very poor[2,7].

In conclusion we believe that coupling between Fe ions is driven by superexchange interaction.

Unfortunately no proper theory of exchange interaction for Fe−DMS (d^6 configuration) was developed so far[50] and thus it is hard to discuss chemical trends in Fe−DMS family, as was done for Mn−DMS[45]. Anyhow, in general we expect that SE interaction (monitored by J_{NN}) increases with increasing hybridization, which is monitored by ion−hole exchange integral ($N_0\beta$)[28,29]. This situation is actually encountered in experiment: stronger interaction between Fe ions and valence band electrons is followed by substantially stronger coupling between Fe ions (cf. Tab.1 and Tab.3).

Table 3
Exchange constants $N_0\beta$ (eV) for Fe− and Mn−DMS

material	Y=Fe	Y=Mn
ZnYSe	-1.74[21,29]	-1.31[52]
CdYSe	-1.6 [29]	-1.11[53]
	-1.45[54]	-1.26[55]
	-1.53[51]	

On the other hand stronger p−d hybridization for Fe ions may be expected since atomic radius is larger for Fe ion than for Mn ion and that should result in stronger overlap of d and p orbitals.

Our knowledge concerning Fe−DMS is still poor in relation to Mn−DMS and further study is necessary to confirm the domination of superexchange interaction. In particular information about location of Fe d levels relatively top of the valence band (resulting from photoemission study)

together with SE theory for d^6 configuration would be very helpful in this task.

CONCLUDING REMARKS

It follows from the data presented above that nonvanishing orbital momentum of Fe ions significantly influences magnetic properties of bulk Fe-DMS crystals. Although basic problems seem to be solved, there are still some fundamental issues to be worked out.

First of all we notice that our knowledge is limited to rather diluted regime (x<0.2) since highly concentrated crystals have not been grown so far in spite of many attempts.

One of the most interesting open magnetic problems is possible existence of spin-glass phase transition for field induced magnetic moments associated with Fe^{++} ions. More experiments are necessary to establish actual situation. Highly concentrated crystals would be very helpful in this challenge. Microscopic mechanism of freezing of field induced moments must also be understood.

Proper Hamiltonian of isolated Fe ion (in particular Jahn-Teller coupling to the lattice) as well as Hamiltonian describing interaction between Fe ions is to be checked. Precise FIR or Raman spectroscopy is necessary.

Information about d-d interaction range would help in confirmation of the interaction nature.

Development of superexchange theory for Fe^{++} (d^6) would enable understanding chemical trends of ion-ion coupling as well as ion-free carrier interaction in Fe-DMS family .

Interesting possibilities of investigating low dimensional magnetism are opened by development of ZnFeSe/ZnSe superlattices and heterostructures.

Finally interaction between field induced magnetic moments and permanent magnetic moments can be studied (Fe^{++}/Fe^{+++} in HgFeSe or Fe^{++}/Mn^{++} in new material like ZnFeMnSe).

We note that understanding of these problems will be crucially helpful in investigation of future Ni-based DMS, which are expected to be another example of system of field induced magnetic moments.

REFERENCES

1.   see review paper: A.Mycielski, J. Appl. Phys. **63**, 3279 (1988)
2.   see review paper: A.Twardowski, J.Appl. Phys. **67**, 5108 (1990)
3.   W.Low and M.Weger, Phys.Rev. **118**, 1119 (1960)
4.   G.A.Slack, S.Roberts and J.T.Vallin, Phys.Rev. **187**, 511 (1969)
5.   R.R.Galazka, S.Nagata , P.H.Keesom, Phys. Rev. **B22**, 3344 (1980)
6.   S.Nagata , R.R.Galazka, D.P.Mullin, H.Arbarzadeh, G.D.Khattak, J.K.Furdyna, P.H.Keesom, Phys. Rev. **B22**, 3331 (1980)
7.   H.J.M.Swagten, A.Twardowski, W.J.M. de Jonge and M.Demianiuk, Phys.Rev. **B39**, 2568 (1989)

8.   H.J.M.Swagten, F.A.Arnouts, A.Twardowski, W.J.M.de Jonge and A.Mycielski, Solid State Commun. **69**, 1047 (1989)
9.   H.J.M.Swagten, A.Twardowski and W.J.M.de Jonge, to be published
     H.J.M.Swagten, A.Twardowski, E.H.H.brentjens, W.J.M.de Jonge, M.Demianiuk and J.A.Gaj, in Proc. of 20 ICPS (Thessaloniki 1990), to be published
10. A.Twardowski, H.J.M.Swagten, W.J.M.de Jonge and M.Demianiuk, Phys.Rev. **B36**, 7013 (1987)
11. see: Semiconductors and Semimetals, **25**, "Diluted Magnetic Semiconductors" ed. J.K.Furdyna and J.Kossut, Academic Press (1988)
     and rewiew paper: Furdyna,J.K., J.Appl.Phys. **64** (4), R29 (1988)
12. A.Twardowski, M.von Ortenberg and M.Demianiuk, J.Cryst.Growth **72**, 401 (1985)
13. C.Testelin, A.Mauger, C.Rigaux, M.Guillot and A.Mycielski, Solid State Commun. in press (1989)
14. D.Heiman, A.Petrou, S.H.Bloom, Y.Shapira, E.D.Isaacs and W.Giriat, Phys. Rev. Letters, **60**, 1876 (1988)
15. D.Scalbert, C.Benoit a la Guillaume, to be published
16. T.Q.Vu, V.Bindilatti, M.V.Kurik, Y.Shapira, A.Twardowski, E.J.McNiff and A.Wold, to be published
17. J.T.Vallin, G.A.Slack and C.C.Bradley, Phys. Rev. **B2**, 4406 (1970)
18. A.Lewicki, J.Spalek and A.Mycielski, J.Phys. C, **20**, 2005 (1987)
19. A.Twardowski, A.Lewicki, M.Arciszewska, W.J.M.de Jonge,H.J.M.Swagten, M.Demianiuk, Phys. Rev. **B38**, 10749 (1988)
20. M.Arciszewska, A.Lenard, T.Diel, W.Plesiewicz, T.Skoskiewicz and W.Dobrowolski, Acta Physica Polonica, **A77**, 155 (1990)
21. A.Twardowski, P.Glod, W.J.M. de Jonge, M.Demianiuk, Solid State Commun. **64**, 63 (1987)
22. see review papers: J.A.Maydosh in "Hyperfine Interactions", Vol.31 of Lecture Notes in Physics (Springer, New York 1986), p.347
     K.Binder and A.P.Young, Rev. Mod. Phys. **56**, 801 (1986)
23. A.Abragam and B.Bleaney "Electron Paramagnetic Resonance of Transition Metal Ions", Clerondon Press, Oxford 1970
24. J.T.Vallin, Phys. Rev **B2**, 2390 (1970)
25. A.Twardowski, H.J.M.Swagten and W.J.M.de Jonge, Phys.Rev. **B42**, no.4 (1990)
26. K.Lebecki, A.Twardowski and Z.Liro, Acta Physica Polonica, in press
27. D.Bloor, G.M.Copland, Rep. Prog. Phys. **35**, 1173 (1972)
     P.M.Levy, Phys. Rev. **177**, 509 (1969)
28. A.Twardowski, K.Pakula, M.Arciszewska and A.Mycielski, Solid State Commun. **73**, 601 (1990)
29. A.Twardowski, K.Pakula, I.Perez, P.Wise and J.E.Crow, Phys. Rev. **B42**, (october 1990)
30. J.Spalek, A.Lewicki, Z.Tarnawski, J.K.Furdyna, R.R.Galazka, Z.Obuszko, Phys. Rev **B33**, 3407 (1986)
31. R.R.Galazka, W.Dobrowolski, J.P.Lascaray, M.Nawrocki, A.Bruno, J.M.Broto, J.C.Ousset, J.of Mag. and Mag. Mat. **72**, 174 (1988)
32. R.L.Aggarwal, S.N.Jasperson, Y.Shapira, S.Foner, T.Sakibara, T.Goto, N.Miura, K.Dwight and A.Wold, Proc. of the 17th Int. Conf. on the Phys. of Semiconductors, San Francisco 1984, p.1419 (ed. J.D.Chadi and

W.A.Harrison) Springer, New York (1985)

Y.Shapira, S.Foner, D.H.Ridgley, K.Dwight and A.Wold, Phys. Rev. **B30**,4021 (1984)

J.P.Lascaray, M.Nawrocki, J.M.Broto, M.Rakoto and M.Demianiuk Solid State Commun. **61**, 401 (1987)

33. B.E.Larson, K.C.Hass, R.L.Aggarwal, Phys.Rev. **B33**, 1789 (1986)

34. T.M.Giebultowicz, J.J.Rhyne, J.K.Furdyna J.Appl.Phys. **61**, 3537 (1987); **61**, 3540 (1987)

35. K.Matho, J. Low Temp. Physics **35**, 165 (1979)
C.J.M.Denissen, H.Nishihara, J.C.van Gool, W.J.M. de Jonge, Phys. Rev. **B 33**, 7637 (1986)

36. A.Twardowski, H.J.M.Swagten, T.F.H.v.d.Wetering and W.J.M.de Jonge, Solid State Commun. **65**, 235 (1988)

37. H.J.M.Swagten, C.E.P.Gerrits, A.Twardowski and W.J.M.de Jonge, Phys.Rev. **B41**, no.10 (1990)

38. G.A.Slack, S.Roberts and F.S.Ham, Phys. Rev. **155**, 170 (1967)

39. A.Twardowski, H.J.M.Swagten and W.J.M. de Jonge, in Proc. of 19th ICPS (Warsaw 1988) ed. W.Zawadzki, Institute of Physics, Polish Academy of Sciences, p.1543 (Wroclaw 1989)

40. A.M.Witowski, A.Twardowski, M.Pohlmann, W.J.M. de Jonge, A.Wieck, A.Mycielski and M.Demianiuk, Solid State Commun. **70**, 27 (1989)

41. M.Hausenblas, C.L.Claessen, A.Wittlin, A.Twardowski, M.von Ortenberg, W.J.M.de Jonge and P.Wyder, Solid State Commun,**72**, 253 (1989)

42. D.Scalbert, J.Cernogora, A.Mauger and C.Benoit a la Guillaume and A.Mycielski, Solid State Commun.**69**, 453 (1989)

43. D.Scalbert, J.A.Gaj, A.Mauger, J.Cernogora C.Benoit a la Guillaume, Phys. Rev. Lett. **62**, 2865 (1989)

44. P.W.Anderson, in Solid State Physics **14**, 99 (1963), ed. F.Seitz and D.Turnbull, Academic Press
G.A.Sawatzky, W.Geertsma and C.Hass, J. of Magn. and Magn. Materials **3**, 37 (1976)
C.E.T.Goncalves da Silva and L.M.Falicov, J.Phys. **C5**, 63

45. H.Ehrenreich, K.C.Hass, B.E.Larson and N.F.Johnson in Material Research Society Fall Meeting (Boston '86), Symposium Proceedings **89**, 159 (1987), Diluted Magnetic (Semimagnetic) Semiconductors, ed. S.von Molnar, R.L.Aggarwal and J.K.Furdyna B.E.Larson, K.C.Hass, H.Ehrenreich and A.E.Carlsson, Solid State Commun. **56**, 347 (1985)

46. J.Spalek, A.Lewicki, Z.Tarnawski, J.K.Furdyna, R.R.Galazka, Z.Obuszko, Phys. Rev **B33**, 3407 (1986)

47. M.A.Ruderman and C.Kittel, Phys. Rev. **96**, 99 (1954)
K.Yoshida, Phys. Rev. **106**,893 (1957)
T.Kasuya, Prog. Theor. Phys. **16**, 45 (1956)

48. N.Blombergen and T.J.Rowland, Phys. Rev. **97**, 1679 (1955)

49. C.Lewiner and G.Bastard, J.Phys. **C13**, 2347 (1980)
C.Lewiner, J.A.Gaj and G.Bastard, J.Phys.(Paris) **41**, C5-289 (1980)

50. K.Hass, this book

51. D.Scalbert, M.Guillot, A.Mauger, J.A.Gaj, J.Cernogora and C.Benoit a la Guillaume, submitted to Solid State Commun. (1990)

52. A.Twardowski, T.Dietl and M.Demianiuk, Solid State Commun. **48**, 945 (1983)
A.Twardowski, M. von Ortenberg, M.Demianiuk,

R.Pauthenet, Solid State Commun. **51**, 849 (1984)

53. R.L.Aggarwal, S.N.Jasperson, J.Stankiewicz, Y.Shapira, S.Foner, B.Khazai and A.Wold, Phys. Rev. **B28**, 6907 (1983)

54. O.W.Shih, R.L.Aggarwal, Y.Shapira, S.H.Bloom, V.Bindilatti, R.Kershaw, K.Dwight and A.Wold, Solid State Commun. **74**, 455 (1990)

55. M.Arciszewska and M.Nawrocki, J.Phys.Chem.Solids **47**, 309 (1986)

INDEX